Eucalyptus Plantations

Research, Management and Development

Proceedings of the International Symposium

Eucalyptus Plantations

Research, Management and Development

uangzhou, China 1 – 6 September 2002

edited by

Run-Peng Wei
Sino-Forest Corporation, Hong Kong

Daping Xu
Research Institute of Tropical Forestry, China

World Scientific
Singapore • New Jersey • London • Hong Kong

Published by

World Scientific Publishing Co. Pte. Ltd.
5 Toh Tuck Link, Singapore 596224
USA office: 27 Warren Street, Suite 401-402, Hackensack, NJ 07601
UK office: 57 Shelton Street, Covent Garden, London WC2H 9HE

British Library Cataloguing-in-Publication Data
A catalogue record for this book is available from the British Library.

First published 2003
Reprinted 2015

EUCALYPTUS PLANTATIONS
RESEARCH, MANAGEMENT AND DEVELOPMENT
Proceedings of the International Symposium

ISBN-13 978-981-238-557-4
ISBN-10 981-238-557-6

PREFACE

As a result of increasing population growth, forestlands getting converted for farming and other purposes, and pressure on logging bans of natural forests, wood supply is becoming an increasingly incisive issue. Also affecting the issue are other factors, such as economic development, environment conservation, etc. It is now widely accepted that tree plantations can play an important role in meeting demand for wood supply, promoting rural development and conserving natural ecosystems and environment. With sufficient input, technical innovations and intensive management, tree plantations can no doubt be objectively cultured to target a high yield of wood with desirable characteristics and supply all kinds of woods and wood products. The intention is the same as those of the high-yield crop plantations that launched the "green revolution" tide in the last several decades, and have continuously been providing food supply for an increasing population. Tree plantations, which usually have to be developed at a large scale in rural and remote regions, create plenty of jobs not only for agricultural labors but also for industrial workers through the downstream product processing and utilization. Besides the inherent ecological function as a forest ecosystem, tree plantations are also unique in producing substitutes for natural woods, and are the most effective sinks for sequestering greenhouse gas, a topic of global interest. The best and most successful instances are probably those countries that produce most of the wood supply with high-yield plantations comprising less than 5% of the total forest area.

As one of the most extensively cultivated forest genus, eucalypts have been widely planted in tropical and subtropical areas all over the world. Since the 1980s in particular, eucalypt plantations have witnessed a rapid development in terms of the planting area and production yield, thanks to research and technical innovations in breeding, silviculture operations, nutrient management, stand management, and particularly mass vegetative propagation technique. Cloning propagation drives breeding to the result that all present genetic variations can be explored for maximizing yield. It is the systematized management and silvicultural practices of eucalypt plantations with infusion of continuous technical innovations that create the yield peak of wood production, which is forestry's own "green revolution". With production of a great amount of logs every year, eucalypt wood products, through processing, are further developed and diversified, from lumber, wood chip, wood pulp, panels, solid wood board to furniture. Currently, many plantation companies also focus their R&D works on high value eucalypt products. The role of eucalypt plantations in conserving nature and the environment is evident as mentioned above. Having brought about great economic, social and environmental benefits, eucalypt plantation management is now becoming not only a unique business, but also the fastest growing one in tree-planting.

Yet, there still exists huge room for eucalypt planters to improve their plantation yield and quality, particularly in countries and regions that have relatively low growth rates and inefficient management systems. Striking contrast between the experimental trials and existing stands reported, between stands with different inputs, or between the yield levels of different countries may easily exhibit such room and potential to improve. On the other hand, the intensive management and the further extension of eucalypt plantations have incurred ecological and environmental concerns or criticisms, some of which may be of scientific grounds while others may be more or less subjective prejudgments. Indeed, fast-growing eucalypt plantations deplete much of available resources such as water and nutrition, facing issues relating to sustainable management. However, it is the same problem for most fast-growing trees (even crops species) plantations aimed at high yield. Moreover, those raised concerns, for eucalypt researchers and planters, point out a research direction rather than a futureless end. Opportunities and challenges always coexist. Continuous R&D works on new technology and knowledge, and their exchange and extension are therefore vital for increasing production yield, sustaining further development and improving or mitigating impacts on the environments of eucalypt plantations.

The theme of the symposium, "*Eucalyptus* plantations", was selected to facilitate the exchange of new research developments, successful experiences and lessons in planting and utilizing eucalypts, among individuals, research institutes and production sectors. Thirty-three papers presented at the symposium were collected and grouped in the present proceedings book with six chapters: I. Development and Ecological/Social Impact, II. Breeding Methods, III. Genetic Testing and Improvement, IV. Nutrition and Site Management, V. Plantation Management, and VI. Wood Processing and Utilization. Considering the book may be particularly of interest to Chinese eucalypt planters and researchers who do not have easy access to outside information, we included a Chinese translation of the table of contents, as well as the abstract at the end of each paper. We hope this proceedings book may be of value for all interested parties in their endeavors to make eucalypt plantations achieve sustainability, high productivity and environmental friendliness.

The symposium was co-organized in Guangzhou/Zhaoqing, China, on 1-6 September 2002 by Research Institute of Tropical Forestry (RITF) and Sino-Forest Corporation (Canada), respectively leading in research and management of eucalypt plantation in China. The symposium organization committee consisted of Allen Chan, Bai Jiayu, Kat Kit Poon, Qi Shuxiong, Xu Daping and Hou Yuanzhao; the program committee consisted of Xu Daping, Run-Peng Wei, Li Mei, Wu Zhongmin and Jayco Fung. Sino-Forest Corporation (Canada) solely sponsored the symposium and the publication of this proceeding book, and RITF was in charge of organizing the symposium. We thank Jayco Fung, Simon Poon, Miranda Lin and Li Mei for their assistances in editing this proceedings book. Many others provided help during the symposium and the post-symposium field trips. We also thank them all.

Run-Peng Wei
Daping Xu

May 2003

CONTENTS

内 容

I. Development and Ecological/Social Impact

EUCALYPTUS: A STRATEGIC FOREST TREE

BERNARD MARTIN

Chargé de mission à l'ENGREF,
14 rue Girardet, 54042 Nancy cedex, France

For the last thirty years, the eucalypts plantations have experienced an important development in the subtropical and tropical zones. Successful genetic improvement amplified by vegetative propagation and more recently by promising biotechnology, has added to the natural qualities of the tree: fast growing, excellent fibre, and relatively high wood density. This has favoured *Eucalyptus* as one of the best tree species for ligniculture.

The high productivity and short rotation, along with the uniformity of clonal plantations and improvement of wood quality, attracted private investments, especially from the pulp industry that continues to invest in R&D to reduce the agronomic, pathological and ecological risks of this crop. *Eucalyptus* has also started to attract financial groups who have joined multinational pulp and paper companies in the investment of its plantation development.

After long time criticism by ecologists, this ligniculture is now better appreciated because it is usually developed on non-forested lands, contributing in fact to reduce the economical pressure on natural forests. This ecological interest is now enhanced by the recent progress in high quality timber production. It is a great new challenge to reinforce the economic and ecologic interests that could increasingly contribute to protect the natural forests. Last, but not least, *Eucalyptus* could become one of the best carbonsinks species, and so, one of the best ecological and economical tools for internalising the global externalities induced by the burning of fossil fuel. This new green gold of the pulp industry can now be considered a "strategic forest tree" as well.

1 Introduction

Although trees provide an important contribution to firewood production and various services in less developed countries, development and research of industrial plantations of *Eucalyptus* escape the public sector's responsibility more and more; it is about a very intensive forest system integrated within the pulp industry.

The concentration of private investment on this tree contributes to very significant progress in terms of not only tree genetics and physiology, but also silviculture and wood technology. This intensification has resulted in greater productivity and much shorter rotations. *Eucalyptus* fibre is now recognized as the international standard of quality for short fibres and it has penetrated all the markets. For pulp industry, *Eucalyptus* has become a strategic tree, a "new green gold".

2 A Successful Domestication

2.1 *Fast Genetic Improvement*

Eucalyptus as an exotic species has a long story (Metro 1955; Eldridge et al. 1994). Since the first *E. camaldulensis* provenances trial, organized by FAO in 1966, immense progresses have been made in this field. Many species were tested in a great number of sites, including the wet tropical and cold climates. Here we should recognize the important contribution of CSIRO and the Indonesian Government that took a very active part to diffuse this invaluable genetic resource from their natural area. We would also like to pay homage to Professor L.D. Pryor, great contributor to the *Eucalyptus* domestication. For all the important species and many others, multi-sites provenances trials were installed, widening considerably the local genetic bases. These trials permit to select plastic populations (vigorous and stable), particularly interestingly for resistance to climatic fluctuations. These populations induced frequently *in situ* and *ex situ* conservation programmes and constituted the starting point of a lot of local programmes of genetic improvement by recurrent way (RW), leading to important gain for many traits such as adaptation, strength, form, resistances against adversities, wood quality, etc. This development occurred in spite of numerous impediments (Haines 1995).

2.2 *Natural and Artificial Hybrids*

Beyond the traditional recurrent way for genetic improvement of the pure species, eucalypts enter quickly into inter-specific hybridization programmes by natural or artificial reciprocal recurrent way (RRW) because *Eucalyptus* genus is very poly-specific, and its species are genetically close and are frequently crossable. This resulted in a strong propensity to natural hybridization in the arboreta or collections of species, and contributed to the emergence of vigorous and well-adapted land races, often forming hybrids swarms in which exceptional individuals are frequently found. In countries with long tradition introduction and cultivation of eucalypts, the very broad development of land races (in space and time) and the transfer of know-how in cloning, often allowed fast development of hybrid clonal plantations with high performances without preliminary sophisticated programmes of improvement by RW (such as cases of Congo, Brazil, Morocco, Venezuela, etc.). Subsequent programmes by RRW were generally integrated into following generations. However, the recombination of natural hybrids is also a simpler way to proceed, with genetic marking being of an invaluable help.

2.3 *Creation of New Varieties*

Grafting practice (which remains relatively difficult with *Eucalyptus*), flowering induction (very effective with paclobutrazole), and controlled crossings (generally very easy but still improvable) are the current practices, and allow accelerated development of improvement programmes. Furthermore, these practices also open the way to the creation of new varieties, combining in the same genotype many advantageous characteristics naturally dispersed in the species.

New local varieties appear increasingly powerful. For example, it is much faster to create hybrids resistant to dryness than to select, inside a significant species, trees resistant to *Phoracanta* that attacks trees suffering from dryness in marginal zone. In the same way, to improve quality of wood, the inter-specific hybrids can give very rapid gains, some of which are even impossible to obtain by selection within the pure species.

2.4 *Increasing Control of Cloning*

In 1956, the first rooted cuttings were obtained with *E. camaldulensis* in Morocco and Tunisia (Franclet 1956, 1970), but not until 1973, the main experiments using coppices of adult trees were conducted in Congo (Martin and Quillet 1974). In 1975, the Congolese technique was transposed at large scale by Aracruz S.A. in Brazil (Marcus Wallenberg Foundation 1984), where "Green Revolution" for *Eucalyptus* appeared in the seventies. Thereafter, many improvements of the cutting system (Chaperon 1983) and development of the *in vitro* culture made cloning a common practice for *Eucalyptus*. Since the mass production of hybrid seeds is unpredictable, difficult and not always possible, the cloning of eucalypt inter-specific hybrids becomes evidently important for plantation development. The technique was also developed in China in 1987, thanks to a FAO Technical Cooperation Project (Campinhos 1987).

In vitro culture, registered with the programmes of many laboratories since the eighties, was quickly obtained for a majority of species, even those more rebellious of the herbaceous stem-cuttings system. On the other hand, for economic reasons, this practice remains limited to experimental scale; the practice with mobilization and rejuvenation of the trees and installation of the clonal hedges still predominantly prevails.

More recently, significant progress in term of rooting potential, root system quality and lowered cost has been made for vegetative propagation. The "mini cutting" system (de Assis 2001) integrates *in vitro* and *in vivo* multiplication in a "super intensive system" for cloning eucalypts at commercial scale. This system uses the axillary's sprouts that were collected on *vitro*-plantlets and cultivated by indoor hydroponics mini-hedges. Much similar to *vitro*-culture, this system is highly sophisticated and generates well rejuvenated plants. It can also enlarge the vegetative propagation of the species that are difficult to root (e.g. *E. globulus*, *E. nitens*, *E. citriodora*, *E. maculata*, etc.).

2.5 *Crossings on a Large Scale*

The programmes that use the best crossings on a large scale are rare. However, this practice is not without interest for clonal tests much delay the selection of clonal varieties. Thus, the new hybrids may by more advantageous than the clones used. In Morocco, the annual production of x *E. camaldulensis* and x *E. grandis* is about 100 to 200,000 artificial hybrid seedlings. In Chile, the FAMASA produces about 4 kg/yr of x *E. globulus* and x *E. nitens* seeds by control pollination, which covers 40% of the need with high genetic gain (Sanhueza and Griffin 2001). The new plantations resulting from controlled seeds, in addition to increase production, become a very important source for selection of new clones, especially in comparison with small-scale progeny tests. In eucalypts, crossing on a large scale is now "a practical reality" (Harbard et al. 1999) facilitated by flowering induction control (Cauvin 1991) and possible pollination on cut stigma not bagged but more or less shortened (Cauvin 1988) and protected by a small tube, a method called "one stop pollination" (OSP).

2.6 *Promising Biotechnology*

Today, biotechnology is very promising, and may greatly affect eucalypts. The species benefits already from an advanced domestication and are now the subjects of research very active in all the biotechnological branches:

- Genetic architecture,
- Marker-assisted selection, and
- GMO Engineering.

Genome prospecting is now possible with new molecular markers: RFLP, RAPD, micro-satellites (Haines and Martin 1995). Bio-technicians are able to answer many questions for tree breeders, from population level (provenances and progenies characterisation and genetic distances estimation) to individual level (clonal identity, consanguinity level, crossing errors, affiliation, genealogy, etc.).

Marker-assisted selection (MAS) is not yet operational with *Eucalyptus* but important research programmes exist (Grattapaglia et al. 1995). Because of the species' inherent characteristics, in particular long-time generation, optimising the crossing is of the first priority. The possibility of choosing the best parents "on genetic map" would be a considerable advantage for heterosis prevision (Vaillancourt et al. 1995). It requires, however, a good advance in the genetic mapping technology so that the prevision is based on the segregation of QTL between parents and hybrid progeny (Myburg et al. 2001).

GMO engineering in *Eucalyptus* is already a reality but many tests are still necessary. It will only involve exceptional clones and will relate to some extrinsic traits like glyphosate tolerance, worst parasitic resistance, or intrinsic traits like low lignin wood, which already exists in poplar at the experimental stage (Chiang 2001). Because these modified clones present big advantages but at a high cost, propagation methods are very sophisticated and should possibly be automated vitro-culture (Teasdale 1995).

The general fear of the public, under influences of the media, delays GMO utilisation. This conjuncture is reinforced by the position of certain major forest certification systems, like FSC, which reject the use of GMO in trees. It is certain that all the aspects have to be examined before using GMO at a large scale, without negative environmental impacts, but we have to dread the "business of the certification" which can, for political raison, exclude GMO and its advantages (Coventry 2001). Wood is not food, and the environmental risks may be reduced with eucalypts. Apart from the natural distribution zones of the species (mainly Australia), the *Myrtaceae* family is very poor in the local floras of the rest of the world.

3 Taking Risks into Account

3.1 *Agronomic Risks*

For forest plantations, the main risk is the soil fertility reduction, getting worse with the low quality of lands used generally unsuitable for good agriculture. These soils have a poor fertility, which often contain very little phosphorus and have a strong nitrogen deficit. This problem is treated as in agriculture, by applying compensatory fertilization, at planting, in the course of growth, and after each harvesting.

In comparison with agricultural productions, the nutrient uptake by the wood is relatively low; however, because of the very high productivity of *Eucalyptus*, the maintenance of the mineral reserves of the soil at an acceptable level is still of primary importance. Research programmes on fertilization are in general very developed, and many fertilization formulas are used, including at the clonal level. The question of nitrogen requirement, poorly satisfied in the tropical regions (Bouillet et al. 2001), is actually the subject of the reinforced studies on, in particular, the utilization of nitrogen fixing species. The mixed-row-species plantation seems an excellent system, especially with slower-growing *Acacia* (Bauhus et al. 2000) or *Albizia* (Binkley et al. 1992). The proportion of nitrogen-fixing species (1 row for 2 to 5 rows of *Eucalyptus*) can depend on the soil quality, the required productivity level and the quality of desired pulp as well. Some downstream products integrated companies could find interest in the harvesting of the two types of wood in the mixture.

3.2 *Sanitary Risks*

It is really a big risk, accentuated by the clonal aspect of plantations. Although *Eucalyptus* are exotics trees for most of the world, parasite occurrence is an increasingly problem, coming from natural areas or from local origin, and causes more or less severe damages. We brief the main pests and diseases as follows.

3.2.1 Insects

a. On eucalypts as exotics
   - *Phoracanta semipunctata*, Coleoptera that attack trees suffering from dryness (*E. globulus* in Mediterranean zone).
   - *Gonipterus scutellatus*, Coleopteran *curculionidae* defoliator on commercial species in South Africa and Spain (mainly *E. viminalis* and species of *E. globulus* group).
   - *Helopeltis schoutedeni*, sucking and stinging Hemipterae, which causes severe damage on the young plantations in Congo.

b. On eucalypts in natives
   - *Mnesampela privata*, (autumn gum moth) Lepidoptera defoliator at larval stage in South Australia and Tasmania (mainly on *E. globulus* and *E. nitens*).
   - *Chrysophtharta bimaculata*, (Tasmanian eucalypt leaf beetle) Coleoptera causes severe damage in Tasmania (*E. regnans* and *E. nitens*).

3.2.2 Fungus

a. *Cryphonectria cubensis* (ex *Diaporthe cubensis*) causes trunk cankers and gummosis (mainly in trees in hot and wet situations).

b. *Mycosphaerella juvenis* causes very severe damage on the youthful sheets of a great number of species and especially *E. globulus*.
c. *Puccinia psidii*, *Cylindrocladium* spp. and *Coniothirium* spp., are rust on *E. grandis* in Argentina and Uruguay (very bad in nursery but rare in plantation).

All the controlling strategies may be used, including chemical, genetic, and biological, which often have very positive results. The cloning technique, although increasing the risks by reducing the genetic base, appears to be a powerful mean of fast selection of the most resistant clones. It has been the case of Congo with the rapid selection of clones relatively resistant to *Helopeltis* (Martin 1987), and also the case of South America with *Puccinia psidii* where molecular markers are very efficient because only a couple of alleles control the susceptible or resistant clones (Junghans 2000). The fungus like *Mycosphaerella*, which is more severe on the hybrids of *E. globulus x E. nitens* than on the parents species (Dungey et al. 1997), seems more dangerous than the insects for which the biological control can be very effective (for example, introduction of *Anaphes nitens* as parasite of *Gonipterus* eggs).

It is necessary to intensify the vigilance, and researchers must redouble their efforts to combine the best strategies. These include molecular markers, essential oil traits (Martin 1987; Shepherd et al. 1995) and GMO creation, particularly with Bt toxin production responsible gene (Harcourt et al. 1995). This should start from the best clones in order to create new genotypes that are resistant to the most serious parasites in plantations, provided the environmental concerns are negligible.

3.3 *Ecological Risks*

The ecologists held the eucalypts accountable for soil sterilization, erosion and loss of biodiversity. In fact, this may be part of a bad war against industrial plantations, and could mask direct attacks against multinational companies. However, the tree growers must remain vigilant on various fronts:

3.3.1 Soils Sterilization

Despite the oil content of leaves and slow decomposition of litters, allelopathy is rare with eucalypts and the mechanism is not very clear (Poore and Fries 1985). On the contrary, *Eucalyptus* often suffers from herbaceous initial invasion, and weeding must be very intensive during the first 2 or 3 years. Good commercial plantations with intensive weeding are usually vigorous and homogeneous, having no understory. When plantations have lower density and longer rotation, trees take longer time to have close canopy, which favour understory vegetation growth. Leaving tree litters on the sites, and mixing *Acacia* in stands, should allow a much better evolution and a fast enrichment of soils out of organic matters.

3.3.2 Erosion

Soil erosion under *Eucalyptus* stands is possible and can be avoided with anti-erosive design for sites of fragile soils and on slopes. On the other hand, eucalypt stand can constitute an excellent reinforcement of the soil, thanks to its very strong rooting systems. However, it is necessary to follow some principles such as planting at low density and along the contour lines, mixing with nitrogen fixing trees or shrubs, etc. For intensively managed plantations on sandy soils, erosions generally always come from use of tracks and firebreaks, which could also be mitigated in a way.

3.3.3 Loss of Biodiversity

Like that of agriculture, the objective of eucalypt plantations is not to increase the biodiversity but to produce a maximum volume of wood per hectare. The growers must avoid amalgam between ligniculture and silviculture. Ligniculture is closer to agriculture and very far from silviculture for natural forests. Although not physically replacing that for the natural forests, ligniculture, concentrated in savannahs or on lands abandoned by agriculture, greatly reduces the increasingly human pressures on the natural forests (Martin 1991) which are today subject of a great interest known as "alternative demand" (environmental and social). Thus, the ecologists cannot simply reproach the plantations of eucalypts for destroying the natural forests and not being able to play the same role.

However, planted on open areas for economical reasons (too expensive and problematical stumping), eucalypts, thanks to their tall and light cover, often support natural vegetation recovery. In the situations of being abandoned for production objective, the old eucalypt stands will not return to savannahs (except if they are frequently traversed by fires), but will quickly be invaded by natural vegetation (favoured especially by the birds) and can be even transformed into reserves of fauna (such as the case of the Loudima plantations in Congo). They could then become excellent carbonsinks (Wright et al. 2001).

3.4 *Research and Development Assistance*

The management of eucalypt plantation sectors should be consistently assisted by active R&D units, as part of the strategy of the companies, and integrated in a vast scientific and open network: public world network allowing information exchanges such as at IUFRO meetings, more restrictive organizations like CAMCORE offering a direct help to each member, or more confidential local networks between regional companies. The need for research is important because the intensively managed eucalypts plantations are subjected to many constraints and are based on the principle of continuous improvement. It is an absolute priority, to assure a good evolution at genetic material, to benefit for a long time from genetic gain, to take up new challenges of the companies, and to adapt to the environmental variations. R&D requires high investments but is guarantee for good results.

4 Best Tree for Ligniculture

4.1 *Clonal Forestry*

The vegetative multiplication was the essential complement giving access to all the genetic gain. Moreover, a true synergy exists between genetic improvement and cloning feasibility. The success of the industrial stem-cuttings was a decisive element for many companies oriented now to hybrids utilization, and complete clonal plantations (similar to *Populus*). New projections are profiled already, by mini-cuttings system for best multiplication application to more species, and for enhanced approach to genetic improvement programmes.

4.2 *Constant Progress in Silviculture*

In intensive forestry management, many constraints exist. In addition to reducing costs, it is necessary to make efficient all the successive operations: nursery management, site preparation, soil cultivation, spacing, fertilization, weeding, and coppices management. Progress depends mainly on good decisions for each of these successive operations.

4.2.1 Mycorrhization

For eucalypts, mycorrhization is not a current operation in nurseries. Generally, mycorrizal fungi appear naturally in the new plantations (*Pisolithus* is very frequent). However, the optimisation of symbioses is rather important (especially in poor soils), which could lead to significant advantages during the first 5 to 8 years, as was widely demonstrated in South China (Zhong et al. 2001).

4.2.2 Nurseries

Classical propagation (rooting stem-cuttings) comprises of 3 successive phases:

a. production of small sprouts in clonal hedges of mother plants.
b. preparation of cuttings and rooting under fog.
c. education of rooted cuttings and root hardening with "root buds" formation (dryness adaptation of roots).

Mini-cuttings system requires *vitro* culture upstream and comprises the 3 phases also, but the first one is more sophisticated with hydroponics culture and very precise hedging operations.

In all the cases, the questions of containers and substrate are permanent. It is crucial to avoid roots deformations, which unfortunately still frequently occurs. In general, balls or pellets (without vertical wall) are much better than other containers, but are very rare in nurseries.

4.2.3 Fertilization

Research about fertilization was systematic and gave birth to many tests. Among the numerous results, often of local interest, we can mention the principal tendencies: basic fertilization, initial fertilization with slow release of nitrogen, and fertilization during the growth and after each harvesting (compensation). The post plantation fertilization is to be preferably made before the fourth year because during the first 3 years, trees make an important stock of mineral elements. Thereafter, it can then use its own reserves and the nutriments accumulated in the litter (Suarez et al. 2001), reducing the effect of the fertilizations during the growth. The studies of mineral uptakes by clones (specific uptake by each clone) led to research of breeding genotypes more responsive to soil nutriments and thus especially less expensive to cultivate (E. Campinhos, pers. comm.).

4.2.4 Spacing and Coppices Management

The spacing and coppicing are important, requiring more research. Rotation for pulp production varies from 6 to 14 years and the adopted density from 400 to 1100/ha. The fear of low production often leads to excessive density with many disadvantages (over cost for planting and harvesting, constraint for mechanical operation, commercial wood loss, excessive stumps mortality, too high competition in coppicing regeneration etc.). Thus, some companies begin to reduce their planting density, following recent research results (Rosales and Rojas 2001).

Concerning the management of the coppices, the stumps should be well balanced with at least two coppices diametrically opposed (reducing the new sprouting coppices after the first coppices thinned), but with two stems per stump, the density is excessive. Coppices thinning should be coupled with stumps thinning, but more researches are necessary in this prospective. However, companies may prefer to replant with new clones or varieties after harvesting, to integrate genetic progress and avoid the following coppice management. However, it is not the case everywhere and genetic gain will perhaps not always be so important. “Know-how” in coppice system seems very weak in comparison with the rest of "high tech know-how".

4.3 *Increasingly Competitive Raw Material*

The improvement in quality of pulpwood continues to progress, so much as regards to methodology (non-destructive microanalyses) and concrete results. Today all the programmes integrate selection for wood density (indirectly estimated in standing trees with pilodyn) and pulp yield related to site fertility, planting density, and stand age.

Mineral contents in wood are an important preoccupation. Macronutrients (N, P, K, Ca, Mg, S, Si) and micronutrients (Fe, Mn, Zn, Cu) depend on genotypes and on site fertility. The nutritional efficiency of different clones is a new concept, which may lead to a very close choice of the best genotypes to the most suitable site (Foelkel and de Assis 1995).

For a long time we lack effective tools to select for pulp yield. The determination of the best genotypes in the families requires numerous and non-destructive analyses (without cutting the tested trees). We have the choice today between micro-pulping method and NIRS analyses (indirect measurements by spectroscopy in the near infra-red), but this last technique, adapted very well to genetic improvement, is in rapid progression.

The micro-pulping analyses need only small wood samples (from some hundreds to a few grams). This analysis is not new, though (Janin 1981). It is based on the separate cooking of wood cores extracted from standing trees, and is a rustic method that gives directly the weight of the pulp obtained from each core. The calibration is made by comparison with the results of the industrial cooking.

The NIRS technique is much more recent and appears very adapted to a genetic improvement programme. After the initial phase of obligatory calibration, this method is rapid in giving the total pulp yield at low cost in comparison with traditional chemical analyses. This method has been used to study within-tree variation (Schimleck and Michell 1987), but remains a tool especially adopted for genetic programmes (Raymond and Schimleck 2002).

4.4 *Solid Hardwood Production Emergence*

In spite of its bad reputation as timber, *Eucalyptus* wood is still oriented to plywood or saw wood by some companies.

In South Africa (tree farms), Argentina and Uruguay, solid wood production with *E. grandis* is not recent but there exist breeding programs aiming at plywood and sawn wood production. For example, TAPEBICUA and COFUSA, etc. aim at timber rather than pulp.

In Chile, *E. globulus* woods are frequently sawn and sold on the international market under the name "Chilean Oak". To diversify production in this country, some new experiments with *E. nitens* and *E. nitens x E. globulus* have been attempted (Medhurst et al. 2001). Compared to the purely pulp plantations, the rotation is longer (20 years) and the silviculture includes one or two heavy thinnings.

Big changes came recently from Brazil, where Aracruz S.A. has begun since year 1999 a new and important production of solid wood with an *E. urophylla x grandis* sawmill (Aracruz 2001). Originally destined for the pulp, clonal plantations have been heavily thinned at 10 years old and harvested at 17. In the future, this rotation can be reduced to 14 years thanks to the genetic improvement. This wood is exported under the registered trademark "Lyptus". The results seem convincing, and the company is now preparing some new important changes with possible timber production accounting for 30% of its total revenue (BOISmag 2002), whereas other potentials (MDF and Laminated wood) are also being investigated.

After a very specialized orientation to single production of fibres, *Eucalyptus* seems now to move towards more traditional forest management with a timber objective, and with fibre taking only a classic by-product place (even if it represents the majority of volume). This new trend should result in a significant increase in profitability, a better competitiveness for the eucalypt pulpwood, and a better integration of pulp industries in the local development.

5 A Great Convergence of Interests

One of the future developments is to tend more and more towards a scission between two complementary types of forestry (Sedjo 2001). While natural forests have a marked conservation objective, intensively managed plantations, particularly in the subtropical and tropical regions, are mainly for wood production. The international forest economic situation seems to confirm this scenario, and many factors contribute to reinforce the intensive culture of eucalypts.

5.1 *Investments Towards the New Forest Potentials*

So far, attracting the private funds for the development of forest plantations is not obvious. Many conditions need to be met:

a. ecologically, sites with relatively hot and wet climates, deep soils but marginal for agriculture.
b. economically, land available and inexpensive, cheap manpower, and proximity of the coast (for exportation).
c. industrially, local infrastructures and industrial environment at a sufficient level.
d. Securely, political, financial and fiscal environment sufficiently sure.

Some countries from the South, thanks to good investment policies, are very successful in their forest plantation development and exportation of the timber and wood products. For *Eucalyptus* planting, there involve mainly six countries (Brazil, Chile, Australia, South Africa, Argentina, and Uruguay). The new "wood fields" show a strategic importance, by their total area (4.7 million ha that is equivalent to one third of all eucalypts planted), their high growth rate (often over 30 m^3/ha/yr), the high degree of technology involved, and their capacity to engender an integrated and competitive industry. This development, even largely extensible, directly threatens the pulp producers' supremacy of the northern hemisphere, which consume expensive raw materials and are in difficulties for competition.

The investments attracted by these "new forest potentials" are increasing. A recent study (Neilson 2000) shows that *Eucalyptus* especially attracts the multinational corporations and is beginning to affect the financial sectors with the participation of important investment groups. The foresters are not accustomed to such a dynamic, which is specific to *Eucalyptus* profitability. The intrinsic low value of wood is compensated by its strategic industrial interest, its high productivity and its short rotation. In addition, locating the plantations around the mills leads to reduction of costs of transportation at a significant level.

5.2 *Tropical Wood in Reduction*

The deforestation in the tropics is a common phenomenon without any immediate homeostatic feedbacks. The combined result of underdevelopment and population growth, exacerbate the problem. Forest companies are the easy targets of the interest groups, but the real causes, which are the migratory agriculture, the "informal sector" and corruption, are seldom denounced. In any case, timber production from tropical forests should be decreased, even if forest regulation could be respected. In such a context, *Eucalyptus* appears to be a true challenger because it can supply forest industry with substitutes of woods from natural forests. The possibility of producing *Eucalyptus* timber with short rotation would be a new alternative, particularly interesting to the companies lacking of raw materials.

5.3 *"Alternative Demand" for Forests of the North*

In the countries with a high standard of living, the increasing urbanization induces a strong recreational demand, and forest spaces are gradually assimilated to as "wild" patrimony for which ecological value starts to surpass the timber value (Martin 1985). Conservation and biodiversity are new objectives, which increasingly constrain forest management and make the production of wood less competitive by considering other factors (higher manpower costs, silviculture constrained by the size of the clear-cut, limited mechanization, high game density, expensive protections of limiting plantations, eco-certification, etc.). In addition, risks due to climatic changes seem to be confirmed (e.g. storms and fires). Finally, all these constraints encourage in fact the intensive management of plantations, and particularly of *Eucalyptus* in the South. The forests close to the great consumption centres in the North, one would have thought, have competence in terms of timber production. In reality, the demand for other goods and services is greater, and thus downgrades the economic interest of the temperate forests for timber production, as we don't know "if we have enough forests" (Wilson 1996).

Big forest companies, which are subjected to the ecological pressures, learn quickly to respond to the challenge. They establish tree plantations in non-forest zones following agricultural rather than silvicultural principles. In this way, they greatly contribute to reduce the pressures on natural forests. Frequently, planted perimeters surround protected zones (galleries forest) and are integrated into the conservation management plan. Importantly, these plantations, durably managed, are already eco-certified. Some companies even signed the HUGUE's charter by which they engage themselves to consider ecology as one of their management priorities.

6 Conclusions

The irresistible development of *Eucalyptus* seems to be confirmed. Economically, combining fast-growing property with a dense wood, *Eucalyptus* is one of the champions of the biomass production. As a very diversified genus, it lends itself particularly well to genetic improvement and biotechnologies, and so to farming intensification. Widely adopted by the international wood products companies, *Eucalyptus* becomes the green gold of the pulp industry in a short time. *Eucalyptus* plantations continue to attract investment for their great profitability. They go probably to other successes, and seem now ready to win the bet of the solid wood, since the production of high quality sawn wood has already started.

Ecologically, responding to the increasing pulp consumption, *Eucalyptus* already reduces the pressure on the natural forests. Moreover, it will probably become one of the fundamental species for creating the future carbonsinks. Indeed, adapted perfectly well to the savannas afforestation and the recolonisation of abandoned lands by agriculture, *Eucalyptus* is a pioneer who, not only very quickly fixes a maximum of carbon, but also stimulates the recovery of natural vegetations. Endowed with longevity, it could, thanks to an adapted silviculture, lead to stable and bio-diversified forests. *Eucalyptus* is thus on the way to become an important ecological tool to tackle global externalities generated by deforestation and greenhouse effect, by releasing the tropical forests from an excessive economic demand and purifying at the same time the atmosphere of its carbon dioxide excess.

References

Aracruz 2001. Ara Lysted Nyse – Full Year 2000 Results. Conference Call, 17 January 2001, http:\\www.aracruz.com.br.

Bauhus, J., Khanna, P.K. and Menden, N. 2000. Aboveground and belowground interactions in mixed plantations of *Eucalyptus globulus* and *Acacia mearnsii*. Can. J. For. Res. 30:1886-1894.

Binkley, D., Dunklin, K.A., DeBell, D. and Ryan, M.G. 1992. Production and nutrient cycling in mixed plantations of *Eucalyptus* and *Albizia* in Hawaii. For. Sci. 38: 393-408.

BOISmag. 2002. Enquête Brésil: "L'*Eucalyptus* dans tous ses états". BOISmag J. (French magazine), April 2002, No.17.

Bouillet, J.P., Safou-Matondo, R., Laclau, J.P., Nzila,, J.D. and Ranger, J. 2001. Changes over successive rotations in NPK fertiliser needs of Congolese clonal *Eucalyptus* plantations - Relationships with changes in soil fertility. In: Proc. of SIMPOSIO IUFRO, Valdivia, Chile, 3 -15 September 2001.

Campinhos, E. 1987. Introduction of new techniques for vegetative propagation of *Eucalyptus* in China - Field document. FAO Technical Cooperation Programme, F0:TCP/CPR/6653.

Cauvin, B. 1988. Pistil treatments for improved fertility in hybridization of *Eucalyptus gunnii* (Hook). In: "Sexual reproduction in higher plants", R. Cresti, P. Gore and E. Pacine (eds), University of Sienna, Italy.

Cauvin, B. 1991. Effet du paclobutrazole sur la mise à fleurs, la croissance de clones juvéniles d'*Eucalyptus*. In: Annales des Recherches Sylvicoles, AFOCEL, p71-88.

Chaperon, H. 1983. Sélection individuelle et clonage de l'*Eucalyptus*. In: Proc. of AFOCEL-IUFRO Conference, Bordeaux, France, 26-30 September 1983, pp535-558.

Chiang V. L. 2001. Regulation of guaiacyl and syringyl monolignol biosynthesis in trees. In: Proc. of International Wood Biotechnology Symposium, Narita, Chiba, Japan, 14 - 17 March 2001.

Coventry, P. 2001. Forest certification and genetically engineered trees: will the two ever be compatible? Oxford Forest Institute, Occasional Papers No.53.

de Assis, T.F. 2001. Evolution of technology for cloning *Eucalyptus* in large scale. In: Proc. of SIMPOSIO IUFRO, Valdivia, Chile, 3 -15 September 2001.

Dungey, H.S., Potts, B.M., Carnegie, A.J. and Ades, P.K. 1997. Mycosphaerella leaf disease: genetic variation in damage to *Eucalyptus nitens*, *Eucalyptus globulus*, and their F1 hybrid. Can. J. For. Res. 27:750-759.

Eldridge, K.G., Davidson, J., Hardwood, C. and van Wyck,G. 1994. *Eucalyptus* domestication and breeding. Oxford Science Publications, Clarendon Press, Oxford, 288p.

Foelkel, C. and de Assis, T.F. 1995. New pulping technology and *Eucalyptus* wood: The role of soil fertility, plant nutrition and wood ion content. In: Proc. of CRCTHF-IUFRO Conference, Hobart, Australia, 19-24 February 1995, pp10-13.

Franclet, A. 1956. Premier travaux d'Amélioration génétique des *Eucalyptus*. Annales de la Station de Recherche et d'Expérimentation Forestière (Rabat, Maroc), pp65-85.

Franclet, A. 1970. Techniques de bouturage des *Eucalyptus camaldulensis*. Tunis, Institut national de recherches forestières. Note technique No.12.

Grattapaglia, D., Bertolucci, F.L., Penchel, R. and Sederoff, R.R. 1995. Advances in genetic mapping of *Eucalyptus grandis*. In: Proc. of CRCTHF-IUFRO Conference, Hobart, Australia, 19-24 February 1995, pp392-397.

Haines, R.J. 1995. Impediments to the generation and capture of genetic gain. In: Proc. of CRCTHF-IUFRO Conference, Hobart, Australia, 19-24 February 1995, pp479-482.

Haines, R.J. and Martin, B. 1995. Biotechnologie et production soutenue des bois tropicaux. Etude sur documents effectuée pour l'OIBT. Conseil International des Bois Tropicaux, Janvier 1995.

Harbard, J.L., Griffin, A.R. and Espejo, J. 1999. Mass controlled pollination of *Eucalyptus globulus*: a practical reality. Can. J. For. Res. 29:1457-1463.

Harcourt, R., Zhu, X., Llewellyn, D., Dennis, E. and Peacock, J. 1995. Genetic engineering for insect resistance in temperate plantation eucalypts. In: Proc. of CRCTHF-IUFRO Conference, Hobart, Australia, 19-24 February 1995, pp406-408.

Janin,G. 1981. Micro-pulping techniques perfecting and precise pulp yield. Annales des Sciences Forestières (INRA, France) 38:107-126.

Junghans, D.T. 2000. Quantificaçao da severidade, herança da resistencia e identificaçao de marcadores RAPD ligados à resistencia à Ferrugem (Puccinia psidii) em *Eucalyptus grandis*. Tese Doctor Scientiae, Viçosa, Minas Gerais, Brasil.

Marcus Wallenberg Fundation 1984. The New Eucalypt Forest. In: Proc. of Marcus Wallenberg Symposium, Falun, Sweden, 14 September 1984.

Martin, B. and Quillet, G. 1974. Bouturage des arbres forestiers au Congo: Résultats des essais effectués à Pointe Noire de 1969 à 1973. Bois et Forêts des Tropiques, No.154, pp41-57; No.155, pp15-33; No.156, pp39-61;No.157, 21-40.

Martin, B. 1985. Impact de l'Homme sur la Forêt, Ecologie ou Economie? In: Revue Forestière Française, XXXVII, 1/1985.

Martin, B. 1987. Amélioration génétique des *Eucalyptus* tropicaux. In: Contribution majeure à la foresterie clonale, ENGREF, Thèse Université PARIS XI (Orsay France) Mai 1987.

Martin, B. 1991. Ecology and economy synergy in *Eucalyptus* plantations in the Congo. In: Proc. of SIMPOSIO IUFRO, Durban, South Africa, 2-6 September 1991, Vol.2, pp1117-1127.

Medhu-st, J.L. Beadle, C.L. and Neilsen,W.A. 2001. Early-age and later-age thinning affects growth, dominance, and intra-specific competition in *Eucalyptus nitens* plantations. Can. J. For. Res. 31:187-197.

Metro, A. 1955. Eucalypts for planting. FAO Forestry and Forest Products Studies II, FAO, Rome.

Myburg, A.A., Griffin, A.R., Sederoff, R.R. and Whetten, R.W. 2001. Comparative QTL mapping of wood and growth properties in a hybrid of *Eucalyptus grandis* and *Eucalyptus globulus*. In: Proc. of SIMPOSIO IUFRO Conference, Valdivia, Chile, 3-15 September 2001.

Neilson, D. 2000. The global *Eucalyptus* resource and some solidwood-panel product development issues. In: Proc. of IUFRO Conference, Launceston, Australia, 20 March 2000.

Poore, M.E.D. and Fried, C. 1985. The ecological effect of *Eucalyptus*. FAO, Rome, FAO Forestry Paper, No.59.

Raymond, C.R. and Schimleck, L.R. 2002. Development of near infrared reflectance analysis calibrations for estimating genetic parameters for cellulose content in *Eucalyptus globulus*. Can. J. For. Res. 32:170-176.

Rosales, L. and Rojas, J.C. 2001. Resultados a 8 anos de un ensayo de espaciamiento con clones del hybrido urograndis en Smurfit Carton de Venezuela. In: Proc. of SIMPOSIO IUFRO, Valdivia, Chile, 3 -15 September 2001.

Sanhueza, R.P. and Griffin, A.R. 2001. FAMASA fiber yield improvement programme (F.Y.I.P.): 10 years experiences breeding *Eucalyptus globulus*. In: Proc. of SIMPOSIO IUFRO, Valdivia, Chile, 3 -15 September 2001.

Shepherd, M., Chaparro, J., Dale, G., Jefferson, L., Duong, H., Vogel, H., Walsh, J., Gibbings, M. and Teasdale, R. 1995. Mapping insect resistance and essential oil traits in a tropical *Eucalyptus* hybrid. In: Proc. of CRCTHF-IUFRO Conference, Hobart, Australia, 19-24 February 1995, pp420-423.

Schimleck, L.R. and Michell, A.J. 1987. Determination of within-tree variation of kraft pulp yield using near-infrared spectroscopy. TAPPI J. 81: 229-236.

Sedjo, R.A. 2001. From foraging to cropping: the plantation forestry, and implications for wood supply and demand. Unasylva, No.204.

Suarez, D.F. and Bonomelli, P.C. 2001. Evolution of demanda de N, P, K, en los tres primeros anos de crecimiento de *E. nitens* Maiden y *E. globulus* Labill en tres ecosistemas de la VIII Region de Chile. In: Proc. of SIMPOSIO IUFRO, Valdivia, Chile, 3 -15 September 2001.

Teasdale, R.D. 1995. Optimising the application of biotechnology to forest plantations. In: Proc. of CRCTHF-IUFRO Conference, Hobart, Australia, 19-24 February 1995, pp427-429.

Vaillancourt, R.E., Potts, B.M., Manson, A., Eldridge, T. and Reid, J.B. 1995. Using RAPDs to detect QTLs in intrespecific F2 hybrid of eucalypts. In: Proc. CRCTHF-IUFRO Conference, Hobart, Australia, 19-24 February 1995, pp 430-433.

Wilson, S. 1996. Do we have enough forests? In: Leader Forest Resources Project, IUFRO Occasional Paper No.5, Laxenburg, Austria.

Wright, J.A., Di Nicola, T. and Gaitan, E. 2001. Carbon sequestration: new objective for eucalypt tree improvement (abstract). In: Proc. of SIMPOSIO IUFRO Conference, Valdivia, Chile, 3-15 September 2001.

Zhong, C., Gong, M., Xu, D., Chen, Y., Wang, F., Chen, Y., Dell, B., Malajczuk, N., Bougher, N. and Brunndrett, M. 2001. Selection and practice of mycorrhizal fungi for *Eucalyptus* plantations in Southern China. In: Proc. of SIMPOSIO IUFRO, Valdivia, Chile, 3 -15 September 2001.

桉树：一种具战略意义的林木

BERNARD MARTIN

Chargé de mission à l'ENGREF
14, rue Girardet, 54042 Nancy cedex -France

众所周知，在过去的三十年，桉树人工林在亚热带和热带区域得到广泛的推广。结合无性繁殖和目前先进的生物技术，成功的遗传改良大大提高这种树种本身固有的自然特征：速生、优质纤维和相对高的木材密度。这些改进了的特征使桉树成为木纤维培植业的最佳树种之一。

桉树高生产力和短轮伐期、无性系人工林的同一性和改进的木材质量，吸引了许多私营，特别是纸浆工业的投资。纸浆工业仍在持续投资有关研究与开发，以期减少耕作、病害和生态各方面的风险。这种树种的种植开发也吸引了许多财团的兴趣，这些财团统合多国纸浆和造纸企业，携手参与其人工林的发展。

饱受长期以来被生态学家的批评，木纤维培植业现在才逐渐受到肯定。木纤维培植主要利用非林业用地，实质上，这有助于缓解天然林所承受的经济压力，利用非林业用地生产高质量木材则进一步说明了木纤维培植业的生态作用。增加木纤维培植业的经济和生态效益，促进天然林的保护，将是一个巨大的新挑战。最后，桉树可能成为最好的碳吸存树种之一，其人工林的经营，可能也是解决由于燃烧矿物燃料导致的全球性客观环境气候问题的一种很好的生态和经济手段。这种纸浆工业的新绿色黄金，现在可以被堪称为“战略性林木”。

INDUSTRIAL ACACIAS IN ASIA: SMALL BROTHER OR BIG COMPETITOR?

STEPHEN J. MIDGLEY, JOHN W. TURNBULL AND KHONGSAK PINYOPUSARERK

CSIRO Forestry and Forest Products, P.O. Box E4008
Kingston, ACT 2604, Australia

Eucalypts are an established part of the landscape in parts of Asia, particularly China, Viet Nam and Thailand. As small, non-formal plantings they meet local needs for fuelwood, poles, lumber, essential leaf oils and tannins, and as large plantations they meet industrial needs for pulp, paper and reconstituted wood products such as MDF and particleboard.

Acacias have similarly being planted in large plantations in Southeast Asia to meet demands for commercial pulpwood. Over 900,000 ha of *Acacia mangium* have been planted in Indonesia and Malaysia, and about 100,000 ha of *A. crassicarpa* plantations have been established on sites with a high watertable and peaty soils on Sumatra, Indonesia.

Eucalypt and tropical acacia pulps are both high quality and short-fibred, with flexible end uses. Before trial shipments were presented to the European market in 1995, there were few opportunities to commercially assess acacia pulp. Currently about 1.9 million tonnes of tropical acacia pulp are produced and used internationally. When commercial plantations in Indonesia become operational in 2010, this supply of acacia pulp will exceed 4 million tonnes annually. Does this present a challenge to eucalypt pulp that shares many of its attractive features?

The capacity to competitively produce acacia pulp has been underpinned by considerable research, particularly into the domestication of these species. This has included characterisation of natural populations, identification of superior natural provenances, breeding, establishment of seed orchards and cloning of fast-growing hybrids. Nutrition research has demonstrated the value of phosphorus fertilisation on most sites and the benefits of inoculation with selected rhizobia and mycorrhizas. Potentially-damaging pathogens and insects have been identified. Development of management practices to maintain site productivity and plantation yield in successive rotations is critical, particularly for peat soils. There remain significant challenges to the ongoing commercial utilisation of tropical acacias but the high quality of their pulps and the extensive established plantation resource will ensure that their domestication will continue.

1 Introduction

The casuarinas, eucalypts and acacias have added an Australian element to the landscape of many countries in SE Asia (Table 1). In addition to small non-formal plantings for fuel, poles, building material and site protection, industrial plantations of eucalypts and acacias are offering a valuable resource base for high quality industrial pulps. Over the past 40 years, eucalypt-based pulps have become major players in global markets for papermaking fibre. Much of this current industrial resource is based upon eucalypt plantations in South America and Europe. The Asian eucalypt resource available to industry has expanded greatly over the past 10 years and significant investments are being considered (and have been made) in the region, based on this resource.

Table 1. Estimated areas of Australian trees planted in Asia[1] (hectares).

Country	Eucalypts	Acacias	Casuarinas
Bangladesh	12,000		
China	800,000[2]	30,000	300,000
India	2,500,000[3]	45,000	500,000
Indonesia	80,000	800,000	
Laos	10,000		
Malaysia	8,000	200,000	
Myanmar	40,000		
Nepal	6,000		
Pakistan	10,000		
Philippines	10,000	10,000	
Sri Lanka	30,000		3,500
Thailand	450,000	30,000	20,000
Viet Nam	350,000[4]	120,000	120,000

[1] Eucalypt figures derived primarily from papers presented in FAO (1995). Blank spaces indicate that no reliable information has been located.
[2] Does not include four-round community plantings equivalent to approx. 1.2 million ha of plantation.
[3] Dewees and Saxena (1995).
[4] Does not include 'scattered trees' equivalent to 700,000 ha of plantation.

Extensive plantations of tropical acacias are reaching maturity in SE Asia, particularly Indonesia, where plantations of *Acacia mangium* and *A. crassicarpa* provide a resource for the pulp industry. Tropical acacia pulps are now on the international market. Will acacia pulps be regarded as cheap eucalypt pulp or will they establish their own niche? How did the resource for these industry pulps so quickly evolve and what possibilities are there for further improvement?

The genus *Acacia* comprises about 1300 species in three subgenera: Acacia, Aculeiferum and Phyllodineae (Maslin 1995). About 1000 species of *Acacia* (almost all in subgenus Phyllodineae) are native to Australia and neighbouring Indonesia, Papua New Guinea and some Pacific island nations. As the great majority of these species are endemic to Australia they are referred to collectively as 'Australian acacias'. About 300 Australian acacias are trees that grow to a height above 5 m. Several species have been planted as exotics in over 70 countries and just a few, principally *A. mangium*, *A. crassicarpa* and *A. auriculiformis* are being planted more widely for industrial and social purposes in SE Asia.

This paper aims to briefly examine the rapid emergence of acacias in the pulp and paper industry in meeting needs for industrial wood. It will offer an overview and history of the domestication and use of two tropical acacias, *Acacia mangium* and *A. crassicarpa*.

2 Pulps from Tropical Acacias

Tropical acacia pulps are typically short-fibred (0.7 mm), of high density (22 million fibres per gram) and compete in international markets for Bleached Hardwood Kraft Pulps (BHKP). They provide a pulp with very high density of very short fibres, which give smoothness and high levels of opacity (Roberts 2002). The pulp is bright and white and suitable for tissue, fine paper, bible paper, directory stock, envelope paper and carbonless base paper.

In 1995, the first trial shipments of acacia pulp were presented to the European market. Since that time the amount of pulp available internationally has increased steadily. In 2002, the large Indonesia producers will all increase production to an estimated 1.9 million tonnes - the Asian Pulp and Paper owned mills, Indah Kiat and Lontar Papyrus expect to produce 600,000 tonnes of pulp (APP data), APRIL expects to produce 500,000 tonnes, PT Tel will produce 450,000 tonnes and Kiani Kertas expects to produce 360,000 tonnes (Roberts 2002). If expansion plans are achieved and the Indonesian pulp sector becomes largely plantation-based, the quantity of acacia pulp on the international market could grow to in excess of 4 million tonnes by 2010. Although this is only a small part of the current annual world pulp production of 179 million tonnes (Pulp and Paper International, July 2002), it represents a significant new commodity for the Asian region.

Current spot prices[1] for tropical acacia pulps (US$480 per tonne at August 2002) would place the current value of tropical acacia pulp at about US$900 million which could conceivably increase to US$1.9 billion by 2010 if production targets are met and prices maintained.

3 Tropical Acacias as Exotics

Tropical acacias are adapted to a wide range of site conditions in the seasonally-dry and lowland humid tropics. The industrial potential of tropical acacias was recognised in the early 1980s. The first species to be planted extensively was *A. auriculiformis* and it has been planted as an exotic in Asia for more than 70 years. Introduced into Malaysia in 1932 and Thailand in 1935 it has since been planted widely. In India it has become a major plantation species and in China over 50,000 ha of plantations and 3,000 km of roadside plantings have been established. Because it is generally a crooked, heavily branched tree, its use has mainly been as an ornamental along roadsides, for fuelwood and small items of furniture or to

---

[1] The Paperloop home page (http://www.paperloop.com) offers an important observation regarding spot and contract prices. Published spot prices should not be confused with contract transactions. These are transactions between buyers and sellers who do business together on a regular basis. In contrast, spot transactions tend to be one-off deals, with no implied understanding between buyer and seller that they will be doing business regularly in the future. Contract prices account for the large majority of tonnage transacted in most grades. The tonnage that is transacted in spot markets varies significantly over time, by region, and by grade. Spot prices can vary dramatically from what most buyers in the market are paying. The movement of spot prices does, however, provide some information as to the future direction of contract prices.

revegetate denuded land following mining (Booth and Turnbull 1994). It was the introduction of *A. mangium* into Sabah, Malaysia, from Australia in 1966 that created the major interest in tropical acacias. Initially introduced as a plant in firebreaks to protect pine plantations, its rapid growth suggested its potential for wood production and it was planted in Malaysia (Yap 1987). Today there are an estimated 900,000 ha of commercial plantations of *A. mangium* in Indonesia and Malaysia (Turnbull et al. 1998; Barr 2000; CSIRO data).

Based on the early introductions and success of a few of the tropical acacias, there was recognition in the 1970s and 1980s that other acacias had potential as highly adaptable, multipurpose trees to provide fuelwood and industrial wood for the tropics (e.g. Boland and Turnbull 1981). This stimulated major exploration of the rich genetic resources of the genus in Australia and adjacent countries. In the early 1980s CSIRO, with the encouragement of FAO and support of various donor agencies (primarily the Australian Centre for International Agricultural Research) and with the cooperation of national research partners and land managers, began comprehensive seed collections of acacias in the humid tropics of Australia, Indonesia and Papua New Guinea (Turnbull et al. 1983; Skelton 1987; Searle 1989; Gunn and Midgley 1991) and a series of large-scale species elimination and provenance trials.

4 Domestication

The challenging process of exploration and manipulation of the wild genetic resource to derive uses and products for maximum social benefit is regarded as domestication. It is an ill-defined process but it has a clear beginning (the wild plant), a middle (human intervention via propagation, selection and manipulation) and an outcome (enhanced human benefit). It begins with the original variability in the natural forest and ends in productive plantations (Libby 1973) and full tree domestication includes identification and characterisation of its germplasm resources; the capture, selection and management of genetic resources; and the regeneration and sustainable cultivation of the species in managed ecosystems (Leakey and Newton 1994). Figure 1 (adapted from Midgley 1995) illustrates the domestication process that is being applied wholly or in part to acacias. It is difficult to identify discrete stages because plant domestication is a continuum and progress through all stages might not be socially necessary or economically viable.

Trees have been planted for thousands of years for food, wood, shelter, medicines and religious purposes. The first woody plants to be cultivated were those yielding food, such as the olive. Trees were moved around the world by the Romans, Greeks, Chinese and others during military conquests. Voyages of discovery by European navigators to the Americas and Asia provided new opportunities to transfer tree germplasm between distant countries. In contrast to that of most agricultural crops, the principal domestication of trees has occurred in the last one hundred years and only a few species are significantly changed from their wild state. Intensive domestication programmes have been applied only to eucalypts, pines or other species with high value for products such as timber and paper.

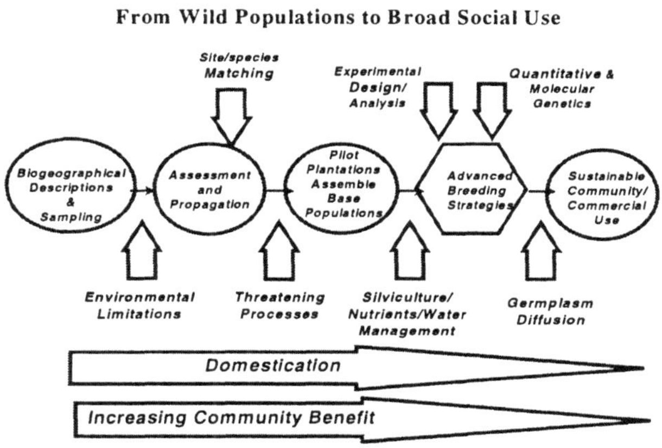

Figure 1. The process of tree domestication.

Domestication of tree species has been a highly variable process. For most multipurpose tree species it may simply consist of identifying good seed sources and developing appropriate propagation and cultural practices. In widely planted and economically important timber species, the full domestication process may involve a systematic sampling and characterization of genetic variation, development of optimal propagation and silvicultural techniques, and intensive breeding, including use of molecular genetics technologies. Domestication seeks to bring out the maximum human benefit within a species as it becomes genetically refined from a wild tree to a cultivated plant. The rapid progress with the domestication of tropical acacias, particularly *A. crassicarpa* that was first planted in species trials in 1984, has been remarkable.

5 Domestication and Use of Acacias: Two Case Studies

5.1 *Acacia mangium Willd.*

This species has been little used in its area of natural occurrence. After an initial introduction to Sabah in 1966, extensive plantations have been established in Malaysia since 1973 and much of the information on silviculture and utilisation of *A. mangium* comes from this area. The initial plantations were established using unselected wild genetic material so the domestication period has been relatively short, but the high value of the plantations and the pulp mills that they feed has resulted in intensive research since about 1980. Much has been learnt about the species in the last 25 years and has resulted in efficient silvicultural practices and the use of selected, highly productive germplasm, including hybrids, sometimes in clonal plantations. Research has focused on the full domestication process involving a systematic sampling and characterisation of genetic variation, intensive breeding, including hybridization and the application of molecular genetics technologies, and development of optimal propagation and silvicultural techniques. Studies have also been made on wood properties and the development of a range of wood products.

A. mangium is a large tree, to 30 m tall, with a straight bole, which may be over half the total tree height but rarely has a diameter over 50 cm. Most natural occurrences are in the coastal, tropical lowlands, usually at an altitude below 300 m. It extends from northern Queensland through the Western Province of Papua New Guinea into the Indonesian provinces of West Papua (Merauke District and Vogelkop Peninsula) and Maluku (Sula, Ceram and the Aru Islands). *A. mangium* grows on the margins of closed forest (rain forest), in open forest and woodland, especially where there is disturbance by fire. It occurs in Papua New Guinea in tall woodland and open forest, frequently in mixed associations with other species of Acacia, Melaleuca and Lophostemon (Skelton 1987). A detailed description of the natural distribution and ecology of *A. mangium* is available in Awang and Taylor (1993).

5.1.1 Sampling and Characterisation of Genetic Variation

Exploration of the natural distribution of *A. mangium* and the systematic collection of seed and botanical material took place mainly between 1980-1990 and is described by Gunn and Midgley (1991). The seed collections were used to establish international provenance trials to examine the patterns of genetic variation in growth rate, stem straightness and frequency of multiple leaders and the extent of genotype x environment interactions (Doran and Skelton 1982). The results of the early trials were reported by Harwood and Williams (1992). There were highly significant differences in performance between experimental sites (19), between provenance regions (5) and among the local provenances within provenance regions. Growth was generally faster at near-equatorial trial sites, with mean annual increment in height around 3-4 m, and slower at sites further from the equator. Papua New Guinea provenances consistently grew fastest, closely followed by the Claudie River provenance from far north Queensland. Significant gains were therefore possible simply by using seed from wild populations in selected localities.

In contrast to large differences in growth performance of families and provenances, allozyme studies indicate low levels of genetic variability within and between populations in this species (Moran et al. 1989a). Harwood and Williams (1992) concluded that allozyme studies examine only a small sub-set of the genome, and therefore do not provide an absolute measure of overall genetic variation, and subsequent studies using restriction fragment length polymorhisms (RFLPs) detected higher levels of variation (Butcher et al. 1998) and high levels of selfing in natural populations (Butcher et al. 1999). Butcher et al. (1996) also used RFLPs to assess genetic variation in the nuclear genome of 10 natural populations and a derived population (an Indonesian seed stand at Subanjeriji, Sumatra). The ranking of populations based on growth performance in provenance trials corresponded to the rankings based on genetic diversity. The low level of diversity in the seed stand reflected the low variation in the Queensland provenances from which it was derived and highlights the danger of establishing breeding populations from a restricted base. Loss of productivity in plantations based on seed from this seed stand has been estimated as about 70% (Turvey 1995). Microsatellite markers have been developed to provide highly variable co-dominant markers for linkage mapping and studies of the breeding system (Butcher and Moran 2000; Butcher et al. 2000). Molecular markers can aid selection at the nursery stage for disease resistance, pulp yield and wood density and provide a means of quality control and monitoring of breeding programs.

5.1.2 Breeding

Since establishment of these trials, further comprehensive seed collections have been undertaken and seedling seed orchards established to produce better quality seed for future planting programs. To support breeding activities, detailed studies of flowering and seeding behaviour have been undertaken (e.g. Sedgley et al. 1992). Flowering and seeding commence at about two years of age under plantation conditions. Pollination is by insects, especially bees, and mature pods can be harvested about seven months after flowering. Natural hybrids between *A. mangium* and *A. auriculiformis* were recorded in Sabah and the potential opportunities and problems noted by Sim (1987). Some hybrids have greater vigour, increased apical dominance, finer branching and greater pulp yields. Research to develop vegetative propagation and controlled seed production technologies began first in Malaysia (Carron and Aken 1992). This work has since gained momentum, especially in Vietnam where outstanding clones have been identified, tested and mass propagated for operational use (Kha 1996, 2001; Nghia 2000). Experimental hybrid seed orchards were established in Indonesia to build up a bank of hybrid material for clonal forestry (Arisman and Havemoller 1994).

5.1.3 Development of Optimal Propagation and Silvicultural Techniques

To fully realize the benefits of the tree breeding and hybridization work an effective and economical means of mass propagation is required. Stem cuttings of young *A. mangium* seedlings are easily rooted when they are treated with hormones and planted in suitable rooting medium. However, the rooting percentage of stem cuttings decreases significantly with older stock plants (see papers in Carron and Aken 1992). There has been significant progress in both micropropagation and conventional root cuttings (e.g. Bon and Montenuuis 1996). The use of cuttings of select trees is now a common feature of commercial plantings of *A. mangium*.

Acacia mangium fixes nitrogen in many tropical soils after nodulating with a range of Rhizobium and Bradyrhizobium strains. It is, however, much more specific in its Rhizobium affinities than some other acacias (Dart et al. 1991). This nitrogen-fixing potential may only be realised in many soils if adequate P and sometimes K fertiliser is applied, but nodule development may be inhibited at higher levels of N. Inoculation with Rhizobium is not usually essential but may be beneficial when seedlings are raised in sterilised media or planted on highly degraded soils or mine spoil. This acacia has associations with both ecto- and endo-mycorrhizal fungi. The ecto-mycorrhizal fungus *Thelephora* spp. forms a beneficial association and *Glomus fasciculatus* and *Gigaspora margarita* have been shown to be effective vesicular arbuscular mycorrhizas (Dela Cruz and Umali-Garcia 1992).

Pests and diseases must be considered as part of the domestication process as they can be major factors in management of plantations. Surveys of insect pests and diseases of *A. mangium* have been made. Foliar diseases, stem cankers, root rots and heart rots have been identified, with root rot widespread and a potentially serious threat to plantations established on recently cut-over forest (Old et al. 2000). There are recent reports that a root rot caused by *Ganoderma* sp. can affect 5-20% of second-rotation *A. mangium* plantations. A sap-sucking bug, *Helopeltis* spp., has been identified as a potentially serious threat in *A. mangium* plantations (Nair and Sumardi 2000).

5.1.4 Utilisation

Acacia mangium is proving to be a very versatile species with a wide range of options for both industrial and local utilisation. Its ecological requirements have been determined and it has proved to be adapted to a wide range of acidic soils (pH 4.0) in the moist tropical lowlands. Its climatic requirements as an exotic are: mean annual temperature 18-28°C, minimum temperature of coldest month 10-24°C, maximum temperature of the hottest month 30-40°C, mean annual precipitation 1000-4000 mm (Booth and Hong 1991).

This acacia has become a major reforestation species in the humid tropical lowlands of Asia and has potential in parts of Africa and South America. Its success is due to its extremely vigorous growth on favourable sites (annual wood volume increment of over 30 m^3/ha); tolerance of acidic soils with low nutrient status; ability to quickly shade out weed competition, such as on Imperata grass; relative freedom from diseases; wood properties which potentially make it acceptable for a wide range of end uses; and ease of establishment. Plantations for chipwood, pulpwood or firewood are not usually thinned or pruned and are harvested at 6-7 years. Saw-log rotations of 15-20 years are envisaged in Malaysia (Mead and Miller 1991).

Basic wood properties have been determined (e.g. Razali and Hamami 1992). Plantation-grown trees have been found to be very promising for the production of unbleached kraft pulp (for bags, wrapping paper, linerboard) and high quality neutral sulphite semichemical pulp (for corrugating, medium and higher-grade products) (Logan 1987; Clark et al. 1994). Most of the large-scale plantations in Malaysia and Indonesia are for the production of paper pulp.

This species produces charcoal of reasonable quality and is suitable for the manufacture of wood pellets and activated carbon (Udarbe and Hepburn 1987). Wood of young trees is not classed as a superior firewood (Gough et al. 1989). The timber can be sawn easily, planed to a smooth surface, polished, drilled and turned. The wood is fairly stable, and it dries without serious defects when suitable kiln schedules are used (Razali and Hamami 1992). It is very durable when exposed to weather but not in contact with the ground. The heartwood is moderately resistant to preservative treatment. It makes excellent particle board, and is suitable for furniture and cabinet-making, light structural works, agricultural tools, boxes and crates. It has been sliced for veneer and was used locally for house construction at one time in Australia.

With its dense foliage, which is retained throughout the year, *A. mangium* makes a useful shade, screening, and soil-cover crop. It has been used experimentally as shade for cocoa in Sabah. Its flowers are a source of bee forage and contribute to honey production.

5.2 *Acacia crassicarpa A.Cunn. ex Benth.*

In 1980 *A. crassicarpa* was almost of insignificant commercial importance in native forests and was best known by botanists and ecologists through its close association with its better-known cousins, *A. mangium* and *A. auriculiformis*. Over the past twenty years, it has gone from a virtually unknown tree in the wilds of north Queensland and New Guinea to a major commercial plantation species for pulp and paper in Southeast Asia. Its natural occurrence in the humid tropics, good bole form and vigour on poorly drained, acidic sites suggested it could find a place in tropical forestry.

It was first tested in Thailand in the mid 1980s and subsequently in neighboring countries such as Indonesia, Malaysia and Vietnam. Over 100,000 ha of commercial plantations have now been established in Sumatra, Indonesia, where it is being grown on highly organic soils with a low pH that may be waterlogged occasionally (CSIRO data; Midgley 2000). Unlike *A. mangium*, the most productive provenances were identified before large-scale planting commenced. Domestication has progressed to the point of establishment of seedling seed orchards but other research has been less intensive than for *A. mangium*. *A. crassicarpa* illustrates the potential for rapid domestication of fast-growing, early-flowering tropical tree species. This has involved substantial R&D work in Australia, Indonesia, Malaysia, Thailand, Vietnam and China. Much of the early work was initially within projects between CSIRO and research partners supported by ACIAR, then under the auspices of the Consultative Group for Research and Development of Acacia (COGREDA), national institutes and most recently in the private pulp and paper sector.

In its natural stands, *A. crassicarpa* is a moderately large tree up to 30 m tall which reaches its best development in the south of Papua New Guinea and West Papua, Indonesia. In north Queensland it is a component of many diverse vegetation associations on a variety of soil types but in Papua New Guinea (PNG) it occurs in woodlands and open-forests commonly fringing seasonally flooded plains. It occurs frequently in mixed associations with *A. auriculiformis*, *A. mangium*, *A. leptocarpa* and species of Melaleuca and Lophostemon (Thomson 1994; Doran and Turnbull 1997).

5.2.1 Sampling and Characterisation of Genetic Variation

The first seed collections of *A. crassicarpa* destined for forestry research were collected for CSIRO in north Queensland in 1981. The FAO Forestry Department and the Danish aid agency, DANIDA, shared an enthusiasm for the potential of tropical acacias for reforestation, and in 1982 supported joint CSIRO/PNG Forest Research Institute collections to complete the sampling of *A. crassicarpa* in

Australia and PNG (Turnbull et al. 1983). Further collections by the CSIRO Australian Tree Seed Centre followed over the next 10 years with donor support from AusAID and ACIAR (Gunn and Midgley 1991). These included the first collections of *A. crassicarpa* in Irian Jaya, Indonesia, in 1990 in collaboration with the Indonesian Ministry of Forestry (Vercoe and McDonald 1991).

Species/provenance trials in Thailand clearly demonstrated the outstanding growth potential of PNG provenances of *A. crassicarpa*, with early height growth rates of 5 m per year being recorded (Pinyopusarerk 1989). Interest in *A. crassicarpa* grew rapidly once high quality, source-identified seed was available from natural provenances. By 1990, the ATSC had sent out over 500 research seedlots, primarily to research partners in South-East Asia and China. By 1993 the species had demonstrated excellent survival and vigour in trials across a range of sites in the humid tropics and was recorded by COGREDA as the third most important tropical acacia in Southeast Asia (Awang and Taylor 1993). Its suitability for shallow sandy soils (e.g. Sim and Gan 1991), peaty soils and the overwhelming superiority of the PNG provenances had been demonstrated (e.g. Chittachumnonk and Sirilak 1991). However, a comparison of the various provenance trials shows there is a strong genotype x environment interaction and not all PNG provenances grow well on all sites (Awang et al. 1998). Queensland provenances appear to be more resistant to cyclone damage than the faster-growing PNG provenances. Provenance variation in economically important wood properties has been demonstrated in Malaysia (Shukor et al. 1998). The same study identified Samlleberr (West Papua) and Olive River (north Queensland) as the most promising seed sources for growth and wood properties for industrial plantations. Seed orchards have been established by CSIRO and research partners in Thailand, Vietnam and Philippines and by commercial companies in Indonesia.

5.2.2 Breeding

The breeding system and genetic diversity were studied by Moran et al. (1989b). Seedling seed orchards were first established in China (Pan 1988) and subsequently in Australia (Harwood et al. 1993), Indonesia (Nirsatmanto 1998), Philippines, Thailand and Vietnam. *A. crassicarpa* is closely related to other commercially important Acacia species (McDonald and Maslin 2000) and there is potential for intra-specific hybrids with *A. auriculiformis*, *A. peregrina*, *A. mangium* and *A. midgleyi* to improve certain attributes or perhaps utilise heterosis for growth as observed in the *A. auriculiformis* x *A. mangium* hybrid in Vietnam. In drier regions, *A. crassicarpa* puts a large amount of photosynthate (and nutrients) into heavy, woody fruit crops after 3-4 years. Creation of sterile lines, e.g. triploids as in the approach of the Pacific Australia Reforestation Company and other companies, may maintain growth rates and wood production in sexually mature trees, as well as increase stress tolerance and minimise any weediness potential.

5.2.3 Development of Optimal Propagation and Silvicultural Techniques

Cuttings can be successfully rooted from trees less than three years old but are difficult to root when taken from older trees (Thomson 1994). Marcotting is effective for vegetative propagation from older trees (Awang et al. 1998).

Maintenance of juvenility in hedge plants is a significant challenge to development of fully-clonal planting programs.

Acacia crassicarpa is susceptible to some foliar disease and root rots but has not yet been damaged by heart rot (Old et al. 2000). Insect pests such as stem borer beetles (*Platypus* spp.) and the stem-girdling beetle (*Sinoxylon* sp.) have been reported (Thomson 1994) but so far damage by pests has been localized.

5.2.4 Utilisation

The utilisation potential of *A. crassicarpa* has yet to be fully explored. It is clearly very useful for pulp. While it has rather similar climatic requirements to *A. mangium* its adaptability to infertile, highly organic soils with a low pH that may be waterlogged occasionally, and to very sandy soils where *A. mangium* does not thrive, give it an advantage in these situations.

It has a dense wood (basic density +/-520 kg/m^3) and the heartwood is golden brown with a reddish cast, giving it potential for furniture making. The wood is strong and durable and has been used for heavy construction, boat building, flooring, hardboard and veneer. In the early 1990s the first report on the pulping and paper-making qualities of *A. crassicarpa* was published (Clark et al. 1991, 1994). Although *A. crassicarpa* has a higher density and a lower pulp yield than the more widely-planted *A. mangium*, its mean annual growth rates are high (>25 m^3/ha) so the yield of pulp per hectare remains acceptable.

The tree has outstanding, as yet largely untapped, potential for other very difficult sites in South-East Asia. These include planting above saline seeps in northeast Thailand for groundwater recharge control, and reforestation of acidic, infertile sands with seasonal waterlogging and shallow hardpan in central Vietnam (Nghia 2000). *A. crassicarpa* also has potential in agroforestry and is now being grown for fuelwood in woodlots by 3,000 farmers in Tabora district, Tanzania, following its introduction in 1988 by the International Centre for Research in Agroforestry (D.J. Boland, pers. comm.).

It has been just 18 years since the first species/provenance trials of *A. crassicarpa* were established. In that short time, the natural variation within the species has been assessed, breeding programs established, molecular marker technologies established, taxonomy clarified, silvicultural studies completed and wood and fibre properties assessed. Such rapid progress in the domestication of a tree species is remarkable. Given the recognised economic importance of the species it is likely that much of the future domestication of *A. crassicarpa* will now be undertaken within the private sector, both by large pulp corporations and smaller growers. The short generation times make the species amenable to rapid tree improvement, for example for better pulp characteristics.

6 Concluding Comments

Tropical acacias have proven to be highly attractive for production of high value, short fibred pulps, and already-extensive plantations are expanding to meet market need.

Do the tropical acacia pulps compete with the better-established eucalypt pulps? Paper and board consumption in Asia is rising, following the established trend of increasing with growing GDPs. China's per capita consumption has increased to 33 kg annually (Jaako Poyry Consulting, pers. comm.) and it stands as the unchallenged market force in the region. By 2010, the market in China alone will be some 70 million tonnes into which the supply of 4 million tones of high quality acacia pulp could find ready markets. Acacia pulps blend well with eucalypts and the indications are that the two resources will be complementary rather than competitive.

The tropical acacias have become an established part of the commercial plantation industry in SE Asia. Their rapid domestication and commercialization has been remarkable. It is likely that several generations of selection, with the starting point of a broad genetic base, will lead to the development of better-adapted land races for specific, difficult or unusual sites.

References

Arisman, H. and Havemoller, P. 1994. Seed supply strategy for a pulpwood plantation project in southern Sumatra. In: Proc. of an International Symposium on "Genetic conservation and production of tropical forest tree seed", R.M. Drysdale and S.E.T. John (eds), Chiang Mai, 14-16 June 1993, ASEAN-Canada Forest Tree Seed Center, Saraburi, Thailand, pp167-169.

Awang, K. and Taylor, D. (eds) 1993. *Acacia mangium* growing and utilization. Winrock International and FAO, Bangkok, MPTS Monograph Series No.3, 280p.

Awang, K., Jamahari, S., Zulkifli, A.A. and Shukor, N.A.A. 1998. Growth, marcottability and photosynthetic rate of *Acacia crassicarpa* provenances at Serdang, Malaysia. In: "Recent developments in *Acacia* planting", J.W. Turnbull, H.R. Crompton and K. Pinyopusarerk (eds), Australian Centre for International Agriculture Research, Canberra, ACIAR Proc. No.82, pp299-304.

Barr, C. 2000. Profits on paper: The political-economy of fibre, finance and the debt in Indonesia's pulp and paper industries. Center for International Forestry Research, Bogor, Indonesia, 52p.

Boland, D.J. and Turnbull, J.W. 1981. Selection of Australian trees other than eucalypts for trials as fuelwood species in developing countries. Aust. For. 44:235-246.

Bon, M.C. and Montenuuis, O. 1996. Biotechnologies au forestieres au Sabah. Bois et Forets des Tropiques No.248, pp31-42.

Booth, T.H. and Hong, Y. 1991. Identifying climatic areas in China suitable for *Acacia mearnsii* and *A. mangium*. In: "Advances in tropical *Acacia* research", J.W. Turnbull (ed), Australian Centre for International Agriculture Research, Canberra, ACIAR Proc. No.35, pp52-56.

Booth, T.H. and Turnbull, J.W. 1994. Domestication of lesser-known tropical tree species: the Australian experience. In: "Tropical trees: The potential for domestication and the rebuilding of forest resources", R.R.B. Leakey and A.C.

Newton (eds), Proc. of ITE symposium No.29, ECTF symposium No.1, Edinburgh, 23-28 August 1992, Her Majesty's Stationery Office, London, pp189-194.

Butcher, P.A. and Moran, G.F. 2000. Genetic linkage mapping in *Acacia mangium* 2. Development of an integrated map from two outbred pedigrees using RFLP and microsatellite loci. Theor. Appl. Genet. 101:594-605.

Butcher, P.A., Moran, G.F. and Perkins, H.D. 1996. Genetic resources and domestication of *Acacia mangium*. In: "Tree improvement for sustainable tropical forestry", M.J. Dieters, A.C. Matheson, D.G. Nikles, C.E. Harwood and S.M. Walker (eds), Proc. of QFRI-IUFRO Conference, Caloundra, Australia, October/November 1996, Queensland Forestry Research Institute, Gympie, Australia, pp467-471.

Butcher, P.A., Moran, G.F. and Perkins, H.D. 1998. RFLP diversity in the nuclear genome of *Acacia mangium*. Heredity 81:205-213.

Butcher, P.A., Moran, G.F. and Bell, R. 2000. Genetic linkage mapping in *Acacia mangium* 1. Evaluation of restriction endonucleases, inheritance of RFLP loci and their conservation across species. Theor. Appl. Genet. 100:576-583.

Butcher, P.A., Swann, K., Morosin, G. and Moran, G.F. 1999. Evidence for the evolution of high rates of selfing in native populations of *Acacia mangium* – implications for domestication. Poster presented at the IUFRO Conference on Forest Biotechnology, Oxford, 10-16 July 1999.

Carron, L.T. and Aken, K.M. (eds) 1992. Breeding technologies for tropical acacias. In: Proc. of an International Workshop held in Tawau, Sabah, Malaysia, 1-4 July 1991, Australian Centre for International Agriculture Research, Canberra, ACIAR Proc. No.37, 132p.

Chittachumnonk, P. and Sirilak, K. 1991. Performance of *Acacia* species in Thailand. In: "Advances in tropical *Acacia* research", J.W. Turnbull (ed), Australian Centre for International Agriculture Research, Canberra, ACIAR Proc. No.35, pp153-158.

Clark, N.B., Balodis, V., Fang, G. and Wang, J. 1991. Pulping properties of tropical acacias. In: "Advances in tropical *Acacia* research", J.W. Turnbull (ed), Australian Centre for International Agriculture Research, Canberra, ACIAR Proc. No.35, pp138-144.

Clark, N.B., Balodis, V., Fang, G. and Wang, J. 1994. Pulping potential of acacias. In: "Australian tree species research in China", A.G. Brown (ed), Australian Centre for International Agriculture Research, Canberra, ACIAR Proc. No.48, pp196-202.

Dart, P., Umali-Garcia, M. and Almendras, A. 1991. Role of symbiotic associations in nutrition of tropical acacias. In: "Advances in tropical *Acacia* research", J.W. Turnbull (ed), Australian Centre for International Agriculture Research, Canberra, ACIAR Proc. No.35, pp13-19.

Dela Cruz, R.E. and Umali-Garcia, M. 1992. Nitrogen fixation and mycorrhizae in acacias on degraded grasslands. In: "Tropical acacias in East Africa and the Pacific", K. Awang and D.A. Taylor (eds), Winrock International Institute for Agricultural Research, Bangkok. pp59-71.

Dewees, P.A. and Saxena, N.C. 1995. Wood product markets as incentives for farmer tree growing. In: "Tree management in farmer strategies: responses to agricultural intensification", J.E.M. Arnold and P.A. Dewees (eds), Oxford University Press, Oxford, pp198-241.

Doran, J.C. and Skelton, D.J. 1982. *Acacia mangium* seed collections for international provenance trials. Rome: FAO, Forest Genetic Resources Information No.11, pp47-53.

Doran, J.C. and Turnbull, J.W. (eds) 1997. Australian trees and shrubs: species for land rehabilitation and farm planting in the tropics. Australian Centre for International Agriculture Research, Canberra, ACIAR Monograph No.24, 384p.

FAO. 1995. Proc. of the Regional Expert Consultation on *Eucalyptus*, 4-8 October 1993, FAO Regional Office for Asia and the Pacific, Bangkok, 196p.

Gough, D.K., Bell, R.E., Ryan, P.A. and Bragg, C.T. 1989. Drying and burning properties of some Australian tree species. In: "Trees for the tropics", D.J. Boland (ed), Australian Centre for International Agriculture Research, Canberra, ACIAR Monograph No.10, pp27-34.

Gunn, B.V. and Midgley, S.J. 1991. Exploring and assessing the genetic resources of four selected tropical acacias. In: "Advances in tropical *Acacia* research", J.W. Turnbull (ed), Australian Centre for International Agriculture Research, Canberra, ACIAR Proc. No.35, pp57-63.

Harwood, C.E. and Williams, E.R. 1992. A review of provenance variation in growth of *Acacia mangium*. In: "Advances in tropical *Acacia* research", J.W. Turnbull (ed), Australian Centre for International Agriculture Research, Canberra, ACIAR Proc. No.35, pp22-30.

Harwood, C.E., Haines, M.W. and Williams, E.R. 1993. Early growth of *Acacia crassicarpa* in a seedling seed orchard at Melville Island, Australia. FAO Forest Genetic Resources Information No.21:46-53.

Kha, L.D. 1996. Studies on natural hybrids of *Acacia mangium* and *A. auriculiformis* in Vietnam. In: "Tree improvement for sustainable tropical forestry", M.J. Dieters, A.C. Matheson, D.G. Nikles, C.E. Harwood and S.M. Walker (eds), Proc. of QFRI-IUFRO Conference, Caloundra, Australia, October/November 1996, Queensland Forestry Research Institute, Gympie, Australia, pp328-332.

Kha, L.D. 2001. Studies on the use of natural hybrids between *Acacia mangium* and *Acacia auriculiformis* in Vietnam. Agricultural Publishing House, Hanoi, Vietnam, 171p.

Leakey, R.R.B. and Newton, A.C. 1994. Domestication of 'Cinderella' species as the start of a woody-plant revolution. In: "Tropical trees: the potential for domestication and the rebuilding of forest resources", R.R.B. Leakey and A.C. Newton (eds), Proc. of ITE Symposium No.29, ECTF Symposium No.1, Edinburgh, 23-28 August 1992, Her Majesty's Stationery Office, London, pp3-15.

Libby, W.J. 1973. Domestication strategies for forest trees. Can. J. For. Res. 3:265-276.

Logan, A.F. 1987. Australian acacias for pulpwood. In: "Australian acacias in developing countries", J.W. Turnbull (ed), Australian Centre for International Agriculture Research, Canberra, ACIAR Proc. No.16, pp89-94.

Maslin, B.R. 1995. Systematics and phytogeography of Australian species of Acacia: an overview. Inst. For. Aust. Newsl. 36:2-5.

McDonald, M.W. and Maslin, B. 2000. A revision of the salwoods: *Acacia aulacocarpa* Cunn. ex Benth. and its allies (Leguminosae: Mimosoideae: section Juliflorae). Aust. Syst. Bot. 13:21-78.

Mead, D.J. and Miller, R.R. 1991. The establishment and tending of *Acacia mangium*. In: "Advances in tropical *Acacia* research", J.W. Turnbull (ed), Australian Centre for International Agriculture Research, Canberra, ACIAR Proc. No.35, pp116-122.

Midgley, S.J. 1995. Seed collection strategies in a changing world. In: "Innovations in tropical tree seed technology", K. Olesen (ed.), Proc. of IUFRO Working Group 2.04, International Tree Seed Symposium, Arusha, Tanzania, 7-10 September 1995, Danida Forest Seed Centre, Humlebaek, Denmark, pp175-186.

Midgley, S.J. 2000. *Acacia crassicarpa*: a tree in the domestication fast lane. CSIRO Forestry and Forest Products, Canberra, Australian Tree Resources News 6:1-2.

Moran, G.F., Muona, O. and Bell, J.C. 1989a. *Acacia mangium*: a tropical forest tree of low genetic diversity. Evolution 43:231-235.

Moran, G.F., Muona, O. and Bell, J.C. 1989b. Breeding systems and genetic diversity in *Acacia auriculiformis* and *A. crassicarpa*. Biotropica 21:250-256.

Nair, K.S.S. and Sumardi, 2000. Insect pests and diseases of major plantation species. In: "Insect pests and diseases of Indonesian forests", K.S.S. Nair (ed), Center for International Forestry Research, Bogor, Indonesia, pp15-38.

Nghia, N.H. 2000. Successful development of Acacia species in Vietnam. APAFRI Publication Series No.5, APAFRI Secretariat, Serdang, Malaysia, 33p.

Nirsatmanto, A. 1998. Growth and performance of *Acacia crassicarpa* seedling seed orchards in South Sumatra, Indonesia. In: "Recent developments in *Acacia* planting", J.W. Turnbull, H.R. Crompton and K. Pinyopusarerk (eds), Proc. of an International Workshop held in Hanoi, Vietnam, 27-30 October 1997, Australian Centre for International Agriculture Research, Canberra, ACIAR Proc. No.82, pp359-362.

Old, K.M, Lee, S.S., Sharma, J.K. and Zi, Q.Y. 2000. A manual of diseases of tropical acacias in Australia, South-East Asia and India. Center for International Forestry Research, Bogor, Indonesia, 104p.

Pan, Z. 1988. Preliminary report on the 3-year-old provenance tests of tropical acacia in China. For. Res. 1:553-558.

Pinyopusarerk, K. 1989. Growth and survival of Australian tree species in field trials in Thailand. In: "Trees for the tropics – growing Australian multipurpose trees and shrubs in developing countries", D.J. Boland (ed), ACIAR Monograph No.10, pp109-127.

Razali, A.K. and Hamami, S.M. 1992. Processing and utilisation of acacias, focussing on *Acacia mangium*. In: "Tropical acacias in East Africa and the

Pacific", K. Awang and D.A. Taylor (eds), Winrock International Institute for Agricultural Research, Bangkok, pp86-91.
Roberts, J. 2002. The case for *Acacia*. Pulp and Paper International (online edition), May 2002, www.paperloop.com.
Searle, S.D. 1989. Seed collections of lesser-known trees and shrubs in Queensland, Australia. In: "Trees for the tropics", D.J. Boland (ed), Australian Centre for International Agriculture Research, Canberra, ACIAR Monograph No.10, pp27-34.
Sedgley, M., Yong, W.C., Newman, V., Harbard, J., Smith, R.M., Ghan, K.K. and Tajuddin, A. 1992. Phenology of *Acacia mangium* and *A. auriculiformis* in Australia and Malaysia. In: "Breeding technologies for tropical acacias", L.T. Carron and K.M. Aken (eds), Australian Centre for International Agriculture Research, Canberra, ACIAR Proc. No.37, pp36-44.
Shukor, N.A.A., Nang, A.N. and Awang, K. 1998. Selected wood properties of *Acacia auriculiformis* and *A. crassicarpa* provenances in Malaysia. In: Recent Developments In: "*Acacia* planting", J.W. Turnbull, H.R. Crompton and K. Pinyopusarerk (eds), Proc. of an International Workshop held in Hanoi, Vietnam, 27-30 October 1997, Australian Centre for International Agriculture Research, Canberra, ACIAR Proc. No.82, pp155-160.
Sim, B.L. 1987. Research on *Acacia mangium* in Sabah: a review. In: "Australian acacias in developing countries", J.W. Turnbull (ed), Australian Centre for International Agriculture Research, Canberra, ACIAR Proc. No.16, pp164-166.
Sim, B.L. and Gan, E. 1991. Performance of *Acacia* species on four sites of Sabah Forest Industries. In: "Advances in tropical *Acacia* research", J.W. Turnbull (ed), Australian Centre for International Agriculture Research, Canberra, ACIAR Proc. No.35, pp159-165.
Skelton, D.J. 1987. Distribution and ecology of Papua New Guinea acacias. In: "Australian acacias in developing countries", J.W. Turnbull (ed), Australian Centre for International Agriculture Research, Canberra, ACIAR Proc. No.16, pp38-44.
Thomson, L.A.J. 1994. *Acacia aulacocarpa*, *A. cincinnata*, *A. crassicarpa* and *A. wetarensis*: an annotated bibliography. Australian Tree Seed Centre, CSIRO, Canberra, 131p.
Turnbull, J.W., Skelton, D.J., Subagyono, M. and Hardiyanto, E.B. 1983. Seed collections of tropical acacias in Indonesia, Papua New Guinea and Australia. Rome: FAO, Forest Genetic Resources Information No.12, pp2-15.
Turnbull, J.W., Midgley, S.J. and Cossalter, C. 1998. Tropical acacias planted in Asia: an overview. In: "Recent developments in *Acacia* planting", J.W. Turnbull, H.R. Crompton and K. Pinyopusarerk (eds), Proc. of an International Workshop held in Hanoi, Vietnam, 27-30 October 1997, Australian Centre for International Agriculture Research, Canberra, ACIAR Proc. No.82, pp14-28.
Turvey, N.D. 1995. Afforestation of *Imperata* grasslands in Indonesia: results of industrial tree plantation trials in Teluk Sirih on Palau Laut, Kalimantan Selatan. Australian Centre for International Agriculture Research, Canberra, ACIAR Technical Report No.16, 43p.

Udarbe, M.P. and Hepburn, A.J. 1987. Development of *Acacia mangium* as a plantation species in Sabah. In: "Australian acacias in developing countries", J.W. Turnbull (ed), Australian Centre for International Agriculture Research, Canberra, ACIAR Proc. No.16, pp157-159.

Vercoe, T.K. and McDonald, M.W. 1991. New introductions of *Eucalyptus* and *Acacia* from Papua New Guinea. FAO, Forest Genetic Resources Information No.19, 5p.

Yap, S.K. 1987. The introduction of *Acacia* species to Peninsular Malaysia. In: "Australian acacias in developing countries", J.W. Turnbull (ed), Australian Centre for International Agriculture Research, Canberra, ACIAR Proc. No.16, pp151-153.

Some Useful References for Tropical Acacias

Awang, K. and Taylor, D.A. (eds) 1992. Tropical acacias in East Asia and the Pacific. In: Proc. of the First Meeting of the Consultative Group for Research and Development of Acacias (COGREDA), held in Phuket, Thailand, 1-3 June 1992, Winrock International, Bangkok, 106p.

Awang, K. and Taylor, D.A. (eds) 1993. Acacias for rural, industrial, and environmental development. In: Proc. of the Second Meeting of the Consultative Group for Research and Development of Acacias (COGREDA), held in Udorn Thani, Thailand, 15-18 February 1993, Winrock International and FAO, Bangkok, 258p.

Awang, K. and Taylor, D.A. (eds) 1993. *Acacia mangium* growing and utilization. Winrock International and FAO, MPTS Monograph Series No. 3, Bangkok, 280p.

CAB International 1998. The forestry compendium – A silvicultural reference. Module 1. 1998. Wallingford, England. CD Rom.

CAB International 2000. The forestry compendium Global Module 2000. Wallingford, England, CD Rom.

Maslin, B.R. and McDonald, M.W. 1996. A key to useful Australian acacias for the seasonally dry tropics. Department of Conservation and Land Management, Bentley, Western Australia and CSIRO Forestry and Forest Products, Canberra, 80p.

Midgley, S.J. and Pinyopusarerk, K. 2002. The Australian Tree Seed Centre and partnerships in Asia. Address to a seminar celebrating the 40th year of activities of the Australian Tree Seed Centre, CSIRO Forestry and Forest Products, 5p.

Midgley, S.J. and Turnbull, J.W. 2003. Domestication and use of Australian acacias: case studies of five important species. Aust. J. Syst. Bot. 16: 89-102.

Pinyopusarerk, K. 1990. *Acacia auriculiformis*: An annotated bibliography. Winrock International (F/FRED) and Australian Centre for International Agriculture Research, Bangkok, 154p.

Thomson, L.A.J., Midgley, S.J., Pinyopusarerk, K. and Kalinganire, A. 2001. Tree domestication: the Australian experience in partnerships with special reference to the Asia-Pacific region. Invited paper to the South-East Asian Workshop on

"Conservation, management and utilization of forest genetic resources", Thailand, 25 February - 10 March 2001.
Turnbull, J.W. (ed) 1987. Australian acacias in developing countries. In: Proc. of an International Workshop held at the Forestry Training Centre, Gympie, Qld., Australia, 4-7 August 1986, Australian Centre for International Agriculture Research, Canberra, ACIAR Proc. No.16, 196p.
Turnbull, J.W. (ed.) 1991. Advances in tropical acacia research. In: Proc. of a Workshop held in Bangkok, Thailand, 11-15 February 1991, Australian Centre for International Agriculture Research, Canberra, ACIAR Proc. No.35, 234p.
Turnbull, J.W., Crompton, H.R. and Pinyopusarerk, K. (eds) 1998. Recent developments in acacia planting. In: Proc. of an International Workshop held in Hanoi, Vietnam, 27-30 October 1997, Australian Centre for International Agriculture Research, Canberra, ACIAR Proc. No.82, 383p.
Wood, H. and Awang, K. (eds) 1998. Acacias for amenity planting and environmental conservation. In: Proc. of the Third Meeting of the Consultative Group for Research and Development of Acacias (COGREDA), Taipei, Taiwan, ROC, 28-29 June 1994, APAFRI Publication Series No.1, 42p.

亚洲的工业用相思：同业友好还是强大竞争对手?

STEPHEN J. MIDGLEY，JOHN W. TURNBULL，
KHONGSAK PINYOPUSARERK

CSIRO Forestry and Forest Products, P.O. Box E4008
Kingston, ACT 2604, Australia

桉树在部分亚洲国家(特别是中国、越南及泰国)已是相当成功的观景树种。如种植规模较小，桉树可以作为当地居民所需要的薪炭材、柱材、木材、桉叶精油和丹宁之用；如种植规模较大，桉树则可以作为纸浆、造纸、纤维板、刨花板等工业生产之原料。

一直以来，亚洲东南部的均有大规模种植相思人工林，以生产商业造纸材为目的。迄今，印度尼西亚和马来西亚两地已种植了超过 900,000 公顷的马占相思，而印度尼西亚苏门答腊岛具有高水位及肥沃泥炭土的地方，则种植了约 100,000 公顷的厚荚相思。

桉树和热带相思均可用作生产高质量短纤维的纸浆，所产纸浆用途广泛。在 1995 年相思纸浆试销欧洲市场前，要从商业上评估相思纸浆并不容易，如今国际市场上已能够生产并耗用差不多 1,900,000 吨热带相思纸浆。当印度尼西亚的工业人工林于 2010 年逐渐成熟可以持续砍伐时，热带相思纸浆的年产量将超过 4 百万吨。这一情况会否对有着同样优势的桉树纸浆业带来挑战?

大量的研究，特别是有关相思树种驯化的研究大大加强了相思纸浆在市场上的竞争力，这些研究包括对其天然群体的特性描述、优良种源的鉴别、种子园的建立以及速生杂种的无性繁殖等。养分研究已显示在大多数立地施用磷肥的价值，并验证选择根瘤生物和菌根来接种的益处。有关研究也鉴别出病虫害可能带来的潜在损失。合适的经营作业对维持多轮伐期的林地生产力和林分产量至关重要，对泥炭土壤尤为如此。虽然现今热带相思在商业应用上仍面对着不断的挑战，但是其高质量的纸浆和已经营造的大量人工林资源则保证相思树种的驯化栽培将不断继续。

BRIEF INTRODUCTION OF EUCALYPT CULTIVATION AND UTILIZATION IN CHINA

SHUXIONG QI

Advisers Committee, Sino-Forest Corporation
Zhangjiang 524022, Guangdong, China

Eucalyptus belongs to the Myrtaceae family, with most of species naturally distributed in mainland Australia, while a few species are from Timor, Indonesia and Papua New Guinea. It was discovered and named in 1770. *Eucalyptus* originated at the end of the Cretaceous period. For plant species with oblique and vertical veining already existed in the Eocene and Miocene age, the primitive *Eucalyptus* had the structural characteristics similar to those of the Mesozoic plants, evolving along with the changing of the geologic ages in Australia. In 1982, the Palaeophytology Group, the Qinhai-Tibet Plateau United Study Team of the Chinese Academy of Science found more than 40 *Eucalyptus* fossils in the Upper Eocene stratum in Litang County, West of Sichuan Province. These fossils included leaves, fruits and buds. The fossil was named as *Eucalyptus reluensis*. It could be assumed there were large areas of eucalypts forest in Tibet 40 to 50 million years ago. The eucalypts gradually vanished in the region due to the weather being colder that was associated with the rising of the Himalayas. The natural distribution of *Eucalyptus* moved down to the south through Malaysia, and Australia now becomes the center of the natural distribution.

For its forest characteristics and economic value, eucalypt have been introduced and cultivated in more than one hundred countries or regions. Several special international symposiums on eucalypt were held since 1950s, for example, those in Rome, 1956, St. Paul, 1961, Canberra, 1967, Lisbon, 1979, Zhanjiang, 1990, Brisbane, 1996 and Brazil, 1998. Knowledge and experiences were exchanged among scientists and planters from different countries at these conferences, which greatly stimulated the development of eucalypt introduction and cultivation all over the world.

Eucalypts were introduced into China in 1890. While several species were brought to Guangzhou, Hong Kong and Macao from Italy, *E. tereticonis* was introduced to Longzhou, Guangxi from France. *E. rudis* was introduced to Fuzhou in 1894, *E. globulus* to Kunming in 1896, and *E. camandulensis* to Sichuan in 1910. *E. robusta* was once planted along Yue-Han Railway from Guangzhou to Henyang. In addition, the records and documents showed early introduction or planting in Santou, Meixian and Zhanjiang in Guangdong, Haikou in Hainan, Nanning, Liuzhou and Beihai in Guangxi, Nanping in Fujian, Wenzhou in Zhejiang, Ganzhou in Jiangxi, and in Chongqin. Zong-Lian Wu, the Chinese Ambassador of the Qing Dynasty in Italy, translated a eucalypt book into Chinese named as "An Pu", which was the first book in China describing the natural habitats, planting and utilization of *Eucalyptus*. Till the end of 1940s, however, eucalypts were only planted in gardens and along roadsides in China.

Leizhou Forest Bureau was set up in Zhanjiang, Guangdong, in the early 1950s, specializing in the planting and extension of eucalypts in large scale. At a national meeting on water and soil conservation in 1957, Zhu De, the former chairman of Chinese People's congress, advocated "widely planting eucalypts in Southern China". Professor Kuan-Zhao Hou of South China Agriculture College published a book named "Eucalypts in South China" from the perspective of botany. Summarizing the experiences and technology of managing large-area of young plantations in Leizhou, Shuxiong Qi wrote "Eucalypts cultivation" in the late 1950s.

To protect road base and supply sleepers, the department of railway had planted eucalypts along roadsides of Jin-Guang, Fu-Nin, Fu-Gan, Xiang-Gui, Bao-Cheng, Cheng-Kun and Xiang-Qian Railway in the middle 1960s. At the same time, more than ten state-own forest farms mainly targeting eucalypt plantations were set up in Qinzhou, Nanning etc. of Guangxi. This was the first wave of eucalypt plantation extension in China. In 1972, right after the Conference of National Tree Breeding Cooperation, the National Eucalypt Science and Technology Cooperative Group were established, and the journal "Eucalypt science and technology" was firstly published. One of the general agricultural science and technology series, "Eucalypts cultivation and utilization", by Shuxiong Qi, was published by China Forestry Publishing House at the end of the1970s. The Southern Eucalypt Research Committee was set up with the members from 12 provinces such as Guangdong, Guangxi, Sichuan, Yunnan, Zejiang, Shanghai, Jiangxi, Anhui and Guizhou etc. to stimulate the development of eucalypt plantation in the regions.

The second wave of planting eucalypts in China started in the mid 1980s with the forest products switching from mine props to wood chips and pulps. So far, more than 300 species have been introduced in more than 600 counties of 16 provinces; among the species introduced, 211 species have been tried in afforestation, but only about 10 species were widely planted. In total, China has about 1,547,000ha of eucalypt plantations, and 1.8 billions of the four-sides trees (equivalent to 354,000ha). Among the countries planting eucalypts, China's total eucalypt plantation area ranks third in the world, just second to Brazil and India. Most of the eucalypt plantations in China are located in the provinces of Guangdong, Guangxi and Hainan. Besides the plantations producing fiber etc, huge number of eucalypt trees was planted in the windbreak forests in the provinces of Guangdong and Hainan, and in the farmland shelter belts and farmland in the Chengdu Flatland of Sichuan Province. To celebrate the hundred anniversary of eucalypt introduction and planting in China in 1990, "Eucalypts in China", edited by Qi Shuxiong, was published, and an "International Eucalypt Symposium" was held in Zhanjiang with the corresponding proceedings published the next year. In 1979, FAO published the Chinese translation of "Eucalypts for planting". "A classification of the eucalypts" was separately translated into Chinese publications by Huoran Wang, Lingying Zeng and Jian Zhong.

If it is said that the planting of eucalypts in 1980s was still conducted by the state-own forest farms or companies, then the private sector and farmers mainly developed most of the eucalypt plantations since 1990s. These plantations with short rotation were managed for producing industrial materials. Among those plantations planted in 1990s, more than 200,000 ha were developed with the loans from the World Bank. Due to the high return, many overseas companies started to invest in managing eucalypt plantations in South China, too. To date, Sino-Forest Corp. has planted more than 50,000 ha and other overseas companies (from Indonesia, Thailand, the United States and Japan) have more than 70,000 ha in total. These plantations are distributed in the provinces of Guangdong, Guangxi, Hainan etc.

The species successfully introduced and widely planted in the early times in China included *E. exserta*, *E. citriodora*, *E. robusta*, *E. camaldulensis*, *E. tereticornis*, *E.* ABL12, *E. globulus*, and *E. maidenii*. During the period of the 6th, 7th, 8th and 9th Five-Year Plan when national research cooperative projects on *Eucalyptus* headed by China Eucalypt Research Center, *E. urophylla*, *E. grandis*, *E. camaldulensis* (cold tolerant) and *E. nitens* were introduced, improved and decided for planting and extension. Many species, provenance, hybrid and clone trials have been established in Guangxi, Guangdong, Yunnan and Fujian, by the Research Institute of Tropical Forest (RITF) and the Forest Research Institute of Chinese Academy of Forestry, the Dongmen State-Own Forest Farm of Guangxi, the China-Australia Cooperative Dongmen Afforestation Program (CACDAP), etc. In particular, it is worthy to mention the RITF and CACDAP projects, which had intensively tested the provenances, families, clones, and hybrid progenies, resulting in the seed orchards of *E. grandis*, *E. urophylla*, *E. camaldulensis* and *E. trerticornis*. The superior selected hybrid clones of *E. grandis* × *E. urophylla*, *E. urophylla* × *E. grandis*, *E. urophylla* × *E. camaldulensis* and *E. urophylla* × *E. trerticornis* have shown great growth potential, and, with help of tissue culture and cutting technology, are now well extended in South China. With support from the Science and Technology Department of the National Forestry Bureau since 1996, Sino-Forest Corporation has invested about 3% of the annual budget for its plantation management in the cooperative research projects on eucalypt breeding and cultivation with Nanjing Forestry University, RITF, and the research institutes of Guangdong, Guangxi and Jiangxi, generating some preliminary but promising results in the perspectives of mountain planting, fertilizing, hybrid breeding, spacing, clone evaluation and tissue culture.

As a non-indigenous species, eucalypts experience much less pest and disease damage in China than in the countries of the natural distribution. Seedling damping-off may happen in nurseries. *E. urophylla* and its hybrids with other species may be infected with *Pseudomonas solanaceaum* at age of 1~2 year old, incurring in some mortality, and, at worst, failure of stands. Damage of insects (e.g. *Eucalyptipsyche spp.*, *Strepsicrates coriariae* and *Phoracantha spp.*) may appear with the leaves and woods of young eucalypt stands. Forest fires do not seriously damage eucalypt trees unless stands are young and short, and have rich under-story vegetations. Seasonal typhoons may lead to loss of the plantation yield or growth by killing, breaking or blowing downs trees in the coastal areas.

Clear cutting is usually applied to eucalypt plantations with rotation depending on the maturity of the designated product, about 15 years for mine-props and 5~6 years for pulp materials and wood chips. Either coppicing or replanting regeneration after harvesting is adopted. Coppicing is commonly used for the clear-cut sites of stands with high yield, survival and sprouting ability, while replanting is employed when new and good planting materials are available, or for the clear-cut sites of stands with low yield, survival and sprouting ability.

Currently, most of the eucalypt plantations are mainly managed for providing industrial materials in China according to the national forest classification management policy. In addition, eucalypts are also planted along the seacoasts and riverbanks, and around the lakes, ponds and agricultural lands for the purposes of protection and sheltering, and planted in the gardens and settlement surroundings and along the roads as the greening species.

In the last decade, the eucalypt wood production has been increasing year by year, reaching 5 millions m^3 annually in the recent years. The processing and utilization industry of the logs and byproducts have also been developing with the extension of eucalypt plantations. Eucalypt wood chips mills were built near the harbors of Zhanjiang, Yangjiang, Haikou, Beihai and Fangcheng, exporting more than one million BDT/year of wood chips to Japan, South Korea and Taiwan. To explore the downstream products, many pulp and artificial board mills are or are being built by state-own or private companies, such as Heda, Liujiang and Fengfang pulp mills in Guangxi, and Jiangmen and Zhujiang pulp mills in Guangdong. The Chinese Government has also planned three pulp mills, one in Zhanjiang with annual production capacity of 500,000 tons, and one in Hainan and one in Guangxi each of 300,000 tons. With eucalypt branches as the raw materials, MDF board mills each with annual capacity of 100,000 tons were separately set up in Zhanjiang, Guangxi, Leshan and Sichuan. Sino-Forest Corporation has built a particleboard mill with annual production of 100,000 m^3 in Gaoyao, and is planning a new artificial board mill with annual capacity of 300,000m^3 in Heyuan, Guangdong. In addition, eucalypt woods are also widely used for construction, furniture and cabinet works etc. Particularly, many small-size factories for producing packing wood boxes and flooring boards are run in the planting areas of eucalypts.

The byproducts of eucalypts are diverse, from the essential oils, tannin and other chemical products extracted from leaves and barks, to charcoals made from stumps, branches and small-size logs. Essential oils are mainly used in medicine and perfume making. Several hundred tons of essential oils were produced each year in the provinces of Yunnan and Sichuan, part of which is exported aboard, and the rest is used to produce value-added products in Shanghai and the other places of China. Some eucalypt chemical products are good materials for making chemical diluents, boiler dirt removers and plant growth or mushroom culture hormones.

To promote the further development of eucalypt plantation and to maximize economic value of the resources in China, processing and utilization should be considered together with plantation management. The existing processing and utilization industry should be restructured and improved, mainly focusing on the production capacity and quality of products and the involvement of plantation management to secure the supply of the raw materials. With the integration of planting and utilization, and supplying and trading of products, eucalypt plantations will bring in greater economic, ecological and social benefits in China.

中国桉树的栽培与利用简介

祁述雄

嘉汉林业国际有限公司顾问委员会
524022 中国广东湛江

中国引种栽培桉树始于 1890 年，从意大利引进多种桉树到广州、香港、澳门；从法国引进细叶桉到广西的龙州。至 20 世纪 40 年代末，只是零星栽培，供作庭院观赏和道路绿化。

中华人民共和国成立后，扩大了桉树的引进栽培。50 年代初，广东省在湛江地区建立雷州林业局，并开始大面积人工林的营造；1957 年，中国人大委员长朱德号召："南方大力种植桉树"；60 年代中央南方局在两广抓桉树基地的建设，推动了南方 12 省、区、市桉树的栽培；80 年代中期，由于桉树栽培目的由坑木改变为短周期工业用材，进一步加速了桉树的发展。

全国共有 16 个省、区、市 600 多个县先后共引进 300 多个种，其中，育苗造林达 211 个，但是，用于大面积造林仅 10 余种。迄今人工林面积达 154.7 万公顷，另有四旁植树 18 亿株（折合面积 35.4 万公顷），在世界营造桉树人工林的国家中，面积次于印度和巴西居第三位。采伐更新一般在工艺成熟时进行，浆粕材 5-6 年，坑柱材 15 年左右。此外，桉树还用于营造防护林、风景林和混交林等。

近年来，每年桉树木材产量增至 500 万立方米以上。桉树木材与副产品的加工利用，遍布产区城乡。一部分木材供加工木片出口日本、韩国，一部分用作国内纸业和人造板工业的原料，而桉叶油等副产品在西南诸省加工为工业和药业原料。

目前，桉树在中国的发展十分迅速，必须在栽培利用上加强科学研究，提高林分单位面积产量和质量，使下游纸和板业产品上档次、上规模，做到营林和加工一体化，产、供、销一条龙，充分发挥桉树在国民经济中的经济、生态和社会三大效益，朝着可持续发展的道路前进。

TAXONOMIC STATUS AND ITS IMPACT ON THE INTRODUCTION AND DOMESTICATION OF THE EUCALYPTS IN CHINA

HUORAN WANG

Research Institute of Forestry
The Chinese Academy of Forestry
Beijing 100091, China

This paper reviews the study of genetic resources and advances in the classification of the eucalypts with emphasis on the new taxonomic system put forward by K.D. Hill and L.A.S. Johnson. The eucalypts comprise three genera, *Angophora* Cav., *Corymbia* K.D. Hill & L.A.S. Johnson and *Eucalyptus* L'Herit., with over 945 taxa totally recognized at present. The effects of the taxonomic changes on domestication and practical forestry with eucalypts are discussed, in particular, with those taxa that have been important plantation species in China. Research direction in eucalypt domestication and tree breeding is also discussed.

1 Introduction

Eucalypts are tall trees, shrubs or mallee, belonging to the Myrtaceae family. The genus, *Eucalyptus* L'Herit., was formally described in 1788, based on the type species, *Eucalyptus obliqua* L'Herit.

It was not until 1986 that the classification, designed by Pryor and Johnson (1971), was introduced to China (Wang 1986a, 1986b). The classification is of great significance, theoretically and practically, for studying eucalypt genetics and cross breeding (Griffin et al. 1988).

This paper attempts to review the advances and changes in the taxonomy of the eucalypts in the last 10 years, with emphasis on the introduction of the current taxonomic status which will impact the practice of tree domestication, improvement and breeding of eucalypts in China. The taxonomy of eucalypts is also important for better understanding of the ecology and biogeography of these essential forest trees in Australia (Pryor 1976; Chippendale and Wolf 1981; Austin et al. 1983; Florence 1981; Gill et al. 1985) as the eucalypts have increasingly become important for plantation forestry worldwide (FAO 1979, 1997). Over 200 species were introduced to China and a number experimental plantings have been set up in southern China for species and provenance testing (Liang et al. 1993; Wang et al. 1992; Wang et al. 1993; Wang et al. 1999), among these species more than 10 have become economically or environmentally important species in China (Wang and Brooker 1991; Wang et al. 1999).

2 Current Taxonomic Status

Since the 1980's, a great number of eucalypt taxa have been described and a very comprehensive revision in classification has been made by Australian botanists

(Boland 1985; Boland et al. 1984; Brooker and Kleinig 1983, 1990, 1994; Carr and Carr 1980, 1985, 1987; Chippendale 1976, 1988; Hill and Johnson 1991a, 1991b, 1991c, 1995; Pryor et al. 1995). As new species were discovered and the genetic variation patterns in genetic diversity were systematically studied (Moran 1992), all of these scientific efforts have resulted in the emersion of new classification of the eucalypts (Brooker 2000; Hill and Johnson 1995).

Hill and Johnson took a cladistic study and phylogenetic analysis of *Corymbia*, a subgenus in the classification system designed by Pryor and Johnson (1971), and logically, a taxonomic revision was thoroughly made. The genus, *Corymbia* was formally established with a type species, *Corymbia gummifera* (Sol. Ex Gaertn.) K.D. Hill & L.A.S. Johnson. *Eucalyptus* L'Herit. is still maintained on the same status as a genus and further divided into 7 subgenera. In contrast to the Pryor-Johnson system, *Nothocalyptus* and *Telocalyptus* are newly designed as subgenera by Hill and Johnson. They emphasize that the eucalypts should, in broad sense, include all species of the three genera, *Angophora* Cav., *Corymbia* K.D. Hill & L.A.S. Johnson and *Eucalyptus* L'Herit. In total, 945 taxa, including 808 species and 137 subspecies or varieties, have been described and accepted by the end of 1998 (Wilcox 1997). Tables 1 and 2 give the detailed information about the systems described by Pryor & Johnson and Hill & Johnson, respectively.

Table 1. Structure of Pryor & Johnson (1971) classification.

Genus/Subgenus (属或亚属)	Code (代码)	Section (组)	Code (代码)	No. of taxa (分类群数)
1. Angophora (杯果木属)	A	Liberia (离被组)	AA	7
2. Blakella (布莱克亚属)	B	Lemuria (鬼桉组)	BA	10
3. Corymbia (伞房亚属)	C	Rufaria (红木组)	CA	30
		Ochraria (褐木组)	CC	9
4. Eudesmia (纹蒴亚属)	E	Quadraria (四萼组)	EA	12
		Apicaria (顶萼组)	EF	5
5. Gaubaea (高伯亚属)	G	Curtisaria (克蒂斯组)	GA	2
6. Idiogenes (昆士兰桉亚属)	I	Gympiaria (昆士兰桉亚)	IA	1
7. Monocalyptus (单蒴盖亚属)	M	Ranantheria (肾药组)	MA	106
8. Symphyomyrtus (双蒴盖亚属)	S	Equatoria (赤道桉组)	SB	3
		Tingleria (丁格组)	SD	1
		Transversaria (横脉组)	SE	17
		Bisectaria (叉形子叶组)	SI	104
		Dumaria (麻利组)	SL	36
		Exsertaria (蜃缘组)	SN	36
		Maidenaria (蓝桉组)	SP	68
		Umbrawarria (山蓝桉组)	SQ	1
		Howittaria (豪威特桉组)	SS	1
		Adnataria (贴药组)	SU	79
		Sebaria (小帽桉组)	SW	1

Table 2. The classification of the eucalypts by K.D. Hill and L.A.S. Johnson with reference to Wilcox (1997, 1998).

	Serial No. (序号)	Species (种)	Ssp./var. (亚种/变种)	Total taxa (分类群总数)
Angophora clade (杯果木进化系)				
***Angophora* Cav. (杯果木属)**	**1-14**	**14**	**0**	**14**
***Corymbia* K.D. Hill & L.A.S. Johnson (伞房属)**	**15-127**	**113**	**23**	**136**
Eucalyptus clade (桉树进化系)				
***Eucalyptus* L'Herit. (桉树属)**	**128-808**	**681**	**114**	**795**
***Eudesmia* (纹蒴亚属)**	**128-149**	**22**	**4**	**26**
***Nothocalyptus* (小帽桉亚属)**	**150**	**1**	**0**	**1**
***Symphyomyrtus* (双蒴盖亚属)**	**151-664**	**514**	**88**	**602**
Tingleria (丁格组)	152	1	0	1
Transversaria (横脉组)	153-174	22	4	26
Michaelianae (麦克桉)	175	1	0	1
Red gums (红桉类)	176-204	29	8	37
Tropical white gums (热带白桉类)	205-220	16	2	18
Maidenaria gums (蓝桉组)	221-300	80	14	94
Boxes (黄杨皮类)	301-363	63	9	72
Ironbarks (铁树皮类)	364-408	45	8	53
Bisectaria (叉形子叶组)	409-602	194	47	241
Dumaria (麻利组)	603-665	63	5	68
***Guabaea* (高伯亚属)**	**666-668**	**2**	**0**	**2**
***Idiogenes* (昆士兰桉亚属)**	**669**	**1**	**0**	**1**
***Monocalyptus* (单蒴盖亚属)**	**668-804**	**137**	**13**	**150**
Rubiginosae (赤褐桉)	668	1	0	1
W Australian Monocalyptus (西澳单蒴盖桉树)	669-698	29	5	34
White mahoganies (白桃花心桉)	699-704	5	0	5
Stringybarks (纤维皮类)	705-743	39	0	39
Blackbutts (弹丸桉组)	744-745	2	0	2
Ashes (王桉组)	746-794	49	5	54
Peppermints (薄荷桉类)	795-804	12	3	15
***Telocalyptus* (赤道桉亚属)**	**805-808**	**4**	**0**	**4**
Total taxa (总计)		**808**	**137**	**945**

3 Impact of the Taxonomic Changes on Eucalypt Domestication in China

3.1 *Changes in Genera and Subgenera*

1) *Angophora* Cav., 14 species, tall tree, occasionally in shrub form. The current sense of eucalypts further indicates that *Angophora* is very closely related to other eucalypts in evolutionary biology which does not affect forestry practice except in botanical study as A. bicostata, and perhaps few others, are only planted in botanical gardens in the southern China.
2) *Corymbia* K.D. Hill & L.A.S. Johnson includes 113 species and 23 subspecies or varieties of ghost gums and bloodwoods, most of which are distributed in the tropical parts of northern Australia. Some species have long been grown in China, for instance, *Corymbia citriodora* (Hook.) K.D. Hill & L.A.S. Johnson is well known to many Chinese foresters.
3) *Eucalyptus* L'Herit. covers all of the remaining taxa. Obviously, it is still rather sophisticated with a great number of species. The genus is further classified into the 7 subgenera. In comparison with the Pryor & Johnson (1971) system, the revised genus has two new subgenera, *Nothocalyptus* and *Telocalyptus*, with the respective *Sebaria* (SW) and *Equatoria* (SB) being distinguished particularly in morphological development (Pryor and Johnson 1971, 1981). The 7 subgenera of *Eucalyptus* in Hill & Johnson system are as follows:
   - Eudesmia (纹蒴亚属)
   - Gaubaea (高伯亚属)
   - Idiogenes (昆士兰桉亚属)
   - Monocalyptus (单蒴盖亚属)
   - Nothocalyptus (小帽桉亚属)
   - Symphyomyrtus (双蒴盖亚属)
   - Telocalyptus (赤道桉亚属)

The subgenus, *Nothocalyptus*, includes only 1 species, *Eucalyptus microcorys* F. Muell., which is unique to the development of operculum. This species can be seen in a few experimental sites in Guangxi and Fujian provinces in China.

A typical species of *Telocalyptus* with economical value is *E. deglupta* Blume, one of the few eucalypts naturally distributed outside Australia. It is found in experimental trials and planted as ornamental trees in Hainan and Guangzhou.

3.2 *Changes at or Below Species Levels*

A number of species were renamed in nomenclature because their taxonomic status had been changed to different genus. Logically, these changes or modifications directly impact on forestry practices, not only on the introduction and utilization of genetic resources but also on tree breeding and botanical study. Examples are given as follows:

1) *Corymbia citriodora* (Hook.) K.D. Hill & L.A.S. Johnson (柠檬桉) used to be a major species for plantation and roadside plantings in the southern China northwards Wenzhou along the eastern coast.

2) *Corymbia maculata* (Hook.) K.D. Hill & L.A.S. Johnson (斑皮桉) was introduced to China in 1980's.
3) *Corymbia papuana* (F. Muell.) K.D. Hill & L.A.S. Johnson (鬼桉), *Corymbia tesselaris* (F. Muell.) K.D. Hill & L.A.S. Johnson (方格皮桉) and *Corymbia torelliana* (F. Muell.) K.D. Hill & L.A.S. Johnson (托里桉), are all beautiful trees for ornamental purposes in South and Southwest China.
4) *E. urophylla* S.T. Blake (尾叶桉) has become a very important plantation species in China since provenance trials were undertaken by China-Australia cooperative research programs in 1980's (Liang et al. 1993). Pryor et al. (1995) studied genetic variations among populations through numerical analyses, which led to the establishment of two new species, *E. orophila* Pryor (高山尾叶桉) and *E. wetarensis* Pryor (维塔尾叶桉). The description of the these two species that are closely related to *E. urophylla* S.T. Blake gives rise to practical effects on tree improvement and breeding of eucalypts in China.
5) *E. nitens* (Deane & Maiden) Maiden (亮果桉) for which discontinued distribution and geographic variation were reported from morphological research and provenance trials (Pryor and Johnson 1971; Pederick 1979; Wang et al. 1993). *E. denticulata* I. Cook & Ladiges (齿叶亮果桉), as new taxon, was separated from *E. nitens* on the basis of morphology of the Errinundra population in Victoria (Cook and Ladiges 1991, 1998). Introduced in 1980's, *E. nitens* is presently doing very well in Yunnan Province and may have extension potential in other regions demanding for cold resistant species.
6) *E. camaldulensis* Dehnh. var. *camaldulensis* (赤桉) and *E. camaldulensis* Dehnh. var. *obtusa* Blakely (钝盖赤桉) can easily be distinguished by operculum shape as two varieties originating respectively from the north and south of Australia (Blakely 1934). However, numerous studies reported that a hybrid swamp existed in the northern Queens land between *E. camaldulensis* and *E. tereticornis* Sm. (e.g. Eldridge et al. 1993). Based on the early findings, *E. camaldulensis* Dehnh. subsp. *simulata* Brooker & Kleinig (昆北赤桉) was formally described, including populations of Petford, Laura and Kennedy River (Brooker and Kleinig 1994). The genetic resources of this subspecies were introduced before 1990's and play a very important role in hybrid breeding in China.
7) *E. saligna/botryoides* complex was taken as an example to explain genetic patterns of eucalypts by Pryor and Johnson (1971). In Bateman's Bay area, these two species are overlapped in natural range that leads to gene introgression. These two species were changed to the status of new combination as subspecies, *E. saligna* Sm. subsp. *saligna* and *E. saligna* Sm. subsp. *botryoides* (Smith) Passioura & Ash, and *E. botryoides* Smith is regarded as the synonym (Passioura and Ash 1993). Both subspecies were introduced to China long ago and can be seen in many places but in large plantations.

8) Based on geographic study, Kirkpatrick (1975) once referred the taxa that are related to Tasmanian blue gum to as subspecies such as *E. globulus* Labill. subsp. *maidenii* (F. Muell.) Kirkpatr. (直干蓝桉), *E. globulus* Labill. subsp. *bicostata* (F. Muell.) Kirkpatr. (双肋蓝桉) and *E. globulus* Labill. subsp. *pseudoglobulus* (F. Muell.) Kirkpatr. (假蓝桉). However, they have been changed back to species status now. *E. globulus* and *E. maidenii* are very important for wood and oil production in the southwestern China, and the corresponding provenance testing have been done in 1980's (Wang et al. 1992).

References

Austin, M.P., Cunningham, R.B. and Wood, J.T. 1983. The subgeneric composition of eucalypt forest stands in a region of southeastern Australia. Aust. J. Bot. 31:63-71.

Blakely, W.F. 1934. A key to the eucalypts. Forest Timber Bureau, Canberra.

Boland, D.J. 1985. Taxonomic revision of *Eucalyptus delegatensis* R.T. Baker (Myrtaceae). Aust. For. Res. 15:173-181.

Boland, D.J., Brooker, M.I.H., Chippendale, G.M., Hall, N., Hyland, B.P.M., Johnston, R.D., Kleinig, D.A. and Turner, J.D. 1984. Forest trees of Australia. 4th ed., Nelson/CSIRO, Melbourne, 687p.

Brooker, M.I.H. 2000. A new classification of the genus *Eucalyptus* L'Her. (Myrtaceae). Aust. Syst. Bot. 13:107.

Brooker, M.I.H. and Kleinig, D.A. 1983. Field guide to eucalypts. Vol.1. Southeastern Australia. Inkata Press, Melbourne, 288p.

Brooker, M.I.H. and Kleinig, D.A. 1990. Field guide to eucalypts. Vol.2. Southwestern and southern Australia. Inkata Press, Melbourne, 428p.

Brooker, M.I.H. and Kleinig, D.A. 1994. Field guide to eucalypts Vol.3. Northern Australia. Inkata Press, Melbourne, 383p.

Carr, D.J. and Carr, S.G.M. 1980. The *Lehmannianae*: a natural group of western Australia eucalypts. Aust. J. Bot. 28:523-550.

Carr, D.J. and Carr, S.G.M. 1985. *Eucalyptus*. I. New or little known species of the *Crymbosae*. Phytoglyph Press, Canberra.

Carr, D.J. and Carr, S.G.M. 1987. *Eucalyptus*. II. The rubber cuticle and other studies of the *Crymbosae*. Phytoglyph Press, Canberra, 372p.

Chippendale, G.M. 1976. *Eucalyptus* nomenclature. Aust. For. Res. 7:69-107.

Chippendale, G.M. 1988. *Eucalyptus, Angophora* (Myrtaceae). In: "Flora of Australia", Vol.19, A.S. George (ed), Australian Government Publishing Service, Canberra, 540p.

Chippendale, G.M. and Wolf, L. 1981. The natural distribution of *Eucalyptus* in Australia. Australian National Parks and Wildlife Service, Special publication No.6, 192p.

Cook, I.O. and Ladiges, P.Y. 1991. Morphological variation within *Eucalyptus nitens* s.lat. and recognition of a new species, *E. denticulata*. Aust. Syst. Bot. 4:375-390.

Cook, I.O. and Ladiges, P.Y. 1998. Isozyme variation in *Eucalyptus nitens* and *E. denticulata*. Aust. J. Bot. 46:35-44.

Eldridge, K., Davidson, J., Harwood, C. and van Wyk, G. 1993. Eucalypt domestication and breeding. Clarendon Press, Oxford, 288p.

FAO. 1979. *Eucalyptus* for planting. FAO, Rome, 783p.

FAO. 1997. State of the World's forests 1997. FAO, Rome, 200p.

Florence, R.G. 1981. The biology of the eucalypt forest. In: "Biology of Australian plants", J.S. Pate and A.J. McComb (eds), Univ. of Western Australia Press, Nedlands, pp147-180.

Gill, A.M., Belbin, L. and Chippendale, G.M. 1985. Phytogeography of *Eucalyptus* in Australia. Australian Government publishing service, Canberra, Australian Flora and Fauna series No.3, 53p.

Griffin, A.R., Burgess, I.P. and Wolf, L. 1988. Patterns of natural and manipulated hybridization in the genus of *Eucalyptus* L'Herit. - a review. Aust. J. Bot. 36:41-66.

Hill, K.D. and Johnson, L.A.S. 1991a. Systematic studies in the eucalypts-3. New taxa and combinations in *Eucalyptus* (Myrtaceae). Telopea 4:223-267.

Hill, K.D. and Johnson, L.A.S. 1991b. Systematic studies in the eucalypts-4. New taxa in *Eucalyptus* (Myrtaceae). Telopea 4:321-349.

Hill, K.D. and Johnson, L.A.S. 1991c. Systematic studies in the eucalypts-5. New taxa and combinations in *Eucalyptus* (Myrtaceae) in western Australia. Telopea 4:561-634.

Hill, K.D. and Johnson, L.A.S. 1995. Systematic studies in the eucalypts-7. A revision of the bloodwoods, genus *Corymbia* (Myrtaceae). Telopea 6:185-504.

Kirkpatrick, J.B. 1975. Geographic variation in *Eucalyptus globulus*. Australian Publishing Service, Canberra, Forestry and Timber Bureau Bulletin 47, 64p.

Moran, G.F. 1992. Patterns of genetic diversity in Australian tree species. In: "Population genetics of forest trees", W.T. Adams, S.H. Strauss, D.L. Copes and A.R. Griffin (eds), Kluwer Academic Publishers, pp49-66.

Liang, K., Zhou, W., Zhong, C. and Yang, C. 1993. Species and provenancec trials in eastern Hainan Island. In: "Research in the broadleaved tree species of Australia". J. Hong and H.R. Wang (eds), China Forestry Publishing House, pp49-57.

Passioura, J.A. and Ash, J.E. 1993. *Eucalyptus saligna* smith subsp. *botryoides* (Smith) Passioura & Ash, Com.nov.: a re-assessment of the *Eucalyptus saligna-Eucalyptus botryoides* complex. Aust. Syst. Bot. 6:181-183.

Pederick, L.A. 1979. Natural variation in shining gum (*Eucalyptus* nitens). Aust. For. Res. 9:41-63.

Pryor, L.D. and Johnson, L.A.S. 1971. A classification of the eucalypts. Australian National University, Canberra, 102p.

Pryor, L.D. 1976. Biology of eucalypts. Edward Arnold (Publishers) Ltd, London, The Institute of Biology's Studies in Biology No.61, 82p.

Pryor, L.D. and Johnson, L.A.S. 1981. *Eucalyptus*, the universal Australian. In: "Ecological biogeography of Australia", A. Keast (ed), The Hague, pp499-536.

Pryor, L.D., Williams, E.R. and Gunn, B.V. 1995. A morphometric analysis of *Eucalyptus urophylla* and some related taxa, with descriptions of two new species. Aust. Syst. Bot. 8:57-70.

Wang, H. 1986a. Review of the current advances and research trend in the taxonomy of *Eucalyptus*. Sci. Silv. Sini. 22: 393-396.

Wang, H. 1986b (Translator). A classification of the eucalypts by L.D. Pryor, and L.A.S. Johnson, Aust. Nat. Univ., 1971, Northeast Forestry University Press, Harbin, 103p.

Wang, H. and Brooker, M.I.H. 1991. A key to eucalypts in China, China Science & Technology Press, Beijing, 74p.

Wang, H., Zheng, Y., Zang, D., Zhang, R. and Wu, X. 1992. Provenances trials of *Eucalyptus globulus* Labill. in China. For. Res. 5:409-416.

Wang, H., Zheng, Y. and Yan, H. 1993. Provenance trials of *Eucalyptus nitens* and its potential in plantation forestry in China. Sci. Silv. Sini. 29:364-371.

Wang, H, Zheng, Y., Zang, D., Zhang, Z. and Huang, Y. 1999. The Introduction of monocalyptus to Southwestern China and its significance in plant geography. Sci. Silv. Sini. 35:2-6.

Wilcox, M.D. 1997. A catalogue of the eucalypts. Groome Poyry Ltd, Auckland, 114p.

Wilcox, M.D. 1998. Supplement to a catalogue of the eucalypts. Groome Poyry Ltd, Auckland, 4p.

桉树分类学研究进展及其对中国桉树引种驯化的影响

王豁然

中国林业科学研究院林业研究所
100091 中国北京

本文论述了桉树属遗传资源和最近 10 年来桉树分类学的进展，重点介绍 Hill & Johnson 新的桉树分类系统及其对中国桉树引种栽培和遗传改良生产实践的影响。桉树属(*Eucalyptus* L' Herit.)属桃金娘科(Myrtaceae)，包括乔木、灌木或丛生桉树(Mallee)种类。桉树遗传资源非常丰富，其中只有 5 种分布于澳大利亚以外的国家和地区，另有 12 种桉树既分布于巴布亚新几内亚又分布在澳大利亚北部，其余的桉树都只分布于澳大利亚。

Pryor & Johnson 的分类从进化的观点出发，在理解种系发生的基础上划分属下组群，进行分类群的排列，旨在体现属以下不同分类群之间的亲缘关系。不同亚属的树种之间不能进行杂交，即未发现自然杂种，也不能通过人工控制授粉产生杂种，杂种化现象只在同一亚属内不同树种之间产生，因此，它不仅对研究桉树属和桃金娘科具有重要的理论意义，并且对鉴定桉树杂种和指导育种改良具有实用价值。最近 10 年以来，许多分类学家相继发表许多新的分类群，森林遗传学家对桉树遗传多样性的变异模式进行了更深入的系统研究。这些植物学新发现和研究成果导致新的分类思想的形成。

1995 年，希尔和约翰逊对 Pryor & Johnson 分类系统中的伞房亚属(*Corymbia*)进行了亲缘研究（Cladistic study）、谱系分析（Phylogenetic analyses）和全面的分类学修订，以伞房花桉[*Corymbia gummifera*（Sol. Ex Gaertn.）K.D. Hill & L.A.S. Johnson]为模式种建立伞房属(*Corymbia* K.D. Hill & L.A.S. Johnson)。他们特别强调，广义来说，桉树应该泛指杯果木属、伞房属和桉树属 3 个属的树种，包括了截至 1998 年底得到承认的桉树 808 种和 137 亚种或变种，总共 945 个分类群。与以前的概念相比，现在的桉树属(*Eucalyptus* L' Herit.)只包括杯果木属和伞房属之外的桉树，仍然划分为 7 个亚属，其中新拟小帽桉亚属(*Nothocalyptus*)和赤道桉亚属(*Telocalyptus*)。Hill &Johnson 分类系统目前已经得到广泛承认和应用，在种和种下水平上，分类地位的变化对生产实践会产生直接影响，特别是对遗传资源选择，如伞房属和尾叶桉(*E. urophylla* S.T. Blake)两个近缘种的命名对于我国今后桉树育种与遗传改良具有实际影响。

笔者曾经多次介绍过桉树遗传资源状况与分类学研究进展。随着大量新种的发现和生物系统学研究的深入，尤其是分子遗传学的发展与应用，进一步推动对桉树遗传资源的发掘和探索，这对遗传学、树木育种与改良、基因资源保存策略和引种驯化等理论研究和林业生产实践产生广泛而深刻的影响。本文旨在使读者认识和了解桉树分类学的历史和发展，把握当前桉树分类的基本框架，作为指导桉树引种驯化、杂交育种和国际学术交流的基本依据，进一步推动我国桉树人工林的发展。

MERGING ECOLOGICAL/ENVIRONMENTAL CONCERNS INTO SUSTAINABLE MANAGEMENT OF SHORT-ROTATION FOREST PLANTATIONS IN SOUTH CHINA: PRACTICES BY SINO-FOREST CORP

RUN-PENG WEI [1, 2]

[1] *Sino-Forest Corporation (Canada), 3129-40,31/F., Sun Hung Kai Centre 30 Harbour Road, Wanchai, Hong Kong*
[2] *Forestry Faculty, South China Agricultural University Guangzhou 510642, China*

In this documentation, ecological/environmental concerns/issues emerging from modern short-rotation plantation management are discussed. Most of these concerns/issues are associated with the intrinsic characteristics of fast-growing and short-rotation plantations. We propose that existing operation procedures are revised or new operation procedures are devised to meet the increasing demands of ecological/environmental conservation and sustainable development. Sino-Forest's practices in South China were used for demonstration. While improving the status of the local/surrounding landscapes, ecosystems, biodiversity, vegetations, and conservation at the present situations in a short time, short-rotation plantations may still worsen sustainable management and slow down the process of ecological/environmental restorations in the regions in a long run. A management system with prescribed techniques or operations could minimize or reverse most of the potential impacts that plantations may have on ecological/environmental conservation.

1 Introduction

Short-rotation plantation management has been becoming a more and more important forest practice in many countries, particularly those lacking land/forest resources and with pressures from increasing populations and demands on forest products. Although high forest yield is an essential end, other aspects of impacts may be huge (Youngquist and Hamilton 2000). A plantation's fast-growth usually also means the fast formation of a forest in a short time. Thus, short-rotation plantations may play a role in the fast restoration of forests and other ecosystems in deforested countries/regions. Forests (even if they are cultivated plantations) are usually the most important ecosystems in the conservation of other ecosystems and local environments. In addition, a good supply of products from short-rotation plantations could greatly reduce the exploration of natural forest or other ecosystems, thus, being seen as an indirect alternative for ecological conservation (Gladstone and Ledig 1990; Sedjo 1999). In China, the legislation on prohibition of cutting natural forest would actually be meaningful only if cultivated plantations provide enough forest products.

Although the positive influences mentioned above are predominant, many concerns or issues about impacts on ecological/environmental conservation and how to maintain the sustainable management of short-rotation plantations have also been raised in recent years, e.g. site productivity degradation, exotic species invasion, ecosystem instability, biodiversity conservation etc. (e.g. Florence 1986; Palmberg 1986; Couto and Betters 1995; FSC 2001). In this paper, we review, summarize and discuss these ecological/environmental concerns/issues. South China is designated as a timber plantation development region and is also a main base for Sino-Forest to develop short-rotation plantations. Considering the status of the local landscapes, ecosystems and environments in this region, Sino-Forest has merged ecological/environmental concerns/issues in managing its short-rotation plantations. We believe that these concerns or issues can be solved or greatly mitigated through suitable techniques/operations at least at the present situation in South China.

2 Main Characteristics of Short-rotation Plantations

Sino-Forest started to plant trees in South China almost 10 years ago. Right now it has become a leading private plantation company in China. Sino-Forest's plantations have the following characteristics:

- Mainly composed of the exotic species, including eucalypts, poplars, acacia, pines etc., although we are also trying other indigenous species like paulownias and bamboos. These exotic species, however, have successfully been introduced and planted at large-scale in Sino-Forest's plantation development or adjacent regions, or similar habitats for a long time, e.g. some eucalypt species for a history of more than 100 years (Qi 2002). There were also piles of publications about the introduction, improvement and silvicultural experiments of eucalypts in these regions and the surrounding areas (e.g. Qi 2002 and references therein).
- Fast early growing and short-rotation. The company's prior development strategy is to produce fiber products with short-rotation plantations for which fast early growing is a prerequisite. According to the growth features and designated products, the rotation is set to 5~7 years for eucalypt/acacia species in Guangdong and Guangxi, 6 years for poplars in Jiangxi and 12 years for pines in South China. Although short-rotation plantation is always the priority, the company does have the plan to diversify the species, plantations and end products in the future. Medium/long rotation species and plantations may be developed. However, breeding and cultivation technology will be imposed on these species to accelerate their early growing and shorten the plantation rotation age.
- With the stand compartments and sub-compartments scattering over landscapes, although Sino-Forest always tries to obtain the tenure of "continuous" and "large-area" lands for its short-rotation plantation development. In most of China-designated areas for commercial plantation development, human settlements and history of exploring natural resources make it difficult to develop continuous and large-area plantations. This means that there exist other ecosystems like water bodies,

national/local reserves, ecological forests, agricultural ecosystems etc around our stands. The landscapes concerned are therefore very different from those with large-area of continuous forest plantations that are often found in the western countries (Kohm and Franklin 1997).

- Intensively managed with high investments. Sino-Forest adopts high-input strategy in achieving *sustainability of its short-rotation plantation management,* which in turn ensures high output. The high-input allows application of new and systematic cultivation technologies in breeding planting stock, site choosing and preparation, fertilizer application, tending, forest protection, harvesting, conservation practice etc (Wei et al. 2001). Basically, optimization principle is followed when we decide our investment in plantation development. That is, the investment aims to achieve maximum profit by solving ecological/environmental issues of concern, and ensuring the sustainability of our plantation management.

3 Ecological/Environmental Issues

Modern tree plantation management considers not only the economic return (growth), but also social development, ecological/environmental conservation etc. In other words, sustainable forest management is an outstanding objective. Due to the characteristics of short-rotation plantations, concerns related to ecological/environmental issues are therefore raised or considered by five classes of parties, that is, the plantation managers themselves (M), academic society (A), local society (L), national/local authorities (N), and other concerning parties (O). Concerning parties include local, national or international organizations, e.g. Green Peaceful Organization (GPO) and Forest Stewardship Council (FSC). Table 1 summarizes these concerns, and analyses which of them could be the potential issues in developing short-rotation plantation at least at the present situations in South China.

Basically, ecological/environmental concerns can be sorted into 3 classes, which need (no) corresponding treatments and provisions:

- Not an issue (codes 0 and 1). Referring to the existing ecological/environmental and social situations, scientific evidence, national/international regulations, and successful management, plantation development will not cause problems, but probably mitigate the problems (Table 1). For example, Sino-Forest has no natural forest conversion problem because its plantations are strictly restricted to abandoned land/barren hills, non-natural stands with low yield, plantation sites etc. Choosing remote and relatively undeveloped regions for plantation development just brings positive influences on local society and cultures (Youngquist and Hamilton 2000). Scientific evidences show that there would not be problems in terms of site adaptability, exotic invasion, plantation restoration/succession, and plantation genetic diversity with genotype mixture (Wei et al. 2001). The fact that long-time deforestation (low forest coverage rate), tree species depression and degradation of ecosystems and environments are actually the driving forces for authorities/society to quickly restore forest coverage and improve environments with (short-rotation) plantations (Deng et al. 2001). Therefore, short-rotation plantation development will play a positive role in bettering local ecosystems and habitats for involved vegetations, species (e.g. RTES), populations and genetic diversity. In fact, modification or improvement of the conventional silvicultural operations without increasing cost would effectively prevent the occurrence of the potential problems (Table 1). For example, the concept of conservation areas is adopted if sites have steep slope and plantations may have impacts on local biodiversity. In a long run, provisions are still needed so that these issues are under proper control.

Table 1. Ecological/environmental concerns raised over short-rotation plantations, and potential issues Sino-Forest may face at the present situation in South China.

Concern	Party#	Potential issue code@ and remark
Site adaptability	G	1, lots of supporting evidences (e.g. Qi 1989; Pan and You 1993; Li et al. 2000); proper treatments and fertilization.
Natural forest conversion	N, O	0, follow national laws, and strictly restricted to certain sites.
Influences on local Society/culture	L, N, O	0, restricted to rural and remote regions without race, religion and culture issues.
Soil and water erosion	G, L	**3**, *a main issue but plantations mitigate the problems, special management operations would further reduce soil erosion.*
Soil compacting	M	**3**, *main issues with controversial views, could be improved by applying enough and suitable fertilizers, leaving litters at sites, controlling nutrient dynamics and balance, enhancing site and soil management, inter-planting green manures (Liu et al. 1999), changing tree species etc.*
Site productivity degradation	G	
Over-consumption of nutrients	M	
Over-consumption of water resources (e.g. Florence 1986)	G	
Water body conservation (e.g. quality)	L, O	2, an issue relevant to local society and concerned by certain organizations (e.g. FSC). Buffer zones and relevant operations are manipulated to control soil erosion and water quality.
Invasion issues of exotic species	N, L, O	0, used species can not be natural-regenerated, no other crossable native species exist in the surrounding areas, etc.
Influences on local ecosystems	N, L, O	0, planted sites and regions with low coverage of forests have long-time history of reforestation and other ecosystem suppression. In general, plantations positively improve environments/habitats for other ecosystems/species.
Habitats for existing flora and fauna	N, L, O	

Table 1 (*continued*). Ecological/environmental concerns raised over short-rotation plantations, and potential issues Sino-Forest may face at the present situation in South China.

Concern	Party#	Potential issue code@ and remark
Plantation restoration and succession	M	0, lots of evidences support that all species used have no problem of plantation regeneration, restoration and succession.
Long-term impacts on landscapes/ecosystems	N, O	2, status: plantations scattered over landscapes, sites and regions with low coverage of forests and long-time history of deforestation and ecosystem suppression. Plantations improve environments for other ecosystems and species, and long-term negative effects can be minimized by adopting proper management and operations. Special treatments can be designed if needed.
Landscape and ecosystem diversity	N, O	
Species diversity	N, L, O	
Rare, threaten and endangered species (RTES)	N, L, O	1, follow authorities' guidelines, no RTES in sites and adjacent areas, and provisions can be given for any new or immigrating RTES, e.g. stand conversion, corridor etc.
Influences on vegetations	N, O	1, plantations restricted to sites with poor vegetations and habitats, and conservation areas and measures can be designated nearby if needed, e.g. forest conversion.
Population dynamics	N, O	
Genetic diversity	N, O	
Plantation's genetic diversity	M	0, controversial issue, but fully considered (Wei et al. 2001).
Conservation sensitive areas	N, O	1, no designated conservation reserves in adjacent areas; buffer zones are set up on the steep slopes and boundaries of the plantations; other provision can be given.
Harmful chemicals application	N, L, O	2, strictly follow national/international regulations.
Other wastes harmful to environments	L, O	1, minimized by adopting proper regulations and operations.
Forest fire and natural damages	L, M	2, impacts on ecosystems, biodiversity and environments could be minimized by properly applying prevention and controlling procedures.
Sustainable development	L, M	**4**, *it is a focus and Sino-Forest is developing sustainable management system for managing its plantations.*

List of the main concerning parties, legend: N=national/local authorities, L=local society, O=concerning organizations, M=plantation managers, G=N+L+O+M+others.

@ code can be changed for different issues over time, legend: 0=not an issue, 2=not an issue with proper conventional management, 4=issue can be solved with new technologies in management, 1=between 0 and 2, 3=between 3 and 4.

- Not an issue with proper management operation (code 2). This type of concerns is repeatedly raised by the developed countries and some international organizations, e.g. water bodies conservation, use of harmful chemicals, influence on landscape, ecosystems, and biodiversity (FSC 2001). Obviously, there is a difference in reference (or basic line) when drawing conclusions for these issues. In forestry-developed countries, natural forests, different ecosystems and environments are much better preserved than in many developing countries (such as the situations described above in South China). Plantations management, therefore, may bring in negative effects in developed countries. In developing countries, however, plantation development would generate positive effects due to long-time deforestation (low forest coverage rate) and degradation of ecosystems and environments (Wei et al. 2001). Efficient management operations will certainly further enhance the positive roles of plantations in these developing countries, at least at present situations. Use of harmful chemicals can be avoided or reduced by employing other alternatives. Forest fires and other natural stresses (disease, pest, drought, wind, frost, flood etc.) may cause serious ecological/environmental problems. Efficient prevention and controlling procedures can reduce, avoid or minimize the resultant impacts or losses.
- Can be major issues (codes 3 and 4). This type of issues may bring about significant negative influences on sustainable short-rotation plantation management, local society development and landscapes/ecosystem management in a long run. Water and soil erosion is particularly popular in developing countries. The most important reasons include long-time deforestation, low forest coverage rate and degradation of ecosystems resulting from over-exploration of different resources. Short-rotation plantation development would quickly recover forest coverage, thus reducing or mitigating water and soil erosion. Issues about soil compacting, site productivity degradation, over-consumption of nutrients and water resources arise with perception that the fast-growing plantations involve frequent disturbance of soil/sites and large amount of nutrients and water uptakes to build up mass production. Indeed, soil compacting and site productivity degradation occurred with eucalypt plantations in Leizhou peninsular (Qi 2002), as the consequences of the improper management and frequent human disturbance, such as use of forest litters and under-story grasses/shrubs as energy, and no appropriate management measures to improve soil structure and nutrient balance. However, the true story probably quite depends on locations of sites, species, species rotation, use of fertilizers, fertilizer type, site management etc. Similar to agronomic crops, plantation (site) productivity may be increased substantially through fertilization, genetic improvement, soil treatments and species rotation (Allen 1987; Powers 1999). Rainfall is rich in the regions where Sino-Forest develops its short-rotation plantations. It can be predicted that water resources would not be an issue in the foreseeable future. Nowadays, sustainable development involving economic, social,

ecological and environmental issues is a priority in planning development at different government levels and for industries. Sino-Forest develops its own sustainable management system, taking all emerging concerns about social, ecological and environmental issues into account when managing its short-rotation plantations.

4 Management System

In this section, we summarize the whole set of the management technologies employed or considered by Sino-Forest, which targets the potential issues listed above, particularly those that may significantly influence the sustainable management of short-rotation plantation and long-term landscape/ecosystem restoration and management (Table 1). These technologies are detailed in the different technical operation guidelines or procedures.

4.1 *Choice of Planting Sites*

We strictly restrict to abandoned (agriculture) lands, degraded lands, barren hills, non-natural regenerated stands with low yield, plantation sites etc. It is not allowed to convert natural forests for plantation purposes. However, we may convert our plantations to other ecosystems with conservation purposes if it is indicated necessary. In addition, the regions of plantation development should have rich rainfalls and reasonable transportation systems. For sites with steep slope (e.g. average slope >25°) or conservation values, conservation areas will be considered, which often appear in hill valleys.

4.2 *Choice of Species and Varieties*

The principle of "site adaptability" is followed. Generally, the existing species, provenances and varieties (families and clones) used by Sino-Forest for short-rotation plantations fully fall in the regions designated as "suitable" (e.g. Pan and You 1994). Sino-Forest is also testing other indigenous and exotic species/varieties with two main purposes: 1) increase species diversity in plantations, and 2) as backup species if site productivity degradation occurs and changing species is needed (Wei et al. 2001). Sino-Forest also invests in tree improvement programs. While growth is targeted, resistance (to disease/pest, drought, frost etc) and adaptability to special sites/locations are also improved. It is expected that the improvement of planting materials would not lead to a worsening of ecological/environmental issues. It should be noted that unsuitable sites might still exist in the designated development regions due to variable forest environmental conditions, micro climates and habitats. In this situation, varieties will be adjusted or species will be replaced with backup species at later practice.

4.3 *Preparation of Site and Planting Holes*

Procedures for preparing sites mainly depend on the slope. Overall soil ploughing is allowed for sites with $\text{slope} \le 15^\circ$ if there are no other considerations. For sites with $15^\circ < \text{slope} \le 25^\circ$, belt soil preparation is adopted with equal horizontal distance between belts (Lima 1990). Belt width is always less than 1 meter, usually the same as the planting holes'. Vegetations between belts are kept. Planting holes with certain size are dug after site preparation. When sites have $\text{slope} > 25^\circ$, only planting holes are dug, and soil/vegetation disturbance should be minimized. Planting rows go along the contour lines with the same vertical distance, but are controlled with same or similar horizontal distance. Non-continuous planting rows may be inserted depending on the variation of slope. Nevertheless, sites with $>25^\circ$ will generally be avoided unless it is shown that soil erosion and other potential issues can be controlled.

4.4 *Road System*

Composed of three levels: general roads, forest roads/auxiliary roads, and in-stand paths. The system with proper drainage function should reasonably be distributed among plantations, and maintained and improved over times. One of the main purposes is to ensure that impacts on ecosystems, biodiversity and environments caused by forest fires and other damages can be minimized by efficiently employing this transportation system. The first two kinds of roads should be available to, at least, each stand compartment and most of sub-compartments. Paths, either permanent or temporary, exist or are planned for each sub-compartment, and should clearly be marked on maps. Accesses of the system to each compartment and sub-compartment can easily be found with help of maps. The drainage function of the road systems avoids and reduces water/soil erosion.

4.5 *Buffer Zones*

A buffer zone is part of a stand or plantation, but has specific silvicultural operations that are different from the other parts. For the boundaries of plantations with roads, water bodies, agriculture lands, buildings etc., *buffer zones* with a width of around 10 meters are set up to prevent soil erosion. This practice is particularly important when trees are planted on steep slopes. Generally, a buffer zone is just needed along the upper side of a body concerned or the lower side of a plantation. Vegetations should always be maintained in the buffer zones, and the operations with minimum vegetation and soil disturbance should be restricted to only digging planting holes (better with small size), competition release, cutting trees etc. Harvesting trees in buffer zones is carefully controlled to reduce impacts on vegetation and soil. The practice of buffer zones is also applied to the ecologically sensitive areas or spots such as steep slopes $>25^\circ$ inside a plantation.

4.6 *Maintenance of Site Productivity*

High site productivity is essential for successful short-rotation plantation management, and long-term stability and health of plantation and other surrounding ecosystems. When soil depression appears, tree growth slows down, vegetation changes, and plantation and surrounding ecosystems become unstable. There are many ways to maintain site productivity, e.g. 1) inoculating seedlings with biological nitrogen-fixation bacteria, 2) applying enough base and topdressing fertilizers at different stages to meet the demand of the fast-growing trees and to keep nutrient balance in soil (Allen 1987; Kohm and Franklin 1997), 3) increasing use of organic and other manures considered "healthy" to soils, 4) maintaining the dead grasses/shrubs and litters of trees in sites for improving the soil structure and nutrient dynamics, 5) inter-planting green manures (Liu et al. 1999; Hauhus et al. 2000), and 6) replacing with backup species if necessary, etc.

4.7 *Spacing Control*

Unlike the common practice using same between-row and between-tree distance, we usually employ relatively wider between-row distance than between-tree distance. For example, instead of 2.45m by 2.45m, a spacing of 4m (row) by 1.5m (tree) or 3m (row) by 2m (tree) is used. One of the main purposes is to have wider continuous vegetation belts along contours so that water and soil erosion are effectively controlled when the disturbed soil surfaces are exposed (Couto and Betters 1995). This practice is particularly efficient for hilly sites. Spacing varies among species depending on planting density.

4.8 *Species and Genetic Diversity with Plantations*

Mixed-species is still not a popular practice, but probably deserves more attentions from Sino-Forest. Much effort has been invested in experiments and generated some favorable evidences. Yet, a more feasible practice is probably to mix species in different neighboring sub-compartments (it is single species or pure stand within a sub-compartment), which will not cause too much difficulty in operations, and, on the other hand, may increase ecosystems' stability and landscapes' appreciation. Genetic diversity is seen as necessary to maintain plantation's stability, and to reduce probability of plantation failure (e.g. Libby 1980) and thus negative impacts on ecological/environment conservation. Theoretically, more than 7 unrelated clones could provide enough genetic diversity to prevent plantation failure (reviewed by Wei et al. 2001). Most of Sino-Forest's plantations either originate from seedling seed orchards with more than 30 half-sib families or are of mixed-clones. Obviously, genetic diversity of plantations with seed orchard seeds is more than enough. For clone-mixtures, 7-10 clones are expected, but usually mixed in nurseries, which create difficulty of identifying clones in plantations. Sino-Forest will strengthen its work on the identification and tracing of varieties (clones or families).

4.9 *Plantation Protection*

An effective system is established at Sino-Forest to prevent forest fires, diseases/pests, and other natural disasters so that influences or damages on ecological/environmental conditions in sites are minimized. We emphasize the importance of prevention and combined-control. Forest fire prevention/control mainly involves the company's stand patrol staffs (full- or part-time) and backup fire-control teams, local communities, township and county fire stations and other relevant groups. Stand patrolling is greatly enhanced at fire-warning seasons. The company does not have normal fire control teams, but is able to recruit enough experienced fire-mans at need. For disease/pest control, resistance is always considered in breeding. Biological and physical controls are preferred to chemical treatments. Local and higher levels of governmental departments or institutes will always help depending on the damage according to Chinese forest laws.

4.10 *Use of Chemicals*

Sino-Forest operations conform to the relevant local, national and international regulations/conventions. Our principle is "no prohibited fungicide, pesticide and herbicides" and "minimum use of harmful chemicals".

4.11 *Harvesting Plan and Operations*

To reduce soil disturbance, exposure period and erosion, it is necessary to have a deliberate harvesting plan and set up a suitable procedure for the relevant operations. As part of a long-term management plan, harvesting plan is prepared at CJV (management unit) level. By following the principle of "sustainability" and avoiding large-scale clearcutting at one location at one time, harvesting is 1) adjusted over rotation years with an allowed variation depending on the company's annual work plan, market situation etc., 2) widely distributed over CJV management areas, and 3) allocated among months in a year. Guidelines are set up for designing harvesting blocks and instructing operations (cutting, stump control, sectioning, collection, loading/unloading, transportation etc.), with one of the main purposes to reduce the influences of harvesting on the ecological/environmental conditions (vegetation and soil disturbance, soil exposure and erosion etc). Special attentions are paid to felling of trees and yarding/collection of logs, particularly in areas of buffer zones, steep slope and sliding soils as described above.

4.12 *In-time Regeneration*

A corresponding regeneration plan is always prepared together with the harvesting plan for cutting permission. From the viewpoint of ecological/environmental conservations, the focus is to have fastest plantation restoration and to minimize the exposure period of the disturbed sites. Either replanting or coppicing can be adopted depending on the valuation of management efficiency. Usually, more attentions should be paid to replanting because of the involvement of other operations (e.g. site preparation). For coppicing, proper selection and thinning of sprouts are necessary for accelerating stand restoration. Some special treatments may also be needed to stimulate stump sprouting and stand restoration in non-growing seasons.

4.13 *Overall Control of Soil Disturbance.*

Besides site preparation and harvesting, tending and other operations (e.g. weed/shrub control, soil de-compacting, intercropping, etc.) may also bring in soil disturbance. The prevention of soil erosion can fully be taken into account by modifying but imposing no difficulties on the relevant operations.

4.14 *Prevention of Dumping Harmful Stuff*

In addition to chemicals used for management purposes, wastes/litters left by the operation staffs and local farmers such as plastic bags/containers, empty cans, etc., although may not be significant at one time (rotation), would accumulate and thus have permanent impacts on ecological/environmental conditions in sites or surrounding areas in a long run. However, this issue can be well controlled through effective operation regulations and agreements with the local communities.

4.15 *Conservation Sensitive Areas*

It is required in our forest management plan that the surrounding regions/areas of conservation values should be identified as part of background information about the plantation sites. The information will help prepare the sound management plan and operations to meet the requirements of local or regional conservation. In addition, unpredicted situations always happen. Some of local ecosystems, species, populations and even individuals may always be added as conservation objects later on. Sensitive areas similar to buffer zones should be set up according to the relevant laws and instructions, which can constitute certain habitats or peculiar animal passage corridors for conservation purposes. Specific operations considering certain requirements will improve the conditions for conservation. By following certain procedure, Sino-Forest's plantations may also be converted for full conservation or other purposes.

4.16 *Monitoring and Feedback of Information about Management*

Sino-Forest is establishing its own forest and management information system, which can also be used to monitor the status/change of the relevant elements listed in Table 1: 1) forest growth/production, 2) site/species adaptability, 3) site productivity (compacting, structure, nutrients etc.), 4) nutrient balance and dynamics, 5) vegetation, 6) species diversity, 7) local landscape/ecosystem composition, 8) soil erosion, 9) local water quality, and 10) other ecological/environmental indicators at sites and surrounding areas. At present, the system includes application of some new technologies such as TRMS (data management system) and 3S (RS, GIS and GPS). Included may also be late on: optimal wood-flow system, resources modeling, climate modeling, forest fire/disease/pest prediction system, cash-flow model etc. Relevant information are systematically collected, processed/analyzed, and, finally, used for: 1) making decisions, 2) revising operation procedures, and 3) improving the existing plantation management system.

5 Conclusion Remarks

Short-rotation plantations can play significant roles in forest/ecosystem restoration and environmental conservation in South China. Site productivity degradation, soil compacting, over-uptakes of nutrients, long-term conservation and other impacts are the main concerns/issues resulting from the inherent characteristic of short-rotation plantations and management operations such as site preparation, harvesting, forest road construction, etc. Long-term sustainable management of short-rotation plantations can be realized when ecological/environmental issues are properly considered in the management and operations. With prescribed operation procedures, site productivity can be maintained, soil erosion caused in operations can be minimized, the conservation ability of short-rotation plantations can be significantly enhanced, but the impacts can be greatly mitigated. Regional conservation demands can easily be considered in forest management plans.

References

Allen, H.L. 1987. Fertilizers: adding nutrients for enhanced forest productivity. J. For. 85:37-46.

Bauhus, J., Khanna, P.K. and Menden, N. 2000. Aboveground and belowground interactions in mixed plantations of *Eucalyptus globulus* and *Acacia mearnsii*. Can. J. For. Res. 30:1886-1894.

Couto, L. and Betters, D.R. 1995. Short-rotation eucalypt plantations in Brazil: Social and environmental issues. Oak Ridge National Laboratory, Oak Ridge, Tennessee, ORNL/TM-12846.

Deng, X., Chen, X. and Peng, S. 2001. Guangdong: investment in forestry is investment in environment. Southern Urban News, 5 June 2001.

Florence, R.G. 1986. Cultural problems of *Eucalyptus* as exotics. Commonw. For. Rev. 65:141-63.

Forest Stewardship Council. 2001. FSC principles and criteria. Forest Stewardship Council, Mexico.

Gladstone, W.T. and Ledig, F.T. 1990. Reducing pressure on natural forests through high-yield forestry. For. Ecol. Mgt. 35:69-78.

Kohm, K.A. and Franklin, J.F. 1997. Creating a forestry for the 21st century. Island Press, Washington, D.C., 475p.

Li, Z., Yang, M., Chen, S. Wang, G., Mo, X. and Yao, D. 2000. Study on regionalization for eucalypt introduction cultural area in China. J. Centr. South For. Coll. 20:1-5.

Libby, W.J. 1980. What is a safe number of clones per plantation? In: "Resistance to diseases and pests in forest trees", H.M. Heybroek, B.R. Stephan and K. von

Weissenberg (eds), Wageningen, Netherlands, Centre for Agricultural Publishing and Documentation, pp342-360.

Lima, W.P. 1990. Overland flow and soil nutrient losses from eucalyptus. IPEF 1:35-44.

Liu, A., Ma, X. and He, Z. 1999. Effects of intercropping green manure crops on degraded soil improvement of Chinese fir plantation after control burning. J. Zhejiang For. Coll. 16:369-374.

Palmberg, C. 1986. Are eucalypts ecologically harmful? Unasylva 38:19-28.

Pan, Z. and You, Y. 1994. Growing exotic trees in China. Beijing Science and Technology Publishing House, Beijing, 758p.

Powers, R.T. 1999. On the sustainable productivity of planted forests. New For. 17:263-306.

Qi, S. (ed) 2002. *Eucalyptus* in China. Chinese Forestry Publishing House, Beijing, 517p.

Sedjo, R.A. 1999. The potential of high-yield plantation forestry for meeting timber needs. New For. 17:339-359.

Wei, R.-P., Yang, Y. and Qi, S. 2001. Role of technology in short-rotation tree plantations: a view from the private sector. In: "Exotic tree species and environment", H. Wang, Y. Zheng and R.-P. Wei (eds), Chinese Environmental Science Publishing House, Beijing, pp15-21.

Youngquist, J.A. and Hamilton, T.E. 2000. A look at the world's timber resources and processing facilities. In: "Forests and society: the role of research", B. Krishnapillay, E. Soepadmo, N.L. Arshad, A.H.H. Wong, S. Appanah, S.W. Chik, N. Manokaran, H.L. Tong and K.K. Choon (eds), XXI IUFRO World Congress, Kuala Lumpur, 7-12 August 2000, pp183-194.

华南地区兼顾生态环境问题之短轮伐期人工林持续经营：嘉汉林业实践例证

魏润鹏[1,2]

[1]加拿大嘉汉林业集团有限公司
香港湾仔港湾道30号新鸿基中心3129-40室
[2]华南农业大学林学院
510642 中国广州

本文主要讨论了现代短轮伐期人工林管理过程中出现的生态/环境问题。这些问题多数与速生短轮伐期人工林的特点相关。现有的操作规程应予修改，或建立新的操作规程以满足不断增长的生态和环境保护需求。嘉汉林业公司在华南地区的实践提供了示范。目前情况下，短期内短轮伐期人工林可以改进当地及周围景观、生态系统、生物多样性、植被状况及环境保护；但从长远看来，亦可能不利持续经营，减慢生态和环境恢复进程。具有规定技术和操作规程的管理系统能减少或转变人工林对生态和环境保护潜在的不利影响。

FORESTRY LAND REGISTRATION IN THE SOUTHERN CHINA

MASATOSHI ENDO [1,2]

[1] *Guangxi Oji Plantation Forest Co., Ltd., 21 Floor Wukuang Building 32 Yuanhu Road, Nanning, Guangxi, China 530022*
[2] *Oji Paper Co., Ltd., Ginza 4-7-5, Chuo-Ku, Tokyo, Japan 104-0061*

Internationally, as the importance of forest plantations increases, investigations to boost forest productivities frequently take place. However, despite its importance, registration of forestry lands is often an overlooked subject in research or in actual administrative processes. This paper first describes experiences of land registration by Oji Paper Co., Ltd.'s overseas forest subsidiaries. Next, it describes the situation of forestry land registration in Southern China, where Guangxi Oji Plantation Forest Co., Ltd., a subsidiary forest plantation company of Oji Paper, operates. In Southern China, forestry land registration is legally defined, but actual registration appears still hampered by several factors. The lack of a registering system is the main physical barrier, while the general lack of understanding for the importance of forestry land registration appears an underlying cause. This paper emphasizes that efficient forestry land registration is one of the important factors in facilitating the development of industrial forest plantations.

1 Background

If one decides to manage commercial forest plantation project by using either "crop sharing" or "leasing"[1], registration of forestry lands becomes an important issue in order to protect the investor's[2] rights. To protect the investor's rights, the following two conditions are important. First, a land provider has clearly defined land ownership[3] and this ownership is registered with the authority. Second, "crop sharing" or "leasing" agreement is clearly registered on the land. Without these two, the landowner can provide the same land to more than one plantation projects in duplicate or can sell the land without acknowledging the existing "crop sharing" or "leasing" agreement.

This paper mainly focuses on how the land is registered when a plantation project secures lands by using "crop sharing" or "leasing". In this paper, "registration of land" ideally refers to a system in which the name of investor and

---

[1] Here, "crop sharing" refers to a land contract in which the investor does not pay land use fee(s), but instead provide land owner(s) with a portion of final product (wood). "Lease" refers to a land contract in which the investor pays land use fee(s) to the land owner(s), but instead the investor has the right to remove all the products. Both terms are not of legal definition.

[2] Refers to forest plantation companies and individuals who establish plantations.

[3] In Australia, China, and some other countries, many forestry lands are owned by the nation, not by individuals and companies. Therefore, in strict sense, it has to be described as "owner of land use rights" and/or "ownership of land use rights". However, for simplicity, "landowner" and/or "land ownership" are used.

his/her rights are registered in the land register of an agent (for example, land office, forestry department, et al.), and register is open to the public. With such a system, a prospective forestry investor can check whether s/he can have legal rights to establish forest plantation before s/he actually starts planting trees on the lands. Some people argue that an approval on a "share crop" or "lease" agreement by an authority is registration. However, the author argues that it is not sufficient because such an approval may not have sufficient effects in preventing the landowner from contracting out the same land twice, for example.

2 Overseas Forest Plantation Projects of Oji Paper Co., Ltd. and the Forest Land Registration

In the case of Japan, seventy percent of the total land is covered with forests. However, because of the low labor productivities due to the steep forest terrains and high labor costs, it is increasingly becoming difficult for Japanese pulp and paper companies to use domestic wood chip. In fact, approximately two-third of all the wood chip that the Japanese pulp and paper industry uses is imported. On the other hand, while it is not widely known, just like oil shortage, chip shortage hit the Japanese pulp and paper industry in 1974, 1980 and 1995. During such periods, the industry suffered from suddenly skyrocketed wood chip prices. In response to such difficult periods, the pulp and paper industry in Japan began considering overseas forest plantation projects in the 1970's. With the rising concerns for environmental protection, many pulp and paper companies in Japan began overseas forestation projects during the 1990's. As for Oji Paper Co., Ltd. (Oji), it adopted an environmental policy to establish a total of 200,000 hectares of overseas forest plantations by 2010. Table 1 summarizes Oji's overseas forest plantation projects (excluding those projects started as pulp projects) and their methods for securing lands. In China, Oji established Guangxi Oji Plantation Forest Co., Ltd. (Guangxi Oji project) in November 2001, starting the first commercial plantation project in China.

Regarding land securing in the Oji's overseas plantation projects, some discussions relevant to the subject of this paper are as follows:

1) Oji projects attempt to use "crop sharing" or "leasing", in contrast to land purchasing, as much as possible to lower initial capital expenditures. Usually, the projects sign "crop sharing" or "leasing" agreements with landowners. The policy in Oji is that these agreements are registered.
2) In Australia, where four Oji projects are located, the land registration system is relatively well established. Oji projects often use the following two kinds of land registration instruments.
   (1) Profit a Prendre

   Profit a Prendre is a legal instrument for forestland registration. It first became popular following enactment of the New Zealand Forestry Rights Registration Act of 1983 in New Zealand, where the forest plantation became widespread much earlier than in Australia. Profit a

Table 1. The list of overseas forest plantation project of Oji with land securing method.

Country	Project Name	Land securing methods
Australia, West Australia	Albany Plantation Forest Company of Australia Pty. Limited	"Crop sharing" or "leasing" using Profit a Prendre, which is based on the Conservation and Land Management Act.
Australia, Victoria	Green Triangle Plantation Forest Company of Australia Pty. Limited East Victoria Plantation Forest Company of Australia Pty. Limited	"Leasing" using Forest Property Agreement, which is based on the Forestry Right Act. Also, purchased lands exist.
Australia, Queensland	Brisbane Plantation Forest Company of Australia Pty. Limited	"Leasing" using Profit a Prendre that is based on the Land Titles Act. Also, purchased lands.
New Zealand	Southland Plantation Forest company of New Zealand Limited	Mainly planting on purchased lands exist.
Vietnam	Quy Nhon Plantation Forest Company of Vietnam Limited	"Leasing" based on the Law on Land and Law on Forest and Forest Development and Protection.
Papua New Guinea	JANT Limited	"Sub-leasing" the land which the government leased from the local farmers.
The People's Republic of China, Guangxi	Guangxi Oji Plantation Forest Co.,	Registering contents of "crop sharing" contracts, according to the Forestry Law and its regulations. This will result in registering the land use rights and the ownership right on timber.

Prendre means taking of profit (thus, wood). In the case of forestry, it refers to the right to take the wood from the land. It is said that this instrument is developed for mining. Using this instrument, an investor can obtain the rights to plant, maintain, harvest trees and take wood out of the land. These rights are registered on the land and show up on the land registry and on the land certificate. One notable feature of this instrument is that the ownership of the trees until cut belongs to the landowner, but the right to take the wood out is as defined in the agreement (usually investor has the right to take out the entire or a portion of wood). This legal instrument is widely used in West Australia, Tasmania, and Queensland.

(2) Forestry Rights Act

Using Profit a Prendre, one could not obtain the ownership right of standing timber. Therefore, Victoria enacted the Forestry Right Act in 1996, enabling the forest investor to have the outright ownership of standing timbers using a contract, called Forest Property Agreement. As in Profit a Prendre, the rights in this agreement is registered on the land and on the forest land certificate. Furthermore, the Forestry Rights (amendment) Act of 2001 enabled that Forestry Property Agreement to define ownership of the carbon right and register carbon right. An interesting observation is for those agreements, signed before the Forestry Rights (amendment) Act of 2001, the carbon right belongs to the owner of trees, but not to the land owner unless specific agreements on the ownership of carbon right exist. In summary, this Victorian act is significant that outright[4] separation of ownership among the land (or land use right), standing timber, and carbon is now possible. This clear separation conceptually favors forestry development, as the investor now can own standing timbers. However, this act differs from a traditional legal view that the trees on the land belong to the landowner. Therefore, it is argued that the legal validity of this Victorian act can be challenged. However, such a possibility is probably slim. As an example in Victoria, Green Triangle Plantation Forest Company of Australia Pty. Limited takes the following steps to register the land and plant trees. In this manner, the land is registered before the land preparation takes place. Investor may have more security if the land can be registered before s/he starts putting the money on the land.

- Negotiation with farmers to lease the land,
- Land survey,
- Sign a lease contract (Forest Property Agreement),
- Register the land (and of agreement) in the land office, and,
- Start land preparation and plant trees.

(3) Land Securing in Vietnam and Papua New Guinea

In Oji's Vietnam project, the project leases the land from the government. Standing timbers on the land automatically become the property of the holder of land use rights. In Papua New Guinean Oji project, supposedly the government leases the land from local farmers and then subleases to the Oji project. However, because of the lack of administrative ability of the New Guinean Government, effectuation

[4] In contrast to profit a prendre.

of sublease tends to take a long time period, being the headache of the project.

3 Forest Land Registration in China[5]

With deteriorating natural environment and shortage of forest resource, China emphasizes forest plantation. According to a recent national plan, China plans to plant total of 13.33 million ha (6.18 million ha of new plantations and 7.15 million ha of plantation renewal) of plantations in the next 15 years. Of this, industrial forest plantation totals 5.86 million ha. In order to achieve this plantation goal, the Chinese government encourages both domestic and foreign companies to invest into forest plantations. As securing lands is an important step in establishing forest plantations, the pattern of forest land ownership in China is explained first. Afterward, forest property rights and their registration process are discussed.

1) In China, ownership and use rights of the forest lands are defined in the Forest Law (People's Republic of China 1984). This is explained in the effective regulations (National Forestry Department 1998), the Land Management Act of the People's Republic of China (1998).
2) The ownership and/or use rights of Forests, standing timber, and forest lands are called the forest property rights.
3) Ownership is categorized as the following
   (1) National Ownership (State Ownership)
   The Chinese Constitution defines that mines, rivers and lakes, forests, ridges, grasslands, barren lands, and sand dunes in mouth of rivers all belong to the nation except where ownership is titled to a certain people's group based on the law. The Forestry Law also defines that all the forests, except forests owned by certain people's groups, belong to the nation.
   (2) Group Ownership
   The Constitution and the Forestry Law have provisions for group ownership of forests. This category includes forests, which became owned by a group due to the land reform, and shelterbelt forests along the road, railroad, and rivers. A people's group can own the standing timber through an agreement with national forest farms.
   (3) Individual Ownership
   The laws can be interpreted that individual (including corporations and institutions) can have the ownership of standing timbers.
4) Use Rights
   In China, as open market economy develops, forms of use rights became more various. Often seen user rights patterns are the following.

[5] The People's Republic of China

(1) A national institution has the use rights of a national forest farm (national lands).
(2) A people's group has the use rights of a national forest farm through a legal process.
(3) A national institution has the use rights of forests of people's group.
(4) An individual, corporation, or institution obtains the use rights of national lands or the forest lands of a people's group, through crop sharing, lease, sale, et al.

5) Forest Land Registration to obtain a forest right certificate

This section is based on an interpretation of Regulation on Wood and Forest Land Tenure Registration (National Forestry Department 2001). However, as far as the author has seen, registration according to this regulation has not yet started.

(1) The people's government registers and issues forest right certificates. A legal owner of the certificate is entitled to legal protection for his/her forest property as specified in the certificate. Other parties cannot infringe his/her forest property rights.
(2) The issuing institution of forest right certificate and certificate form. The people's government, at the level of county or above, can issue forest right certificate according to the relevant regulations of the Forestry Law. Forest right certificate must use a standardized form indicated by the forestry-related-department[6] in the State Council of the Republic of China.
(3) Documents and information necessary for the first time registration of a forest right certificate.
    - Application form for forest property rights registration.
    - Map indicating land boundaries, which matches the actual situation.
    - A description of current forests and/or forest lands. For example, this includes species of trees, the amount of standing timber, area, and the number of trees.
    - Ownership status of standing timber (if applicable).
    - Applicant's certificate (company registration, individual's identification form, et al.) with information about the entity.
    - Evidence that certifies that the applicant has the right to the forest property.
(4) Documents and information necessary when modifying information in a forest right certificate.
    - An existing forest right certificate.
    - Application form for forest property right registration.
    - An explanation of modification.

[6] In practice, referring to State Forestry Administration of the R.P. of China.

- Some documents in (3) immediately above.

(5) Issuing a forest right certificate and time limits. After receiving an application, within 10 days, 30-day public announcement about the application has to be made. If all the below conditions are satisfied, certificate will be issued within 3 months after the application.
- The map and explanation of the current status are all correct, matching the actual status.
- Evidences of forest property rights are adequate and legal.
- There is no counterargument from the public against the application.

4 Some Comments on Forest Land Registration

In this sector, based on experiences of Guangxi Oji project in Guangxi Zhuang minority autonomous region (hereafter, Guangxi), some observational comments on the forest land registration in the southern China are discussed.

1) Forest land registration of foreign companies. In China, registration is mandatory for foreign companies in some cases. For example, the regulations on Foreign Investor's Forest Management in Guangdong Province (Guangdong Forestry Department 1997) defines that a foreign company wishing to plant trees must apply for the registration of forest lands, within 10 days after the company articles are approved, at the country forestry department. The reasons for requiring foreign company to complete forest land registration probably are 1) to protect the rights of the foreign investor and 2) to effectively control forest plantations by foreign companies. In Guangxi, which is located to the west of Guangdong, such regulations for foreign companies have been drafted up (Guangxi Zhuang Autonomous Region's Forestry Department 2000, 2001), however they have not yet in effect.

2) In reality, registration of crop sharing and leasing is not proceeding according to the regulations.

(1) Currently, some foreign companies in the southern China appear to do forest land registration but not according to the forest land registration steps described above. If forest land registration is performed, it usually dealt with a local (county or above) forestry department, using a locally available forest right certificate form.

(2) In some cases, forest land registration with a local county forestry department is not easy or if done, not reliable.

(3) In order to provide a better forest land registration system, for example, the Guangxi government is discussing to adopt a new forest land registration system. In August 2002, the Guangxi government is preparing to do a model registration according to the new regulation in two counties. This involves purchasing of computer systems,

personnel training, and connecting local registration system to the national forest land registration system. If this model registration is successful, the government eventually plans to make such registration system available anywhere in Guangxi.

3) The first forest right certificate of ownership or use rights. Of course, registration of crop sharing and leasing contract on the land is important. However, one can often encounter even more primitive situation, a land provider simply does not have a forest right certificate. In Guangxi, national forest farms usually appear to have the certificate. But many people's groups[7] do not have the certificate for their collective farm lands. Through out China, it is estimated that only 40-50% of people's groups have the forest right certificate, but precise figure is not well known. If the land providers do not have the certificate, it is more difficult to secure the forest property rights through crop sharing or leasing agreement. Put in the other way, if the land providers do not have the certificate, any agreement to use the lands face risk whether the land providers have the legal rights to do so. If such risks prevail, it severely discourages investments.

4) The form of certificate. Currently, there exist several forms of forest right certificates. In Guangxi, forest right certificate is usually printed as Mountain forest right certificate, a term which the Forestry Law and its regulations do not mention. In effect, "Mountain forest right certificate" and "forest right certificate" are the same, but two different names can cause some confusion. Another problem in these certificates is the quality of accompanying maps. In some cases, maps are not clear and confirming the maps against the actual situation can be extremely difficult, if possible at all.

5) Public announcement. According to the regulations, forest land registration will require public announcement for 30 days. For example, even if a registered forest land does not have any tenure disputes, in a process of registering crop sharing or leasing, a third party can freely express counter opinions for the registration, possibly causing unwanted consequences to the land owner. In other words, such public announcement requirement can be a source of new problems, and can be a disincentive for forest land registration.

6) Registration of land and standing timber, can different entities own each? In related regulations in China, it is not clearly defined whether the owner of standing timber can be different from the landowner (or holder of land use rights). However, the most likely interpretation of related regulation and the Law is that the ownership of timber can be different from the land ownership in China. In this aspect, Chinese land registration system is closer to the concept of Forest Rights Act than Profit a Predre system in Australia (Inions 1998).

---

[7] They are usually collective farmers.

5 Some Opinions Expressed by Other Forest Plantation Companies

The author contacted the managers of other forest plantation companies in Guangxi. Some comments are the following.

1) Question: Is forest land registration important?
   - It is considered important.
   - Considered important, but in reality it is not easy to do because the registration system is not yet established.
   - We plant trees on the land of farmers who do not have forest right certificate. In such a situation, registering the land is not easy, and protecting investor's rights is extremely difficult. With some farmers groups, registering alone will not protect the investor, because farmers do not care who has the legal rights.
2) Question: what are the hurdles for registration?
   - Registration system does not exist.
   - Coordinating farmers is not easy. In order to register the land owned by farmers, some two-third of farmers in the group have to sign, but to do this, we have to have a lot of negotiations and treat them to dinners and drinks. Costs add up.
   - Landowners are afraid that after a public announcement, counter opinion on the ownership may appear. Landowners do not want new troubles.
3) Question: what can be done?
   - Hope that the Guangxi Zhuang Autonomous Region's Forestry Department will establish the land registration system soon.

6 Forest Land Registration to Promote Plantation Development

In this section, some tips for establishing a sound forest land registration system, suiting plantation development, are discussed.

1) Forest land registration process is efficient and costs are reasonable. This does not need further elaboration. However, forest land registration is not an area which draws political attentions. Therefore, such system is easy to discuss, but in reality, not easy to establish such a system.
2) Develop a system, which allows registration or pre-registration before land preparation starts. In China, after planting trees, the boundary of plantation became well established, then proceed with forest land registration. This seems prevailing. Of course, this concept has its own merit. However, from the investor's viewpoint, starting land preparation on the land, which may or may not successfully be registered, is a risk. On the other hand, from the landowner's view, registering forest land site before an investor plants trees is a risk because after registering the site, the investor may not actually plant the trees. Therefore, a system which can pre-register land

before planting trees, and can complete registering lands after trees are planted may be ideal. For example, Guangxi Oji project currently takes the following steps to secure lands.

- Discussion with National Forest farms about securing lands.
- Sign a land use contract. Only approximate area is described. Actual area is surveyed after planting trees.
- Land preparation and planting.
- Plantation area survey.
- Land registration.

3) Land survey requirement is not excessively strict, nor too loose, and survey costs are reasonable.
4) Public announcement is important in a sense promoting a fair registration. However, if public announcement becomes subject of malicious attacks by those who do not have legal rights to do so, the merits of announcement are tarnished. Therefore, it may be appropriate to establish fines for malicious counter opinions and/or counter opinions without adequate evidence against a public announcement of land registration.
5) The agreement (e.g. the investor's rights) is carried down to a new owner when the land is sold or succeeded to a next generation. Of course, this is a part of definition of land registration. This is extremely important in order to protect the investor's rights.
6) Open land register. Currently, many land registry in China seems not open to the public scrutiny and the public can only see issued forest right certificate. However, forest right certificate itself can be easily counterfeited. Therefore, in order to take maximum merit of forest land registration, the public must have the right to examine the content of land registry to reconfirm the content of forest right certificate.
7) Clarify that an investor can own the rights of standing timber separate from the land ownership. Such clarification can assist in creating favorable environment for forest plantation development.

However, forest land registration is only a partial answer to solve the forest land issues in China. For example, the author has seen many forest plantation projects in China which used farmers' lands, but resulting in disputes. The main cause of disputes seems the difference in cultures of farmers and people in forest plantation projects. Farmers are not used to long-term contractual agreements. In the mind of some farmers, agreements in a contract may not have significant importance. To them, what agreed in a contract is only valid when it is signed. As time goes by, people's mind changes and thus the content of the agreement must be adjusted, according to their thinking. Currently, reflecting abundant agricultural production in China, the Chinese government encourages some farm lands to be put back into forests. If more farm lands are to be put into forests, ways to better understand the farmers' thinking and to establish real trusts with farmers must be considered. This would help both farmers and forestry projects.

7 Ending Comments

The author is not a legal expert, but a forester and project administrator. Possibly, this paper contains some errors, but the author decided to write this paper because the discussions on this topic are currently very limited. The author wishes that this paper could contribute something to the development of sound forest land registration system.

Acknowledgements

The author thanks Oji Paper Co., Ltd. for permitting the author to publish this paper, Dr. G. Inions for providing profound insights to forest land registration in Australia. Thanks also go to others from Oji's subsidiary overseas forest companies such as Mr. S. Yang, Mr. Y. Huang, Ms. M. Wei, and Mr. Meng for providing information and assistance in preparation of this manuscript.

References

Guangdong Province Forestry Department 1997. Regulations on Foreign Investor's Forest Management in Guangdong Province. 1 August 1 1997, Guangzhou.

Guangxi Zhuang Autonomous Region's Forestry Department 2000. Regulations on Foreign Investor's Forest Management in Guangxi (draft). Nanjing.

Guangxi Zhuang Autonomous Region's Forestry Department (ed) 2001. Training materials for forest land management. Nanning, Guangxi.

Inions, G. 1998. Personal Discussion. East Coast Tree Farms Pty. Ltd., Brisbane, Australia.

National Forestry Department 1998. Regulations on Land Management Practices, 27 December 1998, Beijing.

National Forestry Department 2001. Regulations on Wood and Forest Land Tenure Registration. Beijing.

People's Republic of China 1984. Forestry Law of People's Republic of China. 20 September 1984, Beijing.

People's Republic of China 1998. Land Management Act of the People's Republic of China. Beijing.

关于中国南部造林用地的登记情况

远藤　正俊[1,2]

[1] 广西王子丰产林有限公司
530022 中国广西南宁市园湖路 32 号五矿大厦 21F
[2] 王子制纸株式会社
Ginza 4-7-5, Chuo-Ku, Tokyo, Japan 104-0061

当前，在国际性造林的重要性正在日益凸显的背景下，许多林业研究机构为了提高林业生产量而展开频繁的研究。但是，在植树造林这方面有一个非常重要的问题，却在研究以及实际的行政工作中常常容易被忽视，这就是造林用地的登记。在这份报告中，我首先介绍的就是王子制纸株式会社的海外造林公司在造林用地登记方面所积累的经验，另外，我还将为大家介绍王子制纸株式会社的子公司广西王子丰产林有限公司在中国南部的造林用地登记情况以及面临的问题。在中国的南部，虽然颁布了与造林用地登记手续相关的法律和办法，但是实际登记起来并不是很顺利。登记体系还没有明确的确立起来，是目前我们面临的最大问题。从对这个问题的分析，可以指出，整个社会对造林用地登记的重要性认识不足。在这份报告中，我将说明高效率的造林用地登记体系是促进商业造林健康快速发展的不可或缺的条件。

II. Breeding Method

THE CHALLENGES AND LESSONS FROM BREEDING *EUCALYPTUS*

NUNO M.G. BORRALHO

RAIZ, Forest and Paper Research Institute
Herdade Torre Bela, Apartado 15, Alcoentre

Eucalyptus breeding efforts started relatively late, compared with conifers species and poplar. By 1980, only a few breeding programmes were in place, little was known about the genetics of the species, their reproductive biology and the best deployment schemes, including cloning and seed orchards. It is remarkable and inspiring that by now, some of the most advanced and focused breeding programmes in the world are from *Eucalyptus*. Today, several programs are now well into their 3rd generation of breeding, leading the way in the use of clonal forestry and becoming one of the preferred species in the applications of genomics to forestry.

The paper will briefly review the progress in strategy, technology and outcome over time in eucalypt breeding, in par with the changes in forestry as a whole, and discuss what may be the lessons to take from this co-evolution. The general principles and restrictions underlying *Eucalyptus* breeding programs are presented and put in perspective. Finally, some of the opportunities for developing new and better programs are briefly discussed.

1 Introduction

Systematic *Eucalyptus* breeding efforts started in the late 60's, in Portugal (Dillner et al. 1969), Congo (Delwaulle 1989) and Brazil (see Ferreira and Santos 1997 for a review). It is now being studied and developed in many countries around the world, and constitutes one of the most researched and bred forest species.

Still, by 1980, very little was known about the biology, adaptation and genetics of key traits in all eucalypt species of commercial interest, and there was only a few improved seeds or clones available around the world, mostly in Brazil, South Africa and Portugal. Today, eucalypt breeding programs are amongst the most advanced in the world, several are now in their 3rd cycle of breeding, using large scale clonal deployment schemes and developing state-of-the-art biotechnology tools, including genetic transformation (MacRae and van Staden 1999, Sartoretto et al. 2002) and genomics (Grattapaglia 2001).

Part of this success can be explained by a relatively simple and well defined breeding objective. With few exceptions, *Eucalyptus* breeding programs have been initiated and managed with an interest in establishing new plantations for pulpwood, either as producers or as integrated pulp and paper businesses. A cause for such a sustained interest in improving eucalypts is obviously attributed to the increasing world demand for eucalypt fibre - *Eucalyptus* pulp and paper constitutes now a market reference. The impetus to establish new eucalypt plantations, mostly in areas and countries with no previous eucalypt forestry tradition, has certainly stimulated the initiation of aggressive breeding programmes with a very specific set of goals: high productivity and resistance to major pests or diseases, high pulp yield and reasonably high density.

Today, our knowledge of the basic quantitative genetic parameters (heritabilities and correlations), its stability across environments and changes with age, is reasonable (see Eldridge et al. 1993 and Potts and Wiltshire 1997 for a review). On the other hand, the ability to propagate selected genotypes (clones) instead of seed orchards, has been extensively explored to the point where several programs rely exclusively on clonal forestry to deploy their best material. This included initially a few tropical eucalypt species and hybrids, which are relatively easy to propagate, but has now been applied to temperate eucalypt species such as *E. globulus* (Araújo et al. 1996).

2 The Strategies

Eucalyptus have characteristics which render them unique. The most famous, albeit not unique to the genus, is the ability to produce commercially interesting interspecific hybrids. Eucalypt hybrids are today the basis for most plantations in Brazil, South Africa and Congo, and new hybrids are being developed to broaden this initial generation of crossbreeding. We will not discuss here whether this success is the result of strong heterosis in specific combinations or careful and accurate selection on extremely variable F_1 interspecific progenies. Reported results are conflicting but see Potts and Dungey 2001 for some insights. Eucalypt hybrid breeding and the numerous strategies that emerged, are certainly one of the most challenging and innovative aspect of eucalypt breeding programs.

Even within a single species, and notably with *E. globulus*, *E. grandis* and *E. nitens*, there has also been evidence of important racial (or sub-population) differences. In *E. globulus*, one of the best studied species, large race effects have been found for growth, adaptation and wood characteristics (MacDonald et al. 1997; Dutkowski and Potts 1999; Soria and Borralho 1997). Some of these differences are indeed as large as the genetic variation found within any particular race, and may be large enough to suspect that effects such as outbreeding depression or heterosis may be of significance. As a result, some strategies have been conservative enough to structure their breeding population within such races (e.g. the sublining reported by Woolaston et al. 1991). This would be important to retain the natural racial structure and associated genetic stability.

In terms of breeding strategies, there has been some striking examples of innovation in eucalypts. When the concept of *nucleus* breeding (a smaller elite population plus a larger base population), was extended to forestry, the first reported application was with eucalypts (Cotterill et al. 1989; White 1995). In 1997, when the *rolling front* strategy was enunciated (a continuous, overlapping generation scheme; Borralho and Dutkowski 1998), the described application was also for a *Eucalyptus* program (Araújo et al. 1996). These two examples illustrate well how *Eucalyptus* breeding have been, if not the motive, at least the privileged recipient of innovative breeding schemes in forestry.

Finally, *Eucalyptus* have used a wide range of deployment schemes, including the traditional seed orchard, macro- and mini-cutings and more recently, the possibility of mass controlled or assisted pollination schemes. Most of these processes required considerable effort in understanding the basic reproductive biology of the species (e.g. Eldridge et al. 1993), as well as cost and other considerations (see Borralho 1997 for a review).

3 The Successes and Failures of Breeding

Despite its successful story, eucalypt breeding programs (like with other species, by the way) have and certainly will continue to sustain external pressure. Several programs have been discontinued or are downsized. Most of this has been the result of unavoidable external economic or political cycles. But some of these problems are more permanent and could probably be avoided if simple rules of design are followed. In will review below some of those rules.

Tree breeding is, after all, a combination of activities aiming to improve the performance of individual plants, under a given forest management system. In most cases, such system is a eucalypts plantation, managed to provide wood-chips, or wood to supply directly a mill or group of mills.

The costs and risks associated with growing wood are always substantial. About 15% of the costs of producing one tonne of pulp can be attributed to growing the trees in the first place. Another 30% would be spent in harvesting and transported it to where it can be processed. So, any process that would be able to improve the yield of plantations at reduced expenses or inputs, is a good thing. Tree breeding is one such process, but it will be unwise to think it would be the only one or the best one, or that it will always be justifiable. This needs to be acknowledged and understood by all of those involved in improving plantation performance, particularly by tree breeders. The benefits from breeding should be competitive against other possible developments in the productive process, including better silvicultural practices, better planning of operations and better quality control schemes, which compete against a limited amount of financial resources.

As a simplification, a breeding strategy should achieve three things. Firstly, it should have a measurable and significant effect on the economics of the company or of its members, for example by reducing annually the total wood costs by 2%, increase mill capacity by 5%, or reducing effluent discharges by 10% in 5 years. Secondly, since such changes come at a cost, they must be competitive (financially interesting) compared with alternative R&D investments. Since breeding is replacing old plantations by new improved ones, the internal rate of return of establishing new plantations must be quite high (>10%) to compensate for the relatively high costs and higher risks of replacement. Thirdly, the rate of substitution of old unimproved stocks by new improved ones must be high too (say 5-10% of plantation estate should be replaced every year), otherwise there will be little economies of scale and no justification for a continuous breeding effort of reasonable size. The slow replacement rate is particularly relevant in many *Eucalyptus* plantations, since most species will coppice and there will be a strong pressure to continue to manage the forest in that way, hence slowing down the replacement of old by new stocks.

In dealing with these issues, I will now introduce and discuss the four major domains of breeding (the selection value or merit, the crossing scheme, the testing and measuring and the deployment) and how they will need to be structured.

4 The Four Domains of Breeding

4.1 *Selection Value*

Since breeding is mostly about choosing a few individuals from a large number of candidates, the criteria used to rank them is obviously a major determinant of success. In *Eucalyptus*, as previously said, the aim or the direction of selection has been reasonably well described. The most important variables for those establishing eucalypt plantations for pulpwood today are growth rate and survival, particularly important for forest growers as volume or green weight is how most of the wood is traded, wood density, important for most woodchip overseas exporters, as dry-weight will determine the price, and finally pulp yield, which for most pulp mills would influence the final processing costs, as well as the annual outputs and sales from a given mill. The relative importance of these variables will therefore vary according to the specific needs and characteristics of the business (see Borralho et al. 1993; Fonseca et al. 1995 and Greaves et al. 1997 for a discussion on this), but some sort of selection pressure on each of these three key variables must be considered.

In *Eucalyptus*, growth, wood density and pulp yield have been demonstrated to be genetically independent (see Muneri and Raymond 2000 and Raymond et al. 2001 for recent review on such parameters), so selection will need to carefully weight the three variables, and the merit of each individual tree will be a trade-off between them.

The estimation of genetic merit depends also on the use of appropriate statistical and genetic models for the species and data present. A detailed discussion on this topic is beyond the scope of this paper, but a few considerations will be given below. It is not simple to define what constitutes an appropriate statistical analysis. The choice of the model represents always a simplification of reality, so a compromise must be reached between complexity (which goes as far as we understand the reality) and simplicity (which goes as far as we are willing to reduce that same reality). In the end, the estimates must be reliable (or at least predicted with some appreciation of the errors associated with it) and reasonably unbiased. For that to happen, genetic parameters (variances and co-variances) must be known with rigor, measurements adjusted for the known environmental effect and have very little errors in the data (measurements and pedigree).

In *Eucalyptus*, our understanding of the genetics architecture of the traits studied is restricted to race, additive and dominance genetic effects, so it is not only the models that are simple, but also our understanding of the genetic reality – it remains to be demonstrated that the genetic mode of action is that simple. So, any models used will have necessarily to be simple. In general, for most eucalypt populations, race effects must certainly be included. Additive effects are however the most important, and dominance effects and inbreeding depression should probably be considered if the data includes full-sibs and inbred material. Finally, site and replicate effects, competition between trees (and other plot effects) and within replicate heterogeneity) may need careful consideration.

One unavoidable problem in most eucalypt breeding programs is that with time data become more complex, measurements will cover several ages, different traits will be available in different trees and there is interactions occurring between genotypes and sites. Statistically, this creates unbalance in the data, heterogeneity of variances or difficulty in grouping sites. Given the relative complexity of the data and pedigree, the individual tree mixed model (Best Linear Unbiased Prediction) should be the method of choice in deriving estimates of breeding values in forestry. Borralho (1995) provided an overview on the topic, and Araújo et al. (1996) a simple (probably too simple) example of application to *Eucalyptus* breeding populations. Nevertheless, this remains one of the most challenging and technically important issues in any breeding program.

4.2 *Crossing*

Regardless of the breeding strategy followed (a rigid and discrete sequence of events, or a variable rolling front scheme), crossing must serve the main purpose of recombining the genes in the previously selected group, and create enough variability amongst the plants to be used in the new generation of testing and selection. To do this well, the crossing scheme must choose wisely which particular crosses it creates. It must have enough independent crosses per parent (3 or 4) to allow a good estimate of its breeding value, avoid crosses between relatives and the domination of a few ancestors, prefer crosses which combines well tested and poorly tested selection (compensatory mating), and should create pairs between particularly good parents (assortative mating).

If one looks carefully at these principles, open pollinated schemes are not a good choice at all in eucalypts. Flowering is highly variable amongst genotypes, between 5 and 20% of the progeny will be self pollinated (hence inbred) and some parents will tend to dominate the paternal contribution. Open pollinated schemes are certainly practical and cheap, but in my view they should not be widely used as the basis for future generations of breeding.

The rule to guide the selection and crossing has been reasonably well covered in the recent forestry literature. In short, selection should optimise the aggregate genetic merit minus a penalty of the selected group coancestry. This optimisation has some serious implications and should be carefully considered before a breeding and crossing strategy is adopted (see Brisbane and Gibson 1995 and Wei 1995 for discussion on this). The most important consequence deals with the information used to estimate EBV. If individuals are poorly tested, themselves, but have large families, the contribution to genetic merit will be dominated by information from its relatives, hence being the same for all individuals sharing the same relatives. As a result, trees that are better ranked will have a tendency to the clustered around a few best families. In this case, the restriction imposed on group coancestry will have a large impact on gains. If, on the other hand, families are smaller and each individual tree is well tested (for example by being clonally replicated), then its EBV will be mostly derived from its own phenotypic information and less affected by its relatives' performance. In this case, restriction on group co-ancestry will have a smaller impact on the gains. In other words, the rule of selecting with restriction on group coancestry means the program should try to improve, as much as possible, the accuracy of individual tree selection (or the quality of the information on the phenotype), hence reducing the relative contribution of pedigree information on the EBV.

For a while, it would be possible to reconcile high genetic gains with low group coancestry, but at some stage, a decision will always have to be made between how much coancestry would be allowed to accumulate. This takes us to an important issue. For how long (how many recurrent cycles of breeding and selecting) is the program expected to go? In my view, people have avoided answering this question because they confound two issues. The first one is the expectations of a program (for how long is it justifiable to spend, year after year, resources in improving increasingly better varieties). The second one is the 'evolutionary' question of whether the program will provide society in the future with a established plantation which is sufficiently variable, in a genetic sense, as not to restrict further development decision. Also whether or not it will be useful or sufficient adapted to avoid deteriorating the economic, social and ecological value of the forests now being established.

The answer to the first question is more straightforward (as it can be analysed theoretically) and surely should be no longer than around 5 to 7 cycles of breeding (around 50 years for most *Eucalyptus*). A reasonable target, in this case, would be to accept a final level of coancestry of 0.0625. In other words, clones being deployed after 5 cycles of selection should not exceed, on average, a third-degree relationship (e.g. offspring to great-grandparent or between single first cousins). Even random mating amongst such population would result in an average inbreeding of just 0.125 (which would result, according to estimates in *Eucalyptus*, in a drop of 10% on growth due to inbreeding depression). Such levels of inbreeding are approximately those expected in the seed from orchard with moderate levels of selfing.

This, off course, does not tell us whether or not such populations are sufficiently diverse for the indeterminate future. Levels of genetic diversity are certainly high in trees, and certainly that would be true for *Eucalyptus* (Moran 1992; Potts and Wiltshire 1997), but given the clustered genetic structure of native population, the genetic diversity captured in the first breeding population (a sample of the whole genetic diversity available) can vary considerably (see Dutkowski and Potts 1999 for a discussion).

It is therefore wise to ensure the base population represents as much as possible a reasonable sample of the initial available diversity. In the extreme case of hybrid breeding programs, it is likely that advanced generation populations would in fact be more genetically diverse than any of the existing natural populations of pure species.

4.3 *Testing and Measuring*

The establishment, maintenance and assessment of field trials are the most complicated and costly part of any breeding program. *Eucalyptus* are fast growing trees being managed under short rotations. There has been consistent evidence that age-age correlations are very strong, and genotype by site interactions are low (MacDonald et al. 1997; Wei and Borralho 1998), and even lower for wood traits (MacDonald et al. 1997; Muneri and Raymond 2000; Raymond et al. 2001). These elements strongly suggest a low number of sites required (2 or 3 and certainly less than 5; but see Lindgren 1985 for a theoretical treatment) for reliably testing of a genotype.

The issue of experimental design will depend on local site conditions and the number of genetic entries (but see Williams et al. 2002). However experience tells us that, unless care is taken, a major issue would be to maintain appropriate connectedness between trials.

Finally, a few words on the assessment strategies used in eucalypts. A common paradox of measurements in eucalypt breeding programs is the difference between the economic importance attributed to the various traits, and the amount of information available. Wood characteristics (wood density and pulp yield) are poorly represented in many datasets mostly because of the practical difficulties of sampling the wood, but in many cases because of a conscious decision to measure mostly for growth. In a typical breeding objective for a P&P company, it would be equally valuable to have a clone which grows 20% faster than average, has 10% denser wood or has 3% higher pulp yield. Chances of identifying a tree with such characteristic would mean select 1 out of 20 for pulp yield, 1 out of 40 for volume and 1 out of 70 for density (values were calculated using RAIZ economic values and population parameters for the 3 traits). In conclusion, one would need to screen half as many trees (for the same benefit) when focusing on PY rather than volume. There are other factors influencing the final compromise between trees and traits (cost of assessment being one particularly important). In this regard, it should be referred here the recent developments of simpler and less expensive methods to screen for wood density and pulp yield, namely with the use of Pilodyn for density (cheap but considerable less reliable) and Near Infrared Reflectance (NIR) analysis (see Michell 1995 and Schimleck et al. 1999 for details).

4.4 *Deployment*

The success of any breeding program must be measured by the net benefits derived from the new improved plantations being established. That is the domain of the deployment strategy.

Although there are several technical aspects to be considers, the success in deployment relies more on commercial outcome than on technical know-how. *Eucalyptus* are suitable to a variety of deployment options, however, clonal forestry based on cuttings has been the method of choice. The use of cuttings has been particularly successful with *E. grandis* and some of its tropical hybrids in the Congo, Brazil and South Africa. The development of similar strategies for temperate species such as *E. globulus* and *E. nitens* has been more complicated. These species are difficult to root (Borralho and Wilson 1994; Tibbits et al. 1997) with percentage of genotypes with high potential for cloning being as low as 10 to 20%.

Seed orchards or seed production areas have also been extensively used (and proved effective) in several programs around the world. Also, large scale controlled or assisted pollination has been recently advocated (Leal and Cotterill 1997; Harbard et al. 1999; Williams et al. 1999) and would probably become an interesting option as simple but effective techniques for crossing *Eucalyptus* will continue to improve.

Borralho (1997) compared the various deployment options, under scenario relevant for *Eucalyptus*, and concluded that, under similar circumstances, clonal forestry would only outperform seed production schemes if costs per plant are not €0.50 higher than other seedling alternatives, unless non-additive effects are large. The factors that conditioned the choice are therefore the relative advantage of deploying genotypes (total genetic merit) rather than breeding values (average additive effects) and the plant costs.

5 Concluding Remarks

In the end, a successful scheme is one that achieves the 3 key aims of any breeding program: (i) a significant impact on the economic performance of those investing in the program, (ii) make it very attractive to establish new plantations, such that (iii) the rate of substitution of old by new plants is high.

The need to continuously improve productivity, efficiency and quality will remain a driving force in most activities that use *Eucalyptus* wood. Technological developments, either in the forest, during harvesting and transport, or in the transformation processes further down the chain, will continue to be needed. Tree breeding has particularly elegant features but it is, however, just one of such technological tools. Breeding improves plantation competitiveness by biologically manipulating the productivity of the tree, making it more productive and with better wood characteristics while using the same (or similar) land, water, nutrients and energy inputs. But it also has some drawbacks. Forests, unlike livestock or crops, have a long lifespan and take a long time to be replaced, even when we think of fast growing eucalypt plantations. And replacing an existing plantation by a new more productive one is a long-term (5 to 10 years) investment. So, breeders and foresters in general must continue to work together and find the right balance between the various opportunities to improve the value of eucalypt plantations (after all the major determinant of having a breeding program in the first place). In the end, breeding programs must be designed to serve that balance, a difficult but necessary strategy.

References

Araújo, J.A., Sousa, R., Lemos, L. and Borralho, N.M.G. 1996. Estimates of genetic parameters and prediction of breeding values for growth in *Eucalyptus globulus* combining clonal and full-sib information. Silvae Genet. 45:223-226.

Borralho, N.M.G. 1995. The impact of individual tree mixed model BLUP in tree breeding strategies. In: "Eucalypt plantation: Improving fibre yield and quality", B.M. Potts, N.M.G. Borralho, J.B. Reid, R.N. Cromer, W.N. Tibbits

and C.A. Raymond, Proc. of IUFRO-CRCTHF Conference, CRCTHF, Hobart, Australia, 19-24 February 1995, pp141-145.
Borralho, N.M.G. 1997. Seed orchards or cuttings: which is the best? In: "Silviculture and improvement of eucalypts", Proc. of IUFRO Conference, Salvador, Bahia, Embrapa, Brazil, 24-29 August 1997, pp330-336.
Borralho, N.M.G. and Dutkowski, G.W. 1998. Comparison of rolling front and discrete generation breeding strategies for trees. Can. J. For. Res. 28:987-993.
Borralho, N.M.G. and Wilson, P.J. 1994. Inheritance of initial survival and rooting ability in *Eucalyptus globulus* Labill. stem cuttings. Silvae Genet. 43:238-242.
Borralho, N.M.G., Cotterill, P.P. and Kanowski, P.J. 1993. Breeding objectives for pulp production of *Eucalyptus globulus* under different industrial cost structures. Can. J. For. Res. 23:648-656.
Brisbane, J.R. and Gibson, J.P. 1995. Balancing selection response and rate of inbreeding by including genetic relationships in selection decisions. Theor. Appl. Genet. 91:421-431.
Cotterill, P., Dean, C., Cameron, J. and Brindbergs, M. 1989. Nucleus breeding: a new strategy for rapid improvement under clonal forestry. In: G.L. Gibson, A.R. Griffon and A.C. Matheson (eds), Proc. of IUFRO Conference on "Breeding tropical trees: Population structure and genetic improvement strategies in clonal and seedling forestry", Pattaya, Thailand, OFI, Oxford, UK and Winrock International, Arlington, Virginia, USA, pp39-51.
Delwaulle, J.C. 1989. Plantation clonales au Congo point des recherches sur les choix des clones dix and and après les premiéres plantation. In: G.L. Gibson, A.R. Griffon and A.C. Matheson (eds), Proc. of IUFRO Conference on "Breeding tropical trees: Population structure and genetic improvement strategies in clonal and seedling forestry", Pattaya, Thailand, OFI, Oxford, UK and Winrock International, Arlington, Virginia, USA, pp431-434.
Dillner, B., Ljunger, A., Herud, O.A. and Thune-Larson, E. 1969. The breeding of *Eucalyptus globulus* on the basis of wood density, chemical composition and growth rate. Forskinstiutionen, Billerud, Report 1020e (Mimeographed).
Dutkowski, G. and Potts, B.M. 1999. Geographic patterns of genetic variation in *Eucalyptus globulus ssp. globulus* and a revised racial classification. Aust. J. Bot. 47:237-263.
Eldridge, K., Davidson, J., Harwood, C. and van Wyk, G. 1993. Eucalypt domestication and breeding. Clarendon Press, Oxford, UK.
Ferreira, M. and Santos, P.E.T. 1997. Genetic improvement of *Eucalyptus* in Brazil – brief review and perspectives. In: "Silviculture and improvement of eucalypts, Proc. of IUFRO Conference, Salvador, Bahia, Embrapa, Brazil, 24-29 August 1997, pp14-34.
Fonseca, S.M., Oliveira, R.C. and Silveira, P.N. 1995. Industrial tree selection: procedures, risks and benefits. In: "Eucalypt plantation: Improving fibre yield and quality", B.M. Potts, N.M.G. Borralho, J.B. Reid, R.N. Cromer, W.N. Tibbits and C.A. Raymond, Proc. of IUFRO-CRCTHF Conference, Hobart, CRCTHF, Australia, 19-24 February 1995, pp14-19.
Grattapaglia, D. 2001. Genomic technologies for the development of the eucalypt of the future. In: Proc. of IUFRO Conference Working Group 2.08.03 on

"Developing the eucalypt of the future", INFOR, Valdivia, Chile, 10-15 September 2001, pp17.

Greaves, B.L. Borralho, N.M.G. and Raymond, C.A. 1997. Breeding objective for plantation eucalypts grown for production of kraft pulp. For. Sci. 43:465-472.

Harbard, J.L., Griffin, A.R. and Espejo, J. 1999. Mass controlled pollination of *Eucalyptus globulus* – a practical reality. Can J. For. Res. 29:1457-1463.

Leal, A.M.C. and Cotterill, P.P. 1997. Mass controlled-pollination of *Eucalyptus globulus*. In: Silviculture and Improvement of Eucalypts, Proc. of IUFRO Conference, Salvador, Bahia, Embrapa, Brazil, 24-29 August, Vol.1, pp256-258.

Lindgren, D. 1985. Cost-efficient number of test sites for ranking entries in field trials. Biometrics 41:13-20.

MacDonald, A.C., Borralho, N.M.G. and Potts, B.M. 1997. Genetic variation for growth and wood density in *Eucalyptus globulus* ssp *globulus* in Tasmania (Australia). Silvae Genet. 46:236-241.

MacRae, S. and van Staden, J. 1999. Transgenic *Eucalyptus*. In: "Biotechnology in agriculture and forestry – transgenic trees", Y.P.S. Bajaj (ed), Springer Verlag, Berlin, Vol.44, pp88-114.

Michell, A.J. 1995. Pulpwood quality estimation by near-infrared spectroscopic measurements on *Eucalyptus* wood. Appita J. 48:425-428.

Moran, G.F. 1992. Patterns of genetic diversity in Australian tree species. New For. 6:49-66.

Muneri, A. and Raymond, C.A. 2000. Genetic parameters and genotype-by-environment interactions for basic density, Pilodyn penetration and stem diameter in *Eucalyptus globulus*. For. Genet. 7:317-328.

Potts, B.M. and Dungey, H.S. 2001. Hybridization of *Eucalyptus*: key issues for breeders and geneticists. In: Proc. of IUFRO Conference on "Developing the eucalypt of the future", INFOR, Chile, pp15.

Potts, B.M. and Wiltshire, J.E. 1997. *Eucalyptus* genetics and geneology. In: "Eucalypt ecology, individuals to ecosystems", J. Williams and J. Woinarski (eds), Cambridge University Press, UK, pp56-91.

Raymond, C.A., Schimleck, L.R., Muneri, A. and Michell, A.J. 2001. Genetic parameters and genotype-by-environment interactions for pulp yield predicted using near infrared reflectance analysis and pulp productivity in *Eucalyptus globulus*. For. Genet. 8:213-224.

Sartoretto, L.M., Cid, L.P.B. and Brasileiro, C.M. 2002. Biolistic transformation of *Eucalyptus grandis* x *E. urophylla* callus. Funct. Plant Biol. 29:917-924.

Schimleck, L.R., Michell, A.J., Raymond, C.A. and Muneri, A. 1999. Estimation of basic density of *Eucalyptus globulus* using near-infrared spectroscopy. Can. J. For. Res. 29:194-201.

Soria, F. and Borralho, N.M.G. 1997. The genetics of resistance to *Phoracantha semipunctata* attack in *Eucalyptus globulus* in Spain. Silvae Genet. 46:365-369.

Tibbits, W.N., White, T.L., Hodge, G.R. and Joyce, K.R. 1997. Genetic control of rooting ability of stem cuttings in *Eucalyptus nitens*. Aust. J. Bot. 45:203-210.

Wei, R.-P. 1995. Predicting genetic diversity and optimizing selection in breeding programmes. PhD Thesis, ISBN 91-576-4990-1, Swedish University of Agricultural Sciences, Umeå, Sweden, 64p+12 chapters.
Wei, X. and Borralho, N.M.G. 1998. Genetic control of growth traits of *Eucalyptus urophylla* S.T. Blake in South East China. Silvae Genet. 47:158-165.
White, T.L. 1995. Prototype breeding strategy for clonal forestry: a case study of *Eucalyptus grandis*. In: "Eucalypt plantation: improving fibre yield and quality", B.M. Potts, N.M.G. Borralho, J.B. Reid, R.N. Cromer, W.N. Tibbits and C.A. Raymond (eds), Proc. of IUFRO-CRCTHF Conference, Hobart, CRCTHF, Hobart, Australia, 19-24 February 1995, pp176-183.
Williams, D.R., Potts, B.M. and Black, P.G. 1999. Testing single site visit pollination procedures for *Eucalyptus globulus* and *E. nitens*. Aust. For. 9:346-352.
Williams, E.R. Matheson, A.C. and Haewood, C.E. 2002. Experimental design and analysis for tree improvement, second edition. CSIRO Publishing, Australia.
Woolaston, R., Whiteman, P.H. and Cameron, J.N. 1991. Genetic analysis of four year diameter measures in *Eucalyptus globulus* growing across four sites in Victoria Australia. In: A.P.G. Schonau, Proc. of IUFRO Symposium – "Intensive forestry: the role of eucalypts", Durban, Southern African Institute of Forestry, Pretoria, RSA, 2-6 September 1991, Vol.2, pp371-381.

桉树育种的挑战和教训

NUNO M.G. BORRALHO

RAIZ, Forest and Paper Research Institute
Herdade Torre Bela, Apartado 15, Alcoentre

与许多针叶树种和杨树相比，桉树育种研究起步较晚。截止二十世纪八十年代，才产生了少数育种程序，关于桉树树种的遗传特性、生殖生物学特性、最佳利用方案、包括无性系种子园和实生种子园等方面的知识依然很薄弱。近些年来，桉树育种研究成效显著，振奋人心，许多最先进、最具焦点性的育种程序都来源于桉树树种。现在，有些桉树育种程序已经进入了第三个世代，不仅在无性系林业方面遥遥领先，而且桉树在林木基因组学研究方面也是非常受欢迎的树种。

本文简要介绍了桉树育种研究在育种策略、技术和效益等方面的进展及对林业的贡献，探讨了这种进展可能带来的教训，并展望了未来桉树育种程序的基本原则和规范。最后简要讨论了发展更新和更好育种程序的机遇。

COLD TOLERANT PLANTATION EUCALYPTS FOR SOUTH CENTRAL CHINA

ROGER ARNOLD [1] AND JIANZHONG LUO [2]

[1]*CSIRO Forestry and Forest Products, P.O. Box E 4008*
Kingston, ACT 2604, Australia
[2]*China Eucalypt Research Centre, 30 Mid Renmin Dadao*
Zhanjiang 524022, Guangdong, P.R. China

More than 900,000 hectares of eucalypt plantations have been established to date in China but these are concentrated in warmer coastal regions of southern provinces. Relatively small areas have been planted to date in cooler areas of south central China - areas where some eucalypt species can grow well. These cooler areas include much of China's 'red soils region' which contains large tracts of eroded wasteland; land that is often suitable for forest plantation establishment.

The mean growth of eucalypt species across sites in cooler areas of south central China is summarised. Performance of those species that have been included in trials at three or more sites for each of two distinct environments – the Yunnan Plateau and then lower elevation areas – is reported. The lower elevation sites are spread across Hunan, inland Fujian and northern Guangxi provinces. Eucalypts that have demonstrated the best potential on the Yunnan Plateau include *E. globulus, E. maidenii*, and *E. smithii*. In other south central provinces where the target sites are below 1,000 m elevation, *E. dunnii, E. camaldulensis* and, for milder sites, selected sources of *E. grandis* have shown good potential. *E. nitens* and a range of other cold tolerant eucalypt species warrant further testing and evaluation across the region, but only if this is well coordinated with good communication and collaboration among researchers and institutes.

Significant genetic variation in cold tolerance has been demonstrated within some of the key eucalypts for south central China, providing opportunities to improve their adaptation for cooler sites. However, it emphasized that attention must also be focused on soil, nutritional and silviçultural management for eucalypt plantations to be successful in this region.

1 Introduction

China is one of the world's ten largest importers of forest products. In 2000 its net deficit in timber production was approximately 40 million m^3 and it is estimated this will increase to more than 90 million m^3 in 2010 (Sun and Bean 2001). Rapid annual increases in China's imports have been due to its large annual growth in GDP; increased residential construction; reductions in tariffs and a more open trading system since its entry to the World Trade Organisation; and, sharp reductions in timber harvesting from natural forests after severe floods in 1998. Interestingly, the furniture market in China is one of the strongest growth sectors for forest products, posting approximately 10% growth in 1999 and again in 2000 (Sun and Bean 2001).

To address the deficit in forest production, authorities in China are keen to expand the forest plantation resource in cooler areas of south central provinces[1]. In addition to the national, regional and local economic benefits of such a resource expansion, new plantations in this region could also provide substantial environmental benefits (Xie et al. 1991). Many of the south central provinces lie within the 'red soils region'[2], where the combination of deforestation, erodibility of red soil complexes and inappropriate land uses have resulted in serious sheet erosion of topsoils and, in places, deep gullying into subsoils over the last forty years. Such factors have created substantial areas of largely unproductive degraded lands that now total more than 400,000 km^2 (Xie et al. 1991).

Attempts to establish large areas of productive forest plantations with native species including Chinese fir (*Cunninghamia lanceolata*) and Masson pine (*Pinus massoniana*) in the red soils region have often proved unsuccessful (Xu and Yang 1991). Although such native species can grow well on the more fertile sites with better soils in this region, their growth tends to be poor when planted on the barren eroded lands (Tu 1991). Consequently, a large proportion of the current fir and pine plantations provide little fuelwood or timber, offer little protection against soil erosion, and contribute relatively little to rural economies. Also, many of the natural forest areas across the red soils region in south central China are considered to be severely degraded.

In contrast to the local pine species and Chinese fir, some eucalypt species appear to be well adapted to many degraded sites in China's red soil region (Xu and Yang 1991; Brown 1994). They have demonstrated an ability to sustain relatively fast growth on poorer soils across a wide range of sites in southern China (Turnbull 1981; Bai 1994; Brown 1994; Chen 2002), and their timber is suitable for range of products including pulp, fibreboards, ply and sawn timbers.

More than 900,000 hectares of eucalypt plantations have been established to date in China. However, this resource is concentrated in warmer and/or more coastal regions in Guangdong, Guangxi, Hainan and Yunnan (Chen 2002). In these warmer areas the science and technology of eucalypt species-provenance selection, genetic improvement and silviculture are generally well developed. However, in the cooler regions of south central China where the substantial areas of degraded and largely unproductive lands are located that are imminently suited to forest plantation development are located, the area of eucalypts planted to date is relatively limited. Also, the domestication of cold tolerant eucalypts for these areas is in its relatively early stage. Attempts to directly transfer germplasm from the warmer coastal areas have often met with complete failure – the tropical and sub-tropical eucalypts are ill adapted to southern China's cooler inland areas. For example, more than 2,000 hectares of plantations in Fujian's highlands of *E. urophylla* and *E. urophylla* x *E. grandis* clones introduced from southern Guangdong were killed or severely

[1] South central provinces include Yunnan, Sichuan, Guizhou, Hunan, Jiangxi, Zhejiang, Fujian, and also the northern area of Guangxi province.

[2] South central China's 'red soils' are so named due to characteristic red or yellow subsoil colours. Most are classified as either Red Dermosols or Red Ferrosols (Laffan 2002a). All tend to be acidic in nature (pH 4.5-5.5) and have significant nutrient imbalances.

damaged during severe winter cold (to –8°C) in December 1999 (Qiu Jinqing, Fujian Forestry Department, pers. comm.).

In 1999 researchers from China and Australia commenced a collaborative project, "Development of germplasm and production systems for cold tolerant eucalypts for use in cool regions of southern China and Australia" (the Cold Tolerant Eucalypt Project), to advance the domestication of cold tolerant eucalypts for the cooler regions of south central China. The Chinese partners in this project include the China Eucalypt Research Centre, Zhanjiang; the Yunnan Academy of Forestry, Kunming; Guangxi Academy of Forestry, Nanning; the Seed and Seedling Station of the Fujian Forestry Department, Fuzhou; and, the Hunan Forestry Department, Changsha. The Australian partners include CSIRO Forestry and Forest Products, Canberra; the Centre for Forestry Technology of the Department of Natural Resources and Environment, Melbourne; Forestry Tasmania, Hobart; and, the Australian Centre for International Agricultural Research (ACIAR), Canberra.

The project's research focuses on five key areas: evaluation of existing eucalypt trials in cooler regions of south central China; site evaluation; genetic resources; eucalypt phenology and seed productivity; and, silvicultural practice. This paper presents some general results obtained from the evaluation of existing trials, and discusses important considerations for eucalypt plantations in the target areas.

2 Species Selection from Trial Results

The Cold Tolerant Eucalypt Project has sought to identify superior eucalypt species and provenances for the cooler environments of south central China (including Yunnan province). These target environments are of a "humid mesothermal" climate with mean annual temperatures of 15-24°C, absolute minimum temperatures down to -8° to -10°C and mean annual precipitation generally above 800 mm per annum. Although such conditions appear to be significantly above the lower limit for the many cold hardy eucalypts, some aspects of the region's environment are problematical for adaptability of these species. In terms of heat sums and latitude, much of the region is seemingly subtropical (Zhao 1986). However, this can be misleading. For instance, the mean January temperature (winter) in central Hunan is approximately 4.7°C, frosts occur there on more than 75 days per year and severe cold periods (days of continuous subzero temperatures) can be encountered when masses of cold air flow in from the Gobi Desert and Siberia to the northwest. Also, south central China has a summer rainfall pattern with a drier season over the autumn-winter period which is quite different to that of southern Australia's Mediterranean type climate. Most cold tolerant eucalypts are best adapted to the latter climates with cool moist winters and drier lower humidity summers.

Climatic requirements for development of productive forest plantations of many commercially important eucalypt species adapted to cooler winters, as well as those for other regions, have been well defined (CAB International 2000). Such information is useful in combination with software for developed for species-site climate matching and more recently for predicting frost risks and risk of cold damage to eucalypts in China (Booth et al. 1994; Yan et al. 1995; Yan 2001). However, since climate and especially minimum temperature are strongly influenced by local conditions, and species suitability is influenced by many factors other than climate, such predictions are only broad indicators of species-site suitability. Also, whilst indicating broad adaptability they do not indicate relative productivity of species that are suitable. Thus, the computer software ultimately needs to be accompanied by selection of the most appropriate species and seed sources from trials in the target environment(s) where plantations are to be established.

Through a wide range of projects and independent initiatives, a significant number of eucalypt species, species-provenance, provenance and family trials have been established in cooler regions of south central China since the mid 1980s. However, there has been little integration and coordination of the variety of trials conducted in China so far and a number of these trials had been little known outside their local areas. To help redress such shortcomings, the Cold Tolerant Eucalypt Project has been working to compile a database of known existing eucalypt germplasm trials in cooler areas of Yunnan, Guangxi, Hunan and Fujian provinces that were either accessible to project partners and/or for which adequate published information was available. Of these trials some now offer potential to provide important information and, where appropriate, a number of these have been recently re-assessed.

Two markedly distinct environments are represented in the trials included in the database – Yunnan (17 trials); and, sites in other provinces (18 trials). Yunnan province consists of two distinct regions: – a mountainous canyon region in the west; and, a plateau region in the east (Zhao 1986). The eucalypt trials examined by the Cold Tolerant Eucalypt Project in Yunnan are all located on the plateau area and the elevations of these sites are all above 1,500 m asl with mean annual precipitations of less than 1,000 mm per year. The other trials, with locations in northwest Guangxi, south and central Hunan and inland in Fujian, are all at lower elevations (below 1,000 m asl) with climates having average annual precipitations exceeding 1,400 mm.

On the Yunnan Plateau, *E. globulus, E. maidenii* and *E. smithii* have all performed well across sites (Table 1). Interestingly, there is a long history of cultivation of both *E. globulus* and *E. maidenii* in Yunnan in both plantations and 'four around' plantings (Turnbull 1981). *E. smithii* has more recently gained favour with growers in Yunnan due to its good adaptation, growth and stem form when grown in Yunnan and also for relatively high yields of essential oils from its foliage. A number of other species such as *E. viminalis* and *E. macarthurii* have performed well at one or two sites but poorly elsewhere in Yunnan. Also, there are a large number of eucalypt species which have been included in only one or two trials – too few to enable confident judgment of a species adaptability and productivity. Based on the trials examined to date, several species warrant further monitoring and testing in Yunnan including *E. badjensis, E. macarthurii* and selected provenances of *E. nitens*. *E. badjensis* has shown very good growth and form in a number of trials in southern Australia (Myers 1994; CAB International 2000) and on some higher elevation summer rainfall sites subject to winter frosts in South Africa (Darrow 1996). *E. macarthurii* has proved to be one of the most frost tolerant of all plantation eucalypts in South Africa where it is also favoured for frost prone summer rainfall areas (Darrow 1996; Swain 2001a).

Table 1. Growth of eucalypt species represented in trials at three or more sites on the Yunnan Plateau, P.R. China.

Species	Total No. of sites	Age 1 ½ yrs			Age 3 yrs			Age 4 yrs			Age 12 yrs		
		No. of sites	Ht (m)	DBH (cm)	No. of sites	Ht (m)	DBH (cm)	No. of sites	Ht (m)	DBH (cm)	No. of sites	Ht (m)	DBH (cm)
E. bicostata	5	1	2.6	2.3	3	5.9	5.8	3	10.1	8.7	1	10.1	9.4
E. globules	6	4	2.4	1.8	1	5.2	4.9	4	9.7	7.9	2	13.8	10.8
E. grandis	6	0	*	*	2	4.8	4.4	4	7.2	7.1	1	12.5	12.4
E. macarthurii	4	1	1.4	2.2	3	3.5	2.8	1	5.0	5.4	1	15.1	15.9
E. maidenii	7	3	2.1	1.8	3	6.7	6.4	6	9.6	8.5	2	13.9	10.3
E. nitens	6	2	2.3	1.7	3	5.4	4.9	5	9.1	8.4	1	11.3	12.4
E. smithii	5	2	2.3	2.0	2	6.8	6.7	4	12.0	9.5	1	18.3	15.9
E. viminalis	6	2	2.3	1.5	3	5.4	4.6	5	8.8	7.9	1	13.1	10.9

Note: 1) not all trials were measured at all 4 ages shown; 2) apparent poor growth increments between ages 4 and 13 are due to the fact that early trials were established on very poor quality sites and, the main reason they were still available for measurement was that the trees had not been harvested due their relatively poor growth.

E. nitens showed promise at early ages in a number of the trials on the Yunnan Plateau (e.g. see Wang et al. 1994). However, at later ages (>6 years or so) it often appears poorly adapted. This may in part be due to the area's long dry season (5-6 consecutive months with less than 40 mm rainfall/month), which is at the extreme of that generally tolerated by *E. nitens* (see CAB International 2000). Also, many of the *E. nitens* provenances used in trials have been from Victoria and/or southern New South Wales. In the frost prone summer rainfall areas of South Africa, provenances from the northern New South Wales have proved better adapted and faster growing than those from southern New South Wales and Victoria (Swain 2001b). As northern *E. nitens* provenances are better adapted to summer rainfall climates (CAB International 2000), such as is experienced in Yunnan and other south central provinces, future work there needs to focus on the northern provenances.

At lower elevations elsewhere in cooler parts of south central China, species which have shown the best trial performance across sites are *E. camaldulensis*, *E. dunnii* and, for milder sites, *E. grandis* (Table 2). As an alternate species on selected sites, *E. saligna* has potential, particularly where quality sawn timber is the end product objective (see Turnbull and Pryor 1984). However, only a total of five species had been included in three or more successful trials in this part of China from which information has, so far, been able to be obtained. As is the case in Yunnan, a wide range of other species have also been included in just one or two trials across these south central provinces. This situation highlights the need for better coordination and communication between researchers.

E. dunnii, the species best represented across trials at lower elevations in cooler parts of south central China, has shown good adaptation, growth, stem form and cold tolerance in many plantings in northern Guangxi and, along with *E. camaldulensis*, in Hunan. Both species are now sought after for plantation establishment in those areas. As *E. camaldulensis* is relatively easy to propagate through tissue culture and as rooted cuttings, clones selected for cold tolerance, vigour and stem form are now being propagated and established on a large scale in Hunan and adjoining provinces. The scale of *E. dunnii* plantation establishment is more limited and is being constrained by difficulties of seed supply; it is a shy flowering species and vegetative propagation has very low success rates (CAB International 2000).

3 Potential for Improvement of Cold Tolerance

Eucalypt improvement programs in both Florida and California have successfully demonstrated the effectiveness of within species selection for identifying material with both improved growth and higher frost tolerance and freeze resilience (Franklin and Meskimen 1984; Rockwood 1991; Arnold et al. 1999). During December 1999, a flow of unusually cold air extended southwards over the most part of south China resulting in serious cold damage to extensive areas of eucalypt and acacia plantations, fruit trees, agricultural crops and even some native forest tree species in south central China (Arnold and Midgley 2001). A number of important eucalypt trials in Fujian and also Hunan suffered substantial damage, including provenance-family trials of *E. grandis* and *E. dunnii* located in central Fujian. This provided a timely opportunity to examine genetic variation in cold tolerance in these two species.

Table 2. Growth of eucalypt species represented in trials at three or more sites at lower elevations (<1,000 m asl) in cooler regions of south central China.

Species	Total	Age 2 yrs			Age 3 yrs			Age 5 yrs			Age 8 yrs		
	No. of sites	No. of sites	Ht (m)	DBH (cm)	No. of sites	Ht (m)	DBH (cm)	No. of sites	Ht (m)	DBH (cm)	No. of sites	Ht (m)	DBH (cm)
E. camaldulensis	3	1	5.7	5.1	1	6.9	6.4	0	*	*	1	11.5	15.4
E. dunnii	7	4	6.8	5.9	1	5.8	6.0	6	14.5	12.9	6	16.4	15.1
E. grandis	5	4	5.8	5.1	1	8.6	7.7	0	*	*	0	*	*
E. saligna	4	1	3.1	3.2	3	6.7	6.7	0	*	*	0	*	*
E. tereticornis	3	1	2.7	2.3	2	5.8	5.3	0	*	*	0	*	*

Note: not all trials were measured at all 4 ages shown.

Cold tolerance in both *E. dunnii* and *E. grandis* proved to be under moderately strong genetic control (Table 3). These results in combination with those for growth and form have been used to select superior genotypes for ongoing improvement and for development of grafted clonal seed orchards.

Table 3. Variation in cold tolerance for *E. dunnii* and *E. grandis* in Fujian provenance after severe cold in December 1999.

Species	Trial location	No. of families	Year/ Month planted	Cold event in '99		Height (m)	Survival Rate (%)	h_i^2 for cold tolerance
				min. temp.	days below 0 °C			
E. dunnii	Pushang	216	1998/4	-10°C	10	3.8-8.0	90	0.24±0.10
E. grandis	Yong'an	205	1997/3	-7 °C	8	6.1-12.0	95	0.18±0.02

Sources: Arnold and Midgley (2001); Qiu et al. (2001); and, Luo (2002).

4 Other Factors

Selection and genetic improvement is not, in itself, sufficient to optimise the productivity and sustainability of eucalypt plantations in cooler regions of south central China. In this region great attention must be given to optimising soil, nutritional and silvicultural management.

Although the 'red soils' that dominate south central China generally have very good physical properties for tree growth, there are major differences in other soil properties that have important implications for plantation establishment and management (Laffan 2002a). Successful conversion of large areas of currently degraded land available in the region to sustainable eucalypt plantation use will require more detailed assessments of the properties and distribution of the soil types than are currently undertaken (Laffan 2002b). The 'red soils' with higher contents of iron oxides (Red Ferrosols) are likely to have much higher requirements for phosphatic fertilizers than those with lower levels of iron oxides (Red and Yellow Dermosols). Thus, characterisation and mapping of soils for site selection and appropriate nutritional supplementation will lead to higher plantation productivities.

Quality silvicultural management is essential for the combination of genetic and site potentials to be fully realised in establishment of eucalypt plantations in cooler areas. Development of appropriate tree spacing, pruning and thinnning regimes can increase the yield, quality and hence economic value of forest products (Gerrand et al. 1997; Pinkard 2002; Pinkard and Neilsen 2002). Direct adoption of silvicultural regimes employed in China's tropical and subtropical coastal eucalypt plantations is not appropriate – the species, target end products, sites and market access are widely different. Although useful research on silviculture for cold tolerant eucalypt plantations has already been carried out in Australia and elsewhere, more work will be needed to optimise these systems for the context of south central China. As the most profitable markets for logs from new eucalypt plantations in south central China are likely to be those for higher quality furniture timbers to meet burgeoning domestic demand (Sun and Bean 2001), there is a major need to design and test spacing and stand manipulations more suited to producing such products in the environments in this region.

5 Conclusions

Expansion of eucalypt plantations in south central China, particularly in the red soils region will have substantial multiple benefits. The development and management of the plantations and the timber they produce will increase rural employment and incomes; provide diversification of land use; reduce importation of forest products; and, increase fuelwood availability for households. The expansion of forest cover will help alleviate soil degradation and erosion, ameliorate runoff and enhance biodiversity.

Trials across a wide range of sites have identified a range of eucalypt species well adapted for establishment of such plantations in cooler environments of south central China. Those that have demonstrated the best potential to date include *E. globulus, E. maidenii, E. smithii, E. dunnii* and *E. camaldulensis*, whilst for milder sites selected sources of *E. grandis* can be considered. A range of other eucalypt species warrant further testing and evaluation. It is essential that any such future work is well coordinated with good communications between all researchers and practitioners involved.

Significant genetic variation in cold tolerance has been demonstrated within some of these key eucalypt species, and provides the opportunity to improve their tolerance and extend the range of sites for which they are suitable. However, selection and genetic improvement must not be pursued in isolation. Attention must also be focused on optimising soil, nutritional and silvicultural management for eucalypt plantations to be successful in cooler areas of south central China.

Acknowledgements

Special acknowledgement is due to all the collaborators and dedicated personnel involved in the Cold Tolerant Eucalypt Project who have contributed trial information. This project ("Development of germplasm and production systems for cold tolerant eucalypts for use in cool regions of southern China and Australia") has been supported in part by the Australian Centre for International Agricultural Research (ACIAR project FST/96/125). Special thanks are also due to all the landowners/managers providing significant support of the field trial work.

References

Arnold, R.J. and Midgley, S.J. 2001. Second Annual Report – ACIAR Project FST/95/125: Development of germplasm and production systems for cold tolerant eucalypts in cool regions of southern China and Australia. Client Report 949. Canberra: CSIRO Forestry and Forest Products.

Arnold, R.J., Bacca, M.J. and Clark, N. 1999. Commercial irrigated *E. camaldulensis* plantations – a Californian experience and its applicability to Australia. Aust. For. 62:154-159.

Bai, J. 1994. Genetic improvement of tropical *Eucalyptus* tree species in China. In: "Australian tree species research in China", A.G. Brown (ed), Proc. of an international workshop, Zhangzhou, Fujian Province, P.R. China, 2-5 November 1992, Canberra, ACIAR Proc. No.48, pp32-49.

Booth, T.H., Jovanovic, T. and Yan, H. 1994. Climatic analysis methods to assist choice of Australian species and provenances for frost-affected areas in China. In: "Australian tree species research in China", A.G. Brown (ed), Proc. of an international workshop, Zhangzhou, Fujian Province, P.R. China. 2-5 November 1992, Canberra, ACIAR Proc. No.48, pp26-31.

Brown, A.G. (ed) 1994. Australian Tree Species Research in China. Proc. of an international workshop, Zhangzhou, Fujian Province, P.R. China, 2-5 November 1992, Canberra, ACIAR Proc. No.48.

CAB International 2000. Forestry Compendium – a silvicultural reference. Global Module. Wallingford, UK: CAB International, CD ROM.

Chen, S. 2002. Cultivation and production of eucalypts in the People's Republic of China, with special reference to leaf oils. In: "The Genus *Eucalyptus*", J.J.W. Coppen (ed), London, Taylor and Francis, pp202-215.

Darrow, W.K. 1996. Species trials of cold-tolerant eucalypts in the summer-rainfall zone of South Africa - results at six years of age. Institute for Commercial Forestry Research (ICFR), Pietermaritzburg, South Africa, ICFR Bulletin Series No.9/96.

Franklin, E.C. and Meskimen, G.F. 1984. Choice of species and provenances in cold summer rainfall climates. In: Proc. of IUFRO Colloque international sur les *Eucalyptus* Resistants au Froid. Bordeaux, September 1983, Paris, AFOCEL, pp341-357.

Gerrand, A.M., Neilsen, W.A. and Medhurst, J.L. 1997. Thinning and pruning eucalypt plantations for sawlog production in Tasmania. Tasforests 9:15-34.

Laffan, M. 2002a. Soils, eucalypts and forest practices in southern China. For. Pract. News 4:3-5.

Laffan, M. 2002b. Summary of sites and soils inspected during visits in March/April 2000 and September 2001 to Southern China, and comments on plantation potential. Forestry Tasmania, Hobart, (unpublished report).

Luo, J. 2002. Genetic variation in *Eucalyptus dunnii* in China. MS Thesis, Chinese Academy of Forestry, Beijing.

Myers, B.J. (ed) 1994. Wagga effluent plantation project technical report. CSIRO Division of Forestry, Canberra, September 1994, User Series No.17.

Pinkard, E.A. 2002. Effects of pattern and severity of pruning on growth and branch development of pre-canopy closure of *Eucalyptus nitens*. For. Ecol. Mgt. 157:217-230.

Pinkard, E.A. and Neilsen, W.A. 2003. Crown and stand characteristics of *Eucalyptus nitens* in response to initial spacing: implications for thinning. For. Ecol. Mgt. 172:215-227.

Qiu, J., Lan, H., Xie, G. and Huang, D. 2002. Provenance-family trial of *E. grandis*: growth and cold tolerance to age 3 years. Fujian For. Sci. Techn. (submitted).

Rockwood, D.L. 1991. Genetic improvement of *Eucalyptus* in Florida. California Eucalyptus Grower 6:8-10.

Sun, X. and Bean, R. 2001. Solid wood products annual, Peoples Republic of China. Beijing: USDA Foreign Agricultural Service GAIN Report #CH1032.

Swain, T.L. 2001a. ICFR tree improvement project: status report *Eucalyptus macarthurii*. Institute for Commercial Forestry Research (ICFR), Scottsville, South Africa, *ICFR Bulletin Series* No.6/2001.

Swain, T.L. 2001b. ICFR tree improvement project: status report *Eucalyptus nitens*. Institute for Commercial Forestry Research (ICFR), Scottsville, South Africa, *ICFR Bulletin Series* No.5/2001.

Tu, M. 1991. Forest development in the upland grasslands of southern China. In: "Forages on red soils in China", P.M. Horne, D.A. MacLeod and J.M. Scott (eds), Proc. of a workshop, Lengshuitan, Hunan Province, P.R. China. 22-25 April 1991, ACIAR Proc. No.38, pp86-87.

Turnbull, J.W. 1981. Eucalypts in China. Aust. For. 44:222-234.

Turnbull, J.W. and Pryor, L.D. 1984. Choice of species and seed sources. In: "Eucalypts for wood production", W.E. Hillis and A.G. Brown (eds), Melbourne and Sydney, CSIRO/Academic Press, pp6-65.

Wang, H., Zheng, Y., Yan, H., Zang, D., Zhang, R. and Wu, X. 1994. Introduction and provenance trial of *Eucalyptus nitens* and its potential in plantation forestry in China. In: "Australian tree species research in China", A.G. Brown (ed), Proc. of an international workshop, Zhangzhou, Fujian Province, P.R. China, 2-5 November 1992, ACIAR Proc. No.48, pp50-55.

Xie, K., Zhang, J. and Horne, P.M. 1991. Potential and problems for forage development and animal production in the red soils region of southern China. In: "Forages on red soils in China", P.M. Horne, D.A. MacLeod and J.M. Scott (eds), Proc. of a workshop, Lengshuitan, Hunan Province, P.R. China, 22-25 April 1991, ACIAR Proc. No.38, pp11-14.

Xu, D. and Yang, M. 1991. Soil erosion in the red soil area of southern China. In: "Forages on red soils in China", P.M. Horne, D.A. MacLeod and J.M. Scott (eds), Proc. of a workshop, Lengshuitan, Hunan Province, P.R. China, 22-25 April 1991, ACIAR Proc. No.38, pp82-85.

Yan, H. 2001. Predicting frost risk in China. ACIAR For. Newsl. 26:1-4.

Yan, H., Booth, T.H. and Zuo, H. 1995. GREEN - a climatic mapping program for China and its use in forestry. In: "Matching trees and sites", T.H. Booth (ed), Proc. of an international workshop, Bangkok, Thailand, 27-30 March 1995, ACIAR Proc. No.63, pp24-29.

Zhao, S. 1986. Physical geography of China. Science Press, Beijing and John Wiley and Sons, New York.

适合于华南中部地区生长的抗寒人工林桉树树种

ROGER ARNOLD [1]，罗建中 [2]

[1]*CSIRO Forestry and Forest Products, P.O. Box E 4008*
Kingston, ACT 2604, Australia
[2]国家林业局桉树研究开发中心
524022 中国广东湛江

中国已经有 90 万公顷的桉树人工林，大部分都集中在华南地区，特别是气候比较温暖湿润的沿海各省份。中国中南部气候较冷地区多为红壤区，包括大面积被腐蚀风化的荒地，这些地区比较适合营建人工林，有些桉树树种在这些地方长势较好，但到目前为止，这些较冷地区还只有少数桉树人工林。

本文研究了中国中南部较冷地区桉树树种的平均生长情况，报道了桉树在云南高原和较低海拔地区两种截然不同环境条件下的三个或更多试验林中的生长表现，其中低海拔试验点分布在湖南、福建内地和广西北部地区。试验表明，在云南高原地区最具发展潜力的桉树树种有蓝桉、曼腾桉、谷桉。试验林地点低于海拔 1,000 米的其他中南部省区，具有潜力的有邓恩桉和赤桉，海拔高度在这两者之间的地区，巨尾桉生长表现最好。亮果桉和其他一些抗寒桉树品种还需要在不同的地区进行进一步的试验和评价，这需要不同研究人员和不同研究机构之间的通力合作和交流。

中南部地区一些主要桉树树种之间已经表现出重要的遗传变异，这非常有利于提高它们对较冷气候的适应能力。需要强调的是，要想在这些地区成功营造桉树人工林，还必须注意人工林分土壤和营养的经营管理和采用合适的营林措施。

IMPORTANT TRAITS AND COMBINED EVALUATION OF EUCALYPT CLONES

XIAOYONG MO, SHIYAO PENG, TENG LONG,
WENPING CHEN AND XIAOHONG YANG
Leizhou Forestry Bureau, National Forest Farm Development Company
Zhanjian 524348, Guangdong, China

The difference in and correlation among major characters of eucalypt (species) hybrid clones were studied by investigating and analyzing the performance of 27 different eucalypt clones at 3-year old. Clones significantly differed in the mean annual volume growth and had significant interaction with the conditions of the test sites. A clone's tree form index had significant negative correlation with the yield; the pattern of the correlation between the major characters was different. Tree height and DBH had significant genetic and phenotypic correlation. As expected, DBH was the main factor in determining the plantation yield. Combined evaluation was given to the clones studied by using tree form index and unrestricted index selection. Eighteen superior clones including EC1, CH3 and CH1 were selected for developing eucalypt plantations.

1 Introduction

Eucalyptus is one of the major afforestation species in the tropical and subtropical regions of China. Use and extension of tissue culture and rooting technology have brought the quality and yield of afforestation with superior clones a great progress comparing with seedling plantation. In the tropical and subtropical regions of the other parts of world, the mean annual yield of *Eucalyptus* plantations is more than 20m^3/ha, some commercial plantations in countries such as Brazil, South Africa, Congo and Australian can be as much as 30-50 m^3/ha. In contrast, the mean annual increment in volume of *Eucalyptus* plantations in China is only 5 m^3/ha (Yu et al. 2000), although high yield is also frequently reported for experiment stands. There are two main reasons: 1) clonal plantation is not well extended in afforestation, and 2) the quality of the most clonal plantations is poor (Yu et al. 2000).

Clonal breeding and selection is an important work for the development of high yield *Eucalyptus* plantations in China. Breeding work should target at superior clones that have high yield and desirable performance in other characteristics and are adaptive to different environments. In this paper, we 1) investigate the performance in yield, and the correlation among the growth characteristics, 2) evaluate the overall performance using tree form index and unrestricted index, and 3) discuss the breeding methods of *Eucalyptus* clones.

2 Experimental Sites

Three test sites were arranged at Longmen Forest Farm, Shuixi Forest Farm and Shiling Forest Farm of Leizhou Forestry Bureau. They are located at between

20°16′N-21°55′N, and 109°39′E-110°36′E, with long summer and warm winter. The mean annual temperature is 23.5°C, with no frost all year round; the mean annual rainfall is 1855mm, the evaporation is 1762.9mm; in average, there are four typhoons over class 7 each year, belonging to the tropical monsoon climate.The soil types and the main properties of three sites were: 1) at Longmen Forest Farm, lateritic soil developed from basalt, organic matter 17.27 g/kg, total N 0.72 g/kg, total P 0.33 g/kg, total K 1.63 g/kg, hydrolyzed N 70.80 g/kg, available P 0.75 mg/kg, and available K 31.83 mg/kg, and 2) at Shuixi Forest Farm, lateritic soil developed from shallow-sea deposit, organic matter 14.73 g/kg, total N 0.76 g/kg, total P 0.27 g/kg, total K 1.46 g/kg, hydrolyzed N 59.88 g/kg, available P 3.16 mg/kg, and available K 29.68 mg/kg, and 3) at Shiling Forest Farm, lateritic soil developed from granite, organic matter 13.09 g/kg, total N 0.50 g/kg, total P 0.16 g/kg, total K 3.01 g/kg, hydrolyzed N 44.55 g/kg, available P 1.83 mg/kg, and available K 6.16 mg/kg (Zeng et al. 1995).

3 Materials and Methods

3.1 *Experimental Material*

Twenty-seven *Eucalyptus* clones included in the trial were selected from the superior crosses artificially mated between *E. urophylla*, *E. grandis*, and *E. tereticornis,* and the superior open-pollinated families of *E. urophylla*, *E.12ABL* and *E.leizhouensis* No.1 (table 1).

Table 1. Codes and origins (crossing species) of *Eucalyptus* clones tested.

Clone code	Origin	Clone code	Origin	Clone code	Origin
CH1	*E. grandis* × *E. urophylla*	EC1	*E. grandis* × *E. urophylla*	H1	*E. leizhouensis* No.1
CH3	*E. grandis* × *E. urophylla*	UT1	*E.urophylla* × *E. tereticornis*	H5	*E. leizhouensis* No.1
CH4	*E. urophylla* × *E. grandis*	HU1	*E. urophylla*	L69	*E. leizhouensis* No.1
CH5	*E. urophylla* × *E. grandis*	MLA	*E. urophylla*	MJ41	*E. leizhouensis* No.1
CH6	*E. urophylla* × *E. grandis*	U16	*E. urophylla*	MJ43	*E. leizhouensis* No.1
CH10	*E. urophylla* × *E. grandis*	ZU6	*E. urophylla*	MJ45	*E. leizhouensis* No.1
CH11	*E. urophylla* × *E. grandis*	Wb	*E.* 12ABL	MJ53	*E. leizhouensis* No.1
CH15	*E. urophylla* × *E. grandis*	W2	*E.* 12ABL	MJ57	*E. leizhouensis* No.1
DH32 -32	*E. grandis* × *E. urophylla*	W5	*E.* 12ABL	SH1	*E. leizhouensis* No.1

3.2 *Field Design and Planting*

For all three sites, a randomized block design was used, with 3 replications. Twenty-seven clones were randomly allocated in 60-tree plots (4 rows by 15 trees/row) in each replication. At each site, the planting density was 2892 trees/ha, and the area was about 1.68 ha. Clone W5 was used as the control.

For site preparation, tractor was used to over-all plough to 40 cm depth, and planting ditch was dug to 30 cm depth. The planting spacing was 1 m × 3 m. Before planting in 1997, the planting ditches were applied with 9000 kg of strain mud and 750 kg of calcium super-phosphate per ha. In the first and second years, the young trees were top-dressed with the compound fertilizer at the rate of 900 kg/ha.

3.3 *Measurement and Statistical Analysis*

At 3-year old, height and DBH ($D_{1.3}$) were measured for each tree in the trials. Three trees were randomly sampled to measure bark thickness and test wood density. Two stem form indexes were considered: 1) the breast-height form factor $(F)=V/(hg)$, where V is the tree volume, H the tree height, and g the transect area at breast-height, and 2) the breast-height form quotient $(Q)=D_{0.5}/D_{1.3}$, where $D_{0.5}$ is the diameter at half of tree height (Cai 1989). In each plot, a normal average tree was fell down, and $D_{1.3}$, $D_{0.5}$ and H were accurately measured. Then, all leaves were collected and weighted. By testing the water content of the fresh leaves sampled, the dry weight of the leaves was obtained. The wood yield was converted to the discount dry ton.

Mean stem form index over 3 sites was calculated for each clone. On the basis of the mean values over individuals within a site, difference among the clones was studied, the phenotypic and genetic correlations among the major traits associated with yield were analyzed. Unrestricted combined index was calculated (Moll and Stuber 1974; Nordskog 1978; Ma 1982) and used together with stem form indexes to evaluate the growth performance of each clone at 3-year old.

4 Results and Discussion

4.1 *Difference in Volume*

The results of the analysis were presented in Table 2. Significant difference (1%) was found among sites and clones studied, but no significant difference was found among replications within a site. Obviously, both selection of clones and choice of planting sites are very effective to obtain high-yield plantations. The clones studied had significant interaction with the planting sites, indicating that genotype by environment interaction is an important factor in deciding the yield growth of eucalypt clonal plantations. For specific site conditions, superior clones could be carefully selected to increase yield. The trial at Longmen Forest Farm had the highest mean annual increment in volume (36.25 m^3/ha), followed by Suixi Forest Farm (21.80 m^3/ha), and Shiling Forest Farm (21.45 m^3/ha).

Table 2. Variance analysis for the volume of the 3-site clonal trial.

Source	DF	SS	MS	F
Site	2	11554.96	5777.49	198.19**
Replicate	2	61.15	35.58	1.08
Clone	26	46635.40	1793.67	61.53**
Site × Species	52	16559.72	318.46	10.92**
Error	160	4664.18	29.15	

Note: ** is significant at level of ≤1%.

4.2 *Stem Form Indexes and Correlation with Yield*

Eucalyptus clones differed in the breast-height form quotient (Q) and form factor (F) and yield (Table 3). Normally, Q increased with F, and vice versa. Clone EC1 had minimum Q (0.6322) and F (0.4434) while clone MJ43 had maximum Q (0.85) and F (0.5997). The mean Q and F of 27 clones were 0.7655 and 0.5007. At 3-year old, the yield of a *Eucalyptus* clone was very significantly correlated to Q (r=-0.899, P<0.0001) and F (r=0.844, P<0.0001). The yield drastically dropped as the breast height form quotient and form factor increased.

Table 3. Mean breast-height form quotient (Q) and form factor (F) and yield (t/ha).

Clone code	Q	F	Yield (m^3/ha)	Clone code	Q	F	Yield (m^3/ha)
CH1	0.6711	0.4485	42.223	ZU6	0.7457	0.4677	37.180
CH3	0.6800	0.4530	42.170	Wb	0.7638	0.4885	27.843
CH4	0.7600	0.4802	32.549	W2	0.8019	0.5391	7.347
CH5	0.7272	0.4753	39.266	W5	0.7980	0.5036	14.938
CH6	0.6683	0.4667	33.843	H1	0.7883	0.4932	17.653
CH10	0.7530	0.4776	35.025	H5	0.8055	0.5081	11.039
CH11	0.7500	0.4769	37.230	L69	0.7990	0.5139	9.791
CH15	0.7547	0.4778	31.660	MJ41	0.8010	0.5022	12.009
DH32-32	0.7636	0.4744	34.135	MJ43	0.8500	0.5997	5.310
EC1	0.6322	0.4434	61.796	MJ45	0.8148	0.5839	10.465
UT1	0.7680	0.4803	33.058	MJ53	0.8500	0.5997	5.310
HU1	0.7840	0.5010	27.509	MJ57	0.8210	0.5877	9.191
MLA	0.7830	0.4911	26.724	SH1	0.7590	0.4881	29.753
U16	0.7741	0.4980	27.982				

4.3 *Difference among Clones*

Table 4 presented the mean values and results of variance analysis (SAS Institute Inc. 1995; Lin 1997, 1998) for the difference in 5 traits among clones studied: tree height, DBH, foliage weight, bark thickness and wood density. Difference in 5 traits was very significant among clones studied. As the control, clone W5 averaged 9.46

m height, 6.26 cm DBH, 1.09 kg foliage weight, 0.37 cm bark thickness and 0.427 t/m^3 wood density, all lower than the mean of all clones studied (Table 4).

Table 4. Means and variance analysis for different traits of *Eucalyptus* clones at 3-year old.

Trait	MEAN	DF	SS	MS	F Value
Height (m)	10.90	26	330.920	12.728	9.67**
DBH(cm)	7.39	26	293.021	11.270	13.77**
Foliage weight (kg)	1.948	26	106.333	4.089	61.41**
Bark thickness (cm)	0.43	26	0.799	0.031	10.82**
Wood density (t/m^3)	0.455	26	0.066	0.003	11.96**

Note: ** is significant at level of ≤1%.

Duncan multiple-comparison showed that 17 clones were significantly taller than the mean of 27 clones by the average of 20.3%. The mean superiority in DBH of 17 clones was 18.4%. The minimum height and DBH were 6.49m and 6.14, respectively. Twelve clones significantly produced more foliage than the mean of all clones by the average of 45.11%. The bark of 14 clones was significantly thicker by 19.98%, and the wood density of 10 clones was higher by 18.03% than the respective means of all clones. The extreme differences were, 6.49 m, 6.14 cm, 4.2 kg, 0.34 cm and 0.113 t/m^3 for height, DBH, foliage, bark and wood density, respectively.

4.4 *Correlation between Traits*

Genetic and phenotypic correlation coefficients between 5 traits of 27 clones were presented in Table 5. At 3-year old, height and DBH had significant positive phenotypic and genotypic correlation with the foliage production. The phenotypic correlation was 0.56 and 0.44, respectively, while the genetic correlation was 0.58 and 0.53, respectively. As expected, height had close relationship with DBH, with phenotypic and genetic correlation being 0.92 and 0.93, respectively. No significant phenotypic or genetic correlation was found among other pairs of traits. However, bark thickness tended to increase with DBH and wood density, with phenotypic and genetic correlation close to be significant.

Table 5. Phenotypic (upper triangle) and genotypic (lower triangle) correlation coefficient of clones traits.

Trait	Height	DBH	Foliage weight	Bark thickness	Wood density
Height		0.920**	0.560**	0.050	0.140
DBH	0.930**		0.440*	0.360	0.014
Foliage weight	0.580**	0.520**		0.052	0.026
Bark thickness	0.051	0.340	0.051		0.280
Wood density	0.014	0.015	0.025	0.300	

Note: DF=25, $r_{0.01}$=0.487, and $r_{0.05}$=0.381.

4.5 *Combined Evaluation of 3-Year Old Clones*

At present, *Eucalyptus* plantation is managed mainly for producing wood chips in the region. Therefore, the most important target of *Eucalyptus* clonal selection and breeding is the wood yield or dry weight. The wood yield of *Eucalyptus* clones had significant genetic correlation with height (0.763), diameter (0.82), foliage weight (0.481), bark thickness (0.394) and wood density (0.411). By suing the phenotypic variance and covariance of, and the genetic covariance of wood yield with 5 traits calculated (Table 6), an unrestricted linear selection index involving 5 traits was constructed: I=1.698h+2.28DBH+2.095L-0.680B+1.988WD, where h=height (m), DBH=diameter at breast-height (cm), L=foliage weight (kg), B=bark thickness (cm) and WD=wood density (ton/m^3).

Table 6. Phenotypic variance and covariance and genetic covariance.

	Height	DBH	Foliage weight	Bark thickness	Wood density
Height	4.24	3.67	1.34	1.03×10^{-2}	8.47×10^{-4}
DBH		3.75	0.77	4.61×10^{-3}	8.05×10^{-4}
Foliage weight			1.36	6.16×10^{-3}	8.74×10^{-4}
Bark thickness				1.03×10^{-2}	8.43×10^{-4}
Wood density					8.51×10^{-4}
Wood yield*	18.373	16.397	6.878	3.55×10^{-2}	6.224×10^{-3}

* Genetic covariance with other traits.

Table 7. Unrestricted selection index value (I) of 27 *Eucalyptus* clones.

Clone code	I	Clone code	I	Clone code	I
CH1	52.98	EC1	56.38	H1	35.50
CH3	53.11	UT1	47.04	H5	30.44
CH4	45.43	HU1	44.65	L69	28.52
CH5	47.90	MLA	43.11	MJ41	30.41
CH6	47.35	U16	42.48	MJ43	23.61
CH10	47.07	ZU6	48.77	MJ45	25.61
CH11	47.09	Wb	43.88	MJ53	23.11
CH15	45.81	W2	25.21	MJ57	26.15
DH32-32	45.70	W5	33.22	SH1	44.52

Unrestricted selection index values (I) were calculated for 27 clones (Table 7). The mean (I) of the selection index over 27 clones was 40.19 with standard deviation (σ_I) of 10.14. The 3 best clones (EC1, CH3 and CH1) had the index (I) exceeding $I+\sigma_I$, but seven clones (L69, MJ57, MJ45, W2, MJ43 and MJ53) had index lower than $I-\sigma_I$. By comparing with clone W5 (control) and considering breast-height form indexes (form quotient, Q, and factor, F), 18 clones were selected (Table 8). The 18 selected clones had higher selection indexes (*I*) and lower mean of *Q* and *F* than the control. The mean values of wood yield and 5 traits for 18 selected clones were compared with the control (clone W5) and the population mean in Table 8. All of 18 selected clones had higher wood yield, height, DBH and foliage than the control. The means of wood yield, height, DBH and foliage of the former were 2.3, 1.3, 1.4 and 2.4 times higher than those of the later. All selected clones except H1 also performed better than the mean of 27 clones. Clone H1 was selected because of its superior tree form index. Most of the selected clones had thicker bark but lower density than the control and the population mean. Table 8 also indicated that both higher wood yield and density may be obtained through further selection.

References

Cai, S. 1989. China agriculture encyclopedia. Agriculture Publishing House, Beijing.

Lin, D. 1997. Experimental design and statistical analysis. The postgraduate teaching material of the Tropical Agriculture University of South China.

Lin, D. 1998. SAS tutor of computer programming language. The postgraduate teaching material of the Tropical Agriculture University of South China.

Ma, Y. 1982. The fundamental quantitative genetics of plant breeding. Jiangshu Sci.& Tech Publishing House, Nanjing.

Moll, R.H. and Stuber, C.W. 1974. Quantitative genetics: empirical results relevant to plant breeding. Advances in Agronomy 26:277-313.

Nordskog, A.W. 1978. Some statistical properties of an index of multiple traits. Theor. Appl. Genet. 52:91-94.

SAS Institute Inc. 1995. SAS/STAT Guide for personal computers, Version 6, SAS Institute Inc., Cary, NC, USA.

Yu, X., Xu, D. and Wang, S. 2000. Research on the management of long-term *Eucalyptus* plantation productivity. China Forestry Publishing House, Beijing, pp13-14.

Zeng, T., Liu, Y. and Fu, G. 1995. Studies on the eucalypts ecosystem of short rotation in Leizhou, Guangdong Province. China Forestry Publishing House, Beijing, 241p.

Table 8. The mean values ($\overline{X}$) of 18 selected clones and their superiorities, S1 and S2, over the control (CK) and the population mean ($\overline{P}$), respectively.

Clone	Wood yield			Height			DBH			Foliage weight			Bark thickness			Wood density		
	$\overline{X}$ (t/ha)	S1 (%)	S2 (%)	$\overline{X}$ (m)	S1 (%)	S2 (%)	$\overline{X}$ (cm)	S1 (%)	S2 (%)	$\overline{X}$ (kg)	S1 (%)	S2 (%)	$\overline{X}$ (cm)	S1 (%)	S2 (%)	$\overline{X}$ (t/m^3)	S1 (%)	S2 (%)
EC1	61.80	413.7	238.8	13.72	145.0	125.8	10.14	162.0	137.1	4.41	404.6	226.2	0.44	118.9	102.3	0.51	111.6	112.1
CH1	42.22	282.7	163.1	12.93	136.7	118.6	9.39	150.0	127.0	4.31	395.4	221.0	0.47	127.0	109.3	0.45	99.1	99.6
CH3	42.17	282.3	162.9	13.27	140.3	121.7	10.04	160.4	135.8	3.42	313.8	175.4	0.51	137.8	118.6	0.43	94.5	94.9
CH5	39.27	262.9	151.7	12.30	130.0	112.8	9.12	145.7	123.3	2.67	245.0	136.9	0.50	135.1	116.3	0.48	105.0	105.5
CH11	37.23	249.2	143.9	12.39	131.0	113.6	8.81	140.7	119.2	2.60	238.5	133.3	0.55	148.6	127.9	0.44	96.9	97.4
ZU6	37.18	248.9	143.7	12.33	130.3	113.1	8.89	142.0	120.2	3.38	310.1	173.3	0.49	132.4	114.0	0.41	89.1	89.5
CH10	35.03	234.5	135.3	12.53	132.5	114.9	8.72	139.3	117.9	2.55	233.9	130.8	0.58	156.8	134.9	0.48	104.8	105.3
CH32-32	34.14	228.5	131.9	12.29	129.9	112.7	8.86	141.5	119.8	1.94	178.0	99.5	0.48	129.7	111.6	0.44	96.7	97.1
CH6	33.84	226.6	130.8	12.27	129.7	112.5	8.60	137.4	116.3	3.01	276.1	154.4	0.55	148.6	127.9	0.49	106.6	107.0
UT1	33.06	221.3	127.7	12.68	134.0	116.3	8.70	139.0	117.7	2.45	224.8	125.6	0.54	145.9	125.6	0.45	98.7	99.1
CH4	32.55	217.9	125.8	12.30	130.0	112.8	8.66	138.3	117.1	2.07	189.9	106.2	0.61	164.9	141.9	0.44	96.1	96.5
CH15	31.66	211.9	122.3	12.23	129.3	112.2	8.69	138.8	117.5	2.25	206.4	115.4	0.51	137.8	118.6	0.43	94.3	94.7
SH1	29.25	195.8	113.0	12.42	131.3	113.9	7.90	126.2	106.8	2.25	206.4	115.4	0.29	78.4	67.4	0.45	98.5	98.9
U16	27.98	187.3	108.1	11.77	124.4	108.0	8.17	130.5	110.5	1.58	145.0	81.0	0.37	100.0	86.0	0.40	87.7	88.1
Wb	27.84	186.4	107.6	11.38	120.3	104.4	7.86	125.6	106.3	1.76	161.5	90.3	0.45	121.6	104.7	0.47	102.8	103.3
Hu1	27.51	184.2	106.3	11.64	123.0	106.8	8.04	128.4	108.7	2.87	263.3	147.2	0.57	154.1	132.6	0.46	101.5	102.0
MLA	26.72	178.9	103.3	11.42	120.7	104.7	8.24	131.6	111.4	2.10	192.7	107.7	0.44	118.9	102.3	0.42	90.8	91.2
H1	17.65	118.2	68.2	10.17	107.5	93.3	6.54	104.5	88.5	1.28	117.4	65.6	0.35	94.6	81.4	0.44	96.3	96.7
Mean	34.28	229.5	132.5	12.22	129.2	112.1	8.63	137.9	116.7	2.61	239.4	133.8	0.48	129.7	111.6	0.45	98.5	98.9
CK(W5)	14.94	100.0		9.46	100.0		6.26	100.0		1.09	100.0		0.37	100.0		0.46	100.0	
$\overline{P}$	25.88		100.0	10.90		100.0	7.39		100.0	1.95		99.9	0.43		100.2	0.46		100.0

不同桉树无性系主要性状分析及其综合评价

莫晓勇，彭仕尧，龙腾，陈文平，杨小红
中国国营林场开发总公司雷州林业局
524348 中国广东湛江

通过对 27 个 3 年生不同桉树无性系的几个主要性状进行调查，运用比较、方差分析、相关分析等方法，分析不同无性系主要性状的种间差异，以及各性状间的相互关系。结果表明：不同地点立地条件和桉树无性系之间的年均蓄积量差异极显著，造林地点和无性系存在极显著的交互作用。无性系干形指标与其产量存在极显著的负相关；树高和胸径之间存在显著的表型和遗传相关；通过干形指标和无约束指数选择法，综合评定了各无性系的生长表现，选出 EC1、CH3、CH1 等 18 个表现优良的无性系。

VARIATION IN FERTILITY AND ITS IMPACT ON GENE DIVERSITY IN A SEEDLING SEED ORCHARD OF *EUCALYPTUS TERETICORNIS*

MOHAN VARGHESE [1,2], NAMASIVAYAM RAVI [2], SEOG-GU SON [1,3] AND DAG LINDGREN [1]

[1] *Department of Forest Genetics and Plant Physiology*
Swedish University of Agricultural Sciences
SE 901 83, Umea, Sweden
[2] *Institute of Forest Genetics and Tree Breeding*
Coimbatore - 641002, India
[3] *Korea Forest Research Institute*
Cheongryangri 2-207, Seoul, Korea

Impact of fertility variation in a progeny trial of *Eucalyptus tereticornis* in south India was studied at four years of age with the intention to convert it to a seedling seed orchard. A hypothetical phenotypic selection of 200 trees for tree height resulted in only 18% fertile trees. There was tremendous variation in fertility (sibling coefficient, A=17.4). This would cause excessive representation of the 12 most fertile trees resulting in high genetic drift, inbreeding and low status number (N_s=8.5). Collection of equal quantities of seed per tree would reduce sibling coefficient and almost double the status number in seed crop. Much of the diversity loss occurs in the first generation as indicated by 65% decline in relative status number in the first generation compared to six advanced generations. Retaining 35 additional pollen parents would increase status number and variance effective population size by 100% in the first generation, but reduce gain by 2% compared to the trees selected for genetic merit. Both strategies would reduce inbreeding coefficient in the second generation. Combination of the two - retaining trees for pollination only and limiting the number of seeds harvested from individual trees would reduce impact of fertility variation in seedling seed orchards and be useful as management strategies in domesticating tropical eucalypts.

1 Introduction

Progeny trials of candidate trees selected from natural forests are commonly used as breeding populations in improvement programs for tropical eucalypts (Eldridge et al. 1993). Seedling seed orchards are generally used as production populations in short rotation eucalypts because they help in combining selection based on family and provenance information and seed production. This process also enables estimation of genetic parameters like heritability and correlation of important traits, knowledge of which are essential in domesticating the species.

It is assumed that open pollinated seeds obtained from seed orchards are generated from random mating of genotypes retained after final evaluation of the progeny trial. Selection is often focussed on breeding values and erratic flowering and sexual asymmetry among orchard trees could result in drastic deviation from panmixis leading to low predicted gains (Griffin 1982). The situation could vary between locations and species and a proper understanding of the fertility pattern would be essential in determining the orchard stocking.

The Indian land race of *Eucalyptus tereticornis* (Mysore gum) has evolved from a narrow base (Kaikini 1961) and suffers from inbreeding and hybrid breakdown (Davidson 1998). Domestication of this species has been given a new thrust with fresh introductions to initiate breeding programs for improving the productivity (Doran et al. 1996). Natural seedlots of *E. tereticornis* are however observed to have poor flowering in peninsular India and adjoining Sri Lanka (Arnold 1996). Differences in gamete contribution could result in over representation of the productive provenances or trees leading to increased inbreeding and genetic drift in the seed crop.

Seedling seed orchards are expected to serve as a means of packaging genes and generate improved seed at the culmination of a generation of breeding. Breeding values are normally used for culling the inferior families and individuals to maximise gain. Fertility of trees also has to be considered as it ultimately decides the transfer of the genes to the seed crop. The coancestry of the fertile trees and the effective number of trees that contribute to random mating will give an estimate of the diversity and inbreeding in the subsequent generations. Estimating these factors assume importance in the first generation as it is the point where a species is shifted from its native condition to a modified environment and where maximum gene losses occur (Andersson et al. 1999).

The quality of seed output from an orchard will be judged from the possible gain and the diversity of the deployed crop. Genetic value of seed will be determined by the breeding values and the maternal and paternal gametes produced by the orchard trees. This study aims to investigate fertility variation in a first generation progeny trial of *E. tereticornis* and predict the impact of a hypothetical selection procedure on relatedness and inbreeding in advanced generations. Management options to mitigate diversity loss in the seed crop have also been considered.

2 Material and Methods

2.1 *Seedling Orchard*

An open pollinated progeny trial of *E. tereticornis* was evaluated at 4 years of age for tree growth (height) and reproductive traits. The trial established at Coimbatore (11°N latitude, 76°58′E longitude, 1500 mm rainfall and 300 m altitude) in south India comprised of 42 families from 17 provenances with 24 trees per family in 4 tree plots and 6 replications at spacing of 3 m x 2 m. It was to be converted to a seedling seed orchard after an early evaluation, as part of a breeding program for the species (Doran et al. 1996).

An estimate of breeding value for each tree was obtained from combined index values computed (Varghese et al. 2002b) according to Falconer and Mackay (1996).

2.2 *Assessment of Flowering*

Flower and fruit production in the trees were recorded during the period October 2001 to January 2002. An assembled aluminum ladder with a platform and wheels (Varghese et al. 2000a) was used to go up to the top of the tree and record data. The number of primary, secondary and tertiary branches were counted in each tree and the flowers per tertiary branch recorded for all the trees during two visits in the peak flowering period of October - November. Number of stamens per open flower was recorded in a sample of 10 flowers in each tree. The number of developed fruits per secondary branch was recorded in two visits during the period of seed collection (December - January). Estimates of number of fruits and stamens per tree were obtained by extrapolating the counts made on stamens and fruits (Bila et al. 1999; Kang and Lindgren 1998).

E. tereticornis produces perfect flowers in axillary umbels with 5-7 buds per cluster, normally in new shoots. When the flower blooms, the operculum of the bud opens, revealing large number of creamy white stamens and a single style in the centre. Flowers are mostly outcrossed as they are protandrous, with the stigma becoming receptive after pollen are shed. Stigma is receptive usually about third to fourth day after anthesis when it becomes swollen and sticky. Eucalypts are insect pollinated and honeybees are the major pollinators. Self-fertilisation is not common because of natural biological barriers associated with the protandrous nature of flowers. Flowers open progressively in the tree, which could result in some self-pollination. Bees rarely visit the flowers after second day of anthesis.

2.3 *Variation in Fertility*

The male and female fertility of a tree was considered to be equal to the number of male and female gametes produced by the tree (Gregorius 1989). Gender fertilities were assumed to be equal to the number of reproductive structures (stamens and fruits) expressed as a proportion of all trees. Total fertility of a tree (p_i) was taken as the average of the male (f_i) and female (m_i) fertilities of each tree. When equal number of fruits was collected from each tree, the female fertility was assumed to be constant, equal to $1/N$, where N is the census number of fertile trees in the orchard.

Sibling coefficient, calculated from the number of fertile trees in the orchard (N) and individual fertility (p_i) of each tree was used to describe fertility variation,

$$A = N\sum_{i=1}^{N} p_i^2 \, . \qquad [1]$$

Sibling coefficient is also related to the coefficient of variation ($C.V.$) in fertility. The parental sibling coefficient can be divided into maternal sibling coefficient (A_m) and paternal sibling coefficient (A_f),

$$A_m = N\sum_{i=1}^{N} m_i^2 = CV_m^2 + 1 \text{, and} \qquad [2]$$

$$A_f = N\sum_{i=1}^{N} f_i^2 = CV_f^2 + 1 \, , \qquad [3]$$

where CV_m and CV_f are the coefficients of variation of maternal and paternal fertilities respectively.

The total fertility variation in the orchard was calculated (Kang and Lindgren 1999) as

$$A = CV_t^2 + 1\,, \qquad [4]$$

where CV_t is the average tree fertility of combined male and female fertilities.

When equal amount of seed is collected from each genotype, the maternal fertility is constant and fertility variation will be expressed as a function of paternal fertility,

$$A = 0.25CV_f^2 + 1\,. \qquad [5]$$

2.4 *Coancestry and Effective Population Size*

Group coancestry (Θ) is the probability that two genes chosen at random from a gene pool are identical by descent (Cockerham 1967). If the trees are non related and non inbred, all pair wise coancestries are equal to zero and all self coancestries equal to 0.5 and the group coancestry is calculated as

$$\Theta = 0.5\sum_{i=1}^{N} p_i^2\,. \qquad [6]$$

Depending on the relatedness between the selected trees in the orchard, the group coancestry was calculated according to Lindgren and Mullin (1998),

$$\Theta = \sum_{i=1}^{N}\sum_{j=1}^{N} p_i p_j \theta_{ij}\,, \qquad [7]$$

where p_i and p_j are the probabilities that genes sampled at random from the gamete pool originate from genotypes *i* and *j* respectively and θ_{ij} is the probability that genes sampled from genotypes *i* and *j* are identical by descent (coancestry between *i* and *j*).

If different sexes among parents are considered, the probabilities can be divided into a paternal and maternal component and an interaction component,

$$\Theta = \sum_{i=1}^{N}(m_i + f_i)\sum_{j=1}^{N}(m_j + f_j)\theta_{ij}\,. \qquad [8]$$

Status number is half the inverse of group coancestry (N_s = 0.5/Θ). Provided the trees are unrelated it can be computed as,

$$N_s = \frac{1}{\sum_{i=1}^{N} p_i^2}\,, \qquad [9]$$

where p_i is the contribution from individual genotype *i* to the gamete pool and *N* the census number of trees in the orchard. N_r was used to compare the effective number of trees contributing to random mating with the actual number of trees retained in the orchard,

$$N_r = \frac{N_s}{N}. \quad [10]$$

Variance effective population size ($N_e^{(v)}$) was computed to describe the size of the population that would give the same drift in gene frequencies in the seed crop as the orchard parents (Kang et al. 2001b),

$$N_e^{(v)} = \frac{A}{2\Theta(A-1)}. \quad [11]$$

2.5 *Prediction of Relatedness Over Generations*

The gametes produced by the orchard parents form the gene pool of the seed crop. The group coancestry of the seed crop (Θoffspring) is influenced by the coancestry of the orchard parents (Θgametes),

$$\Theta_{offspring} = \frac{0.5}{N_{offspring}} + (1 - \frac{0.5}{N_{offspring}})\Theta_{gametes}. \quad [12]$$

The group coancesty of the gamete pool is a function of inbreeding (F), fertility variation (A) and number (N) of parent trees (Bila et al. 1999; Kang 2001),

$$\Theta_{gametes} = \frac{0.5(1+F)A}{N} + (1 - \frac{A}{N})\frac{N\Theta - 0.5(1+F)}{N-1}. \quad [13]$$

Under assumption of random mating conditions, group coancestry of the parents becomes the expected inbreeding of the offspring as shown below,

$$F_{offspring} = \Theta_{offspring}. \quad [14]$$

A program developed on excel spread sheet was used to compute the predicted values (Θ, F, N_s, and N_r) for six generations assuming that the fertility variation observed in the parental generation in the orchard is continued over six generations of progenies and that the offspring population size is uniform.

The cumulative contribution of the orchard parents to the seed crop was described using the power function $F_x = x^a$ where x is the relative fertilities of each parent ranked according to the relative proportions. The parameter a, cannot be less than one and is computed from A (Kang and Lindgren 1998),

$$a = A\left(1 + \sqrt{1 - \frac{1}{A}}\right). \quad [15]$$

2.6 *Gene Diversity*

Expected gene diversity (*GD*) is a function of the group coancestry and can be calculated in the orchard relative to a reference population. The reference population, which is the natural forest, has zero group coancestry as it is considered to have an infinite number of unrelated individuals,

$$GD = 1 - \Theta . \qquad [16]$$

Expected average heterozygosity in the orchard crop (He_t) can also be expressed on the same basis as gene diversity (Stoehr and El-Kassaby 1997). The heterozygosity of first generation orchards will be one because the parents are considered to be a sample from the base population,

$$He_t = (1 - \frac{1}{2N_e^{(v)}}) He_{t-1} . \qquad [17]$$

2.7 *Genetic Gain*

Expected genetic can be computed based on the fertility of the orchard parents and the breeding value of the trees (Griffin 1982; Lindgren et al. 1989; Kang 2001),

$$\Delta G = \sum_{i=1}^{N} G_i p_i . \qquad [18]$$

where p_i is the fertility of the *i*th tree, *N* the number of trees, and G_i, the breeding value of the *i*th tree.

3 Results

Breeding value was estimated for each tree. The trees were arranged according to phenotypic values for tree height and 20% of the best trees (200 trees) were truncated based on the deviation of the individual tree value from the mean value of all the trees.

3.1 *Fertility Estimation*

Variation in fertility was high and only 12% of the trees in the progeny trial were fertile. Among the 200 selected trees, only 35 trees produced flowers and fruits. The number of stamens was quite variable, ranging from 183 to 252 in each flower, and 60% of the flowers had around 200 stamens. Thus 200 stamens per flower were used to extrapolate the number of stamens produced per tree. There was no significant correlation (r=0.057) between tree growth and fertility, but there was a high correlation between male and female fertilities (r=0.981). Coancestry was determined for the 35 fertile trees and 35 additional pollen parents by considering their relatedness and male and female fertilities as per equation [7] and [8].

Sibling coefficient (*A*) which takes into account the sum of squared contributions from each tree varied between male and female fertilities indicating that trees varied greater in their seed output than in pollen production (Table 1). When equal quantity of seed was collected from each tree, sibling coefficient reduced to less than half the original value. It was still eight times the value of equal fertility for all trees, which results in the same gene frequencies in seed crop as that of the parent trees. Variation in fertility was further substantiated by the values of parameter *a*, which indicate the magnitude of flowering differences between trees. Values of *a* also indicated the same trend as that of sibling coefficient.

To increase the pollen availability, a hypothetical situation was considered - retaining additional 35 fertile trees, which ranked below the selected trees, as pollen contributing parents only. The differences in male and average tree fertilities were reduced in this option. The female fertility however remained the same; only the initially selected 35 trees were used as seed parents.

Table 1. Fertility variation and gain with selection for genetic value and extra pollen parents.

	Selected trees		70 pollen parents	
Fertility type	*A*	*a*	*A*	*a*
Male	15.8	31.2	10.2	20
Female	19.3	38.2	19.3	38.2
Av tree fertility	17.4	34.3	16.3	32.2
Female constant	8.2	16	6.9	13.3
Equal fertility	1	1	1	1
Gain	46.57		45.65	
Eff. Pop. No.	11.5		16.5	

3.2 *Changes Across Generations*

For a certain pattern of fertility variation (decided by sibling coefficient) predictions were made on the changes in relatedness and inbreeding for six advanced generations. Inbreeding and coancestry would increase considerably, if seed collection were to be continued from the 35 selected trees, with the existing pattern of fertility differences (Table 2). Census number of orchard parents was 200 and assumed to be 200 for next generations. Status number followed an opposite decreasing trend in advanced generations, with a steep decrease in the second generation.

Heterozygosity and gene diversity followed a similar trend as that of status number, decreasing in advanced generations, indicating increase in relatedness in advanced stages of improvement.

Table 2. Measures of coancestry, inbreeding and diversity in orchard parents and six advanced generations.

Fertility type	Generations						
	Gen 1	Gen 2	Gen 3	Gen 4	Gen 5	Gen 7	Gen8
			Varying fertility				
Θ	0.0586	0.097	0.135	0.172	0.207	0.240	0.273
F	0.0	0.059	0.097	0.135	0.172	0.207	0.240
N_s	8.532	5.135	3.694	2.910	2.418	2.080	1.834
N_r	0.043	0.026	0.018	0.015	0.012	0.010	0.009
$N_e^{(v)}$	9.053	-	-	-	-	-	-
H_e	0.945	0.913	0.883	0.857	0.833	0.811	0.791
GD	0.941	0.903	0.865	0.828	0.793	0.760	0.727
% Div loss		4.295	4.393	4.40	4.413	4.425	4.439
			Constant seed collection				
Θ	0.0294	0.049	0.068	0.087	0.106	0.124	0.142
F	0.0	0.029	0.049	0.068	0.087	0.106	0.124
N_s	17.007	10.230	7.334	5.738	4.728	4.031	3.521
N_r	0.085	0.051	0.037	0.029	0.024	0.020	0.018
$N_e^{(v)}$	19.369	-	-	-	-	-	
H_e	0.974	0.958	0.943	0.928	0.914	0.90	0.888
GD	0.971	0.951	0.932	0.913	0.894	0.876	0.858
% Div loss		2.048	2.071	2.076	2.082	2.087	2.093
			Extra pollen parents				
Θ	0.0293	0.068	0.105	0.140	0.174	0.207	0.239
F	0.0	0.029	0.068	0.105	0.140	0.174	0.207
N_s	17.065	7.365	4.779	3.571	2.872	2.416	2.096
N_r	0.085	0.037	0.024	0.018	0.014	0.012	0.010
$N_e^{(v)}$	18.180	-	-	-	-	-	-
H_e	0.972	0.938	0.908	0.881	0.856	0.834	0.813
GD	0.971	0.932	0.895	0.860	0.826	0.793	0.761
% Div loss		4.140	4.104	4.116	4.127	4.139	4.151
			Equal fertility				
Θ	0.00636	0.009	0.011	0.014	0.016	0.019	0.021
F	0.0	0.006	0.009	0.011	0.014	0.016	0.019
N_s	78.616	56.473	44.063	36.124	30.609	26.555	23.450
N_r	0.393	0.282	0.220	0.181	0.153	0.133	0.117
$N_e^{(v)}$	infinity	-	-	-	-	-	-
H_e	infinity	infinity	infinity	infinity	infinity	infinity	infinity
GD	0.994	0.991	0.989	0.986	0.984	0.981	0.979
% Div loss		0.252	0.252	0.253	0.254	0.254	0.255

3.3 *Altering Fertility Status*

If the improvement program were to depend on selected orchard trees, the existing fertility pattern would lead to rapid increase in coancestry (Figure 1) and substantial

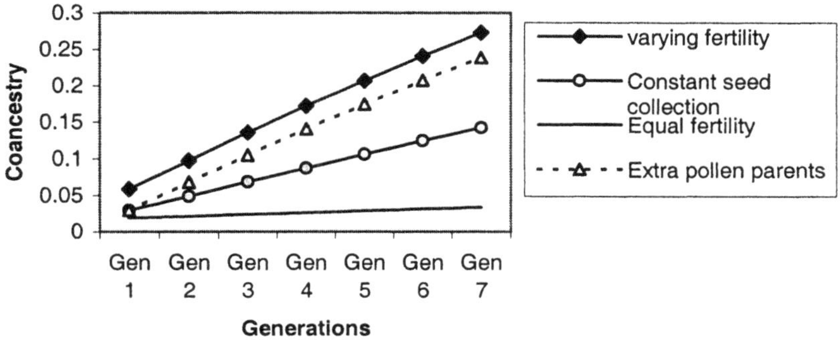

Figure 1. Coancestry variation in different fertility conditions.

reduction in status number. Alternative options tested like constant seed collection from the mother trees and increased pollen contribution would help in improving the trend in later generations. Restricting seed contribution from orchard parents would help in lowering coancestry by 92% and inbreeding by 94% in the seventh generation. Similarly status number would be higher by 92%. Retaining additional pollen parents was found to be a less effective strategy. This enables coancestry and inbreeding values to be reduced by 14 and 16% respectively and status number to be increased by 14% in the seventh generation. Thus the effective population size can be increased in advanced generations if equal seed collection was done with more male parents in the orchard. These strategies however achieve a lower status number than the ideal situation of equal fertility among all trees.

Constant seed collection and maintaining additional pollen parents were also effective in increasing variance effective population size (twice as that of the existing fertility situation when female fertility is constant). $N_e^{(v)}$ and heterozygosity are considered infinite in the ideal situation, when all trees are assumed to have same fertility. A higher level of gene diversity can be maintained in the first generation by these strategies. However with the existing levels of fertility differences, when additional pollen parents are considered, loss is drastic in subsequent generations and in the seventh generation, diversity is 13% lower than the option of equal seed collection from all trees. Constant seed collection from all trees enables minimum loss in gene diversity with each generation (Table 2). When equal fertility is assumed for all 200 selected trees, greater number of related trees (than the fertile 35 trees) results in a higher uniform loss in diversity across all generations. Loss in diversity is fairly uniform across generations as it is assumed that the same fertility differences persist in all generations. There is a reduction of only 4 to 7% in diversity loss in each generation if additional pollen parents are retained, as seed collection is restricted to the selected mother trees with the existing pattern of variation in female fertility.

There would be a reduction in gain of 2.02% when an equal number of fertile trees of lower genetic potential were retained in addition to the trees selected for phenotypic superiority.

4 Discussion

Genetic improvement programs in *Eucalyptus* aim to upgrade genetic value of the crop through recurrent selections made in breeding populations, that are converted to seedling seed orchards (Doran et al. 1996). Genetic testing and thinning are two components used to achieve the anticipated improvement. These steps are completed by fourth year, when the orchard is normally ready for operational seed production (Meskimen and Francis 1990).

4.1 *Selection for Genetic Value*

Maximisation of gain is often the first priority in any improvement program. Combined index selection is effective in maximising gain but poor in conserving diversity (Wei 1995) as it confines to selections from superior families. Phenotypic selection achieves moderate gain but is efficient in maintaining diversity even after several generations (Spanos et al. 1996; Andersson et al. 1998). An improvement program employing phenotypic selection is generally considered to be risk free as it ensures adequate representation of different families and low coancestry level in the orchard. This has prompted breeders to establish unpedigreed orchards thinned purely on phenotypic performance as a routine improvement strategy in eucalypts (Harwood et al. 1996). Fertility of the trees is the other important factor influencing gain in the crop. Different fertility patterns can alter the mating system in the orchard. Fertility status often assumes equally important role as the genetic value of the parent, in determining the quality of seed. Thinning, done to remove genetically inferior families may alter the mating pattern further. A proper analysis of the fertility variation would be essential in proposing a thinning regime.

4.2 *Fertility Variation*

It has been seen in this study that fertility was not correlated with tree growth, and selections made for growth traits would not promote seed production. Differences in fertility rates between individuals would result in corresponding differences in gamete contribution leading to over representation of more productive trees. If the trees had same fertility, each of the 200 selected trees would have contributed 0.005% of the gamete pool. The actual female fertilities varied from 22.9 to 0% and the male fertilities varied from 18.8 to 0%. Twelve most fertile trees produced almost 81% of the total gametes. This causes a big shift in the really contributing trees in relation to the actual number of trees in the orchard. The values of sibling coefficient indicate the impact of fertility differences on later generations, as it is dependent only on the fertility status and not the relatedness of trees (Kang and Lindgren 1999). Gene frequencies would remain the same if trees had the same fertility leading to random mating. Each unit increase in sibling coefficient indicates the rate at which each generation deviates from the previous one, and the extent to which inbreeding and relatedness would build up. Cumulative contribution of the seed orchard genotypes to the seed crop is indicated by the values of parameter a, which is derived from A (Kang and Lindgren 1998). Impact of shift in individual contribution from the ideal, (equal seed contribution) is effected, exponential to the values of a. Only

17.5% of the 200 selected trees produced seed and the rate of build up in relatedness (*A* = 17.4) would be highly biased from ideal panmictic situation. Associated with this would be a chain of events like increase in inbreeding and mating between related trees and a proportionate loss in diversity.

Status number of the selected trees, which was 4.3% of the census number in first generation, would reduce to 0.9% by the seventh generation. Variance effective population size can be used to quantify extent of drift that results from fertility differences (Kjaer and Wellendorf 1998). Unequal gamete contribution has caused a net effect of less than ten equally fertile trees among the selected ones, contributing genes to the next generation. Inbreeding increases to the extent that 12 most fertile trees cause seed production. High sexual symmetry observed in trees could be a cause of concern as there could be within tree mating and mating between related trees apart from the selfing within each flower (which is normally expected to be low). Remedial measures like, collection of equal quantities of seeds and enhancing the male parent contribution (Lindgren and El-Kassaby 1989) have been put forth to mitigate the adverse effects. When the pollen contribution was doubled, the number of contributing parents was increased and hence the drift was reduced (Table 2). Selective seed harvest, genetic thinning and combination of the two are reported to increase gain but adversely affect diversity (Kang et al. 2001a).

Loss of diversity and increase in relatedness are expected in the advanced stages of improvement in a species. Improvement in gain is at the cost of diversity with each advance in generation. It is however important to monitor the consequences of selection and thinning done to enhance gain. As relatedness increase beyond a certain point, depending on the deviation from random mating, much of the desired benefits may not be achieved. Kang et al. (2001b) recommended an effective clone number of at least 10 in an advanced generation clonal seed orchard with an equal number of ramets per clone. There is a steady loss in status number (8.53 to 1.83) from the first to seventh generation, which results in 29% loss of gene diversity and 19% loss in expected heterozygosity. Loss is drastic in the second generation. Low fertility in orchard parents has resulted in 66% loss in status number in the first seed crop. Subsequent losses are comparatively less and emphasis should be made on curtailing losses at the entry, which decides the genes tapped for further improvement in the domestication process.

In a survey on fertility differences in seed orchards and stands Kang (2001) concluded that fertility variation and sibling coefficient was more in tree stands than in seed orchards. The study made on clonal seed orchards of four broad-leaved species (Bila 2000) gives a maximum value of A= 9.32 (average =2.62). The present study reports a higher value (A=17.4), as it pertains to a first generation seedling orchard. A first generation progeny trial of an exotic species like *Eucalyptus* may have greater asynchrony in flowering than a seed stand or a seed orchard, which is often established with selections from plantations and thus several generations into the domestication process. Variation in individual trees in a seedling seed orchard could be more than that in the ramets of different clones in a clonal seed orchard. The drift could be high in poor flowering years compared to abundant and intermediate flowering years (Kjaer 1996). Observations need to be continued in subsequent years and in different locations to determine the optimum age for seed collection in a seed lot for a particular location.

4.3 *Impact of Modifying Fertility Status*

As found in the present study, fertility plays a very important role in converting the genetic value of the selected trees to gain in the next generation. It would be better to have a fertile tree of slightly lower genetic value than to have a superior tree that does not contribute to seed production. Unequal fertility can also cause deviation in the mating pattern. Thus the major concern would be to have random mating of maximum number of trees of known genetic potential. This would be especially true in a first generation seedling seed orchard, where the species is initiated into domestication process. Equal seed collection helps to have uniform female contribution and remove impact of excessive fertility of few individual trees. Increasing the number of male parents enhances the effective number (Lindgren and El-Kassaby 1989), which however lowers the genetic value of the crop, as the additional male parents selected are genetically inferior to the trees selected for merit. A combination of the two strategies would serve to achieve both objectives, as seed collection is limited to the superior trees and higher diversity levels are maintained by contribution from males. The impact of modifying fertility status can be monitored by the reduction in values of sibling coefficient and the parameter '*a*'. This strategy ensures reduced drift and inbreeding in the seed crop, which will decide the constitution of advanced generations.

4.4 *Implications in Eucalypt Domestication*

Modification of a species to suit an exotic site involves adaptability to survive and perform well in the new location. Eucalypts are mostly used, as pulpwood in India and the prime objective of an improvement program would be to produce more volume of bole wood of good form. Extensive provenance testing has been done since early 1980s (Rao 1984) and sufficient information is available on the suitable populations for different regions of the country. But there has been no serious effort to quantify the flowering potential of selected trees. Venkatesh and Kedharnath (1965) who detected inbreeding in eucalypts gave guidelines for seed collection to improve genetic purity and quality of seed crop. The problems are acute in first generation introductions where maximum drift and loss in diversity occur. Sufficient care has to be taken in the first generation seed orchard and suitable strategies employed to reduce genetic erosion. Lack of awareness of this factor in many eucalypt growing countries of Asia Pacific region has been highlighted by Davidson (1998) who recommended suitable measures (Varghese et al. 2000b) to increase genetic base of plantations.

Balancing gain and genetic diversity is important in eucalypts because of the fast turnover resulting from short rotations. Much emphasis is placed on gain, that clonal plantations are becoming extremely popular (Lal et al. 1993). However, seed still remains the major planting material in developing countries and clonal option remains a short-term strategy in most breeding programs (Varghese et al. 2002a). Selection can be made in every advanced generation, for deployment as clones. In such a situation it is desirable to have large breeding populations in each generation that get converted to seedling seed orchards which provide fair diversity in seed output, enhanced gain when necessary in the form of clones, and information on progeny performance of parents in the previous generation. Some of the clonal trials could be suitably thinned and converted to clonal seed orchards with proper knowledge of their fertility for greater gain. The seedling orchard provides opportunities for correcting fertility problems and producing seed for different requirements. Additional pollen parents could be retained until the non-flowering trees selected for genetic merit contribute seed. Paclobutrazol application is used to improve flowering in orchards (Hetherington 1994).

Effective number of contributing trees may increase with generations as seeds are collected from fertile trees, but the status number may decrease as inbreeding increases. Levels of inbreeding and drift appear to be high in the present seed orchard. This is commonly encountered in first generation seedling seed orchards, which calls for steps to improve flowering and reduce the impact of low fertility. The effective number of mothers can intentionally be forced to become more similar to the real number by extracting seeds tree wise and discarding excess seeds from mothers contributing most, or the prolific seed bearers may not be completely harvested. This principle of equal seed contribution is widely used in establishing what are known as extensive seedling seed orchards (Nikles et al. 1984; Varghese et al. 2000a) in eucalypts.

Acknowledgements

This study was conducted when Mohan Varghese was a visiting scientist at the Department of Forest Genetics and Plant Physiology, Swedish University of Agricultural Sciences, Umea. International Tropical Timber Organisation, Japan and Kempe Foundation, Sweden provided financial support. Seog-Gu Son was supported by postdoctoral fellowship program from Korea Science & Engineering Foundation (KOSEF), the Korea Forest Research Institute (KFRI) and the Swedish Association for Forest Tree Breeding (Föreningen Skogsträdsförädling).

References

Andersson, E.W., Sanchez Rodriguez, L. and Andersson, B. 1999. Group coancestry controlled selection in *Pinus sylvestris* L. breeding program. Theor. Appl. Genet. 99:73-80.

Andersson, E.W., Spanos, K.A., Mullin, T.J. and Lindgren, D. 1998. Phenotypic selection compared to restricted combined index selection for many generations. Silva Fen. 32:111-120.

Arnold, R.J. 1996. *Eucalyptus* genetic improvement strategies – prepared for Sri Lanka Forest Department (unpublished report). Australian Tree Seed Centre, CSIRO, Canberra.

Bila, A.D. 2000. Fertility variation and its effects on gene diversity in forest tree populations. PhD Thesis, Department of Forest Genetics and Plant Physiology, Swedish University of Agricultural Sciences, Umea.

Bila, A.D., Lindgren, D. and Mullin, T.J. 1999. Fertility variation and its effect on diversity over generations in a teak plantation (*Tectona grandis* L.f). Silvae Genet. 48:109-114.

Cockerham, C.C. 1967. Group inbreeding and coancestry. Genetics 56:89-104.

Davidson, J. 1998. Domestication and breeding programme for *Eucalyptus* in the Asia-Pacific region. UNDP/FAO Regional Project on "Improved productivity of man-made forests through application of technological advances in tree breeding and propagation (FORTIP)", Philippines, 252p.

Doran, J.C., Pinyopusarerk, K., Arnold, R. and Harwood, C.E. 1996. Breeding plan for *Eucalyptus camaldulensis* in Tamil nadu. UNDP/FAO Regional Project on "Improved productivity of man-made forests through application of technical advances in tree breeding and propagation" (RAS/91/004 - FORTIP), FAO, Los Banos, Phillipines, 40p.

Eldridge, K.G., Davidson, J., Harwood, C.E. and van Wyk, G. 1993. Eucalypt domestication and breeding. Clarendon Press, Oxford.

Falconer, D.S. and Mackay, T.F.C. 1996. Introduction to quantitative genetics, forth edition. Longman, Malaysia, 464p.

Gregorius, H.R. 1989. Characterisation and analysis of mating system. Ekopan Verlag, Germany, 158p.

Griffin, A.R. 1982. Clonal variation in radiata pine seed orchards. 1. Some flowering, cone and seed production traits. Aust. For. Res. 12:295-302.

Harwood, C.E., Nikles, D.G., Pomeroy, P. and Robson, K.A. 1996. Impact of thinning via phenotypic selection on the genetic base of planted seed production areas. In: "Tree improvement for sustainable tropical forestry", M.J. Dieters, A.C. Matheson, D.G. Nikles, C.E. Harwood and S.M. Walker (eds), Proc. of QFRI-IUFRO Conference, Caloundra, 27 October - 1 November 1996, Vol.2, pp148-153.

Hetherington, S. 1994. Management of eucalypt seed orchard with pacloburtrazol. In: Proc. of Australian Forest Growers Conference on "Faces of farm forestry", Launceston, Tasmania, pp235-238.

Kaikini, N.S. 1961. *Eucalyptus* in Mysore state. In: Proc. of the Tenth All India Silvicultural Conference, Dehra Dun, pp546-553.

Kang, K.S. 2001. Genetic gain and gene diversity of seed orchard crops. PhD Thesis, Department of Forest Genetics and Plant Physiology, Swedish University of Agricultural Sciences, Umea.

Kang, K.S. and Lindgren, D. 1998. Fertility variation and its effect on the relatedness of seeds in *Pinus densiflora*, *P. thunbergii* and *P. koraiensis* clonal seed orchards. Silvae Genet. 47:196-201.

Kang, K.S. and Lindgren, D. 1999. Fertility variation among clones of Korean pine (*Pinus koraiensis*) and its implications on seed orchard management. For. Genet. 6(3):191-200.

Kang, K.S., Lindgren, D. and Mullin, T.J. 2001a. Prediction of genetic gain and gene diversity in seed orchard crops under alternative management strategies. Theor. Appl. Genet. 103:1099-1107.

Kang, K.S., Harju, A.M. Lindgren, D., Nikkanen, T., Almqvist, A. and Suh, G.U. 2001b. Variation in effective number of clones in seed orchards. New For. 21:17-33.

Kjaer, E.D. 1996. Estimation of effective population number in a *Picea abies* (Karst.) seed orchard based on flower assessment. Scand. J. For. Res. 11:111-121.

Kjaer, E.D. and Wellendorf, H. 1998. Variation in flowering and reproductive success in a Danish *Picea abies* (Karst.) seed orchard. For. Genet. 4:181-188.

Lal, P., Kulkarni, H.D. and Srinivas, K. 1993. *Eucalyptus* improvement programme of ITC Bhadrachalam Paper boards Ltd. In: K. Vivekanandan, K.N. Subramanian, N.Q. Zabala and K. Gurumurthi (eds), Proc. of Workshop on "Production of improved planting material for afforestation programmes", Coimbatore, India, 18-25 June 1993, UNDP/FAO Regional Project on "Improved productivity of man-made forests through application of technological advances in tree breeding and propagation (FORTIP)", Philippines.

Lindgren, D. and Mullin, T.J. 1998. Relatedness and status number in seed orchard crops. Can. J. For. Res. 28:276-283.

Lindgren, D. and El-Kassaby, Y.A. 1989. Genetic consequences of combining selective cone harvesting and genetic thinning in clonal seed orchards. Silvae Genet. 38:65-70.

Lindgren, D. Libby, W.S. and Bondesson, F.L. 1989. Deployment to plantations of numbers and proportions of clones with special emphasis on maximising gain at a constant diversity. Theor. Appl. Genet. 77:825-831.

Meskimen, G. and Francis, J.K. 1990. Rose gum *Eucalyptus*. In: "Silvics of North America", Hardwoods, R.M. Burns and B.H. Honkala (eds), USDA Forest Service, Washington DC, Agricultural Handbook 654, Vol. 2., 877p.

Nikles, D.G., Spidy, T., Bowyer, P.C. and Newton, R.S. 1984. Seedling seed sources of *Pinus caribaea var. hondurensis* in Queensland. In: R.D. Barnes and G.L. Gibson (eds), Proc. of IUFRO Conference on "Provenance and genetic improvement strategies in tropical forest trees", Mutare, Zimbabwe, 9-14 April 1984, pp 579-581.

Rao, D.V. 1984. Provenance trials of *Eucalyptus*. Indian For. 110:28-34.

Spanos, K., Andersson, E.W. and Lindgren, D. 1996. Multi-generation comparison between restricted phenotypic and combined index selection for genetic gain and diversity. In: "Conservation of forest genetic resources", M. Kurm and Y. Tamm (eds), Proc. of Nordic Group for Forest Genetics and Tree Breeding Meeting, Estonia, Estonian Agricultural University, Tartu, 3-7 June 1996, ISBN 9985-830-11-3.

Stoehr, M.U. and El-Kassaby, Y.A. 1997. Levels of genetic diversity at different stages of the domestication cycle of interior spruce in British Columbia. Theor. Appl. Genet. 94:83-90.

Varghese, M., Nicodemus, A., Nagarajan, B., Siddappa, K.R.S., Bennet, S.S.R. and Subramanian, K. 2000a. Seedling seed orchards for breeding tropical trees.

Scroll Press, Institute of Forest Genetics and Tree Breeding, Coimbatore, India, 126p.

Varghese, M., Nicodemus, A., Nagarajan, B. and Subramanian, K. 2000b. Hybrid breakdown in Mysore gum and need for genetic improvement of *Eucalyptus camaldulensis* and *E. tereticornis*. In: Proc. of QFRI/CRC-SPF Conference on "Hybrid breeding and genetics of forest trees", Noosa, Australia, 9-14 April 2000, pp519-525.

Varghese, M., Nicodemus, A., Nagarajan, B., Ravi, N. and Hegde, R. 2002a. A breeding programme for improving productivity of *Eucalyptus camaldulensis* and *E.tereticornis* in India. In: "Recent eucalypt research in India", S.K. Bagchi, M. Varghese and Siddappa (eds), Institute of Forest Genetics and Tree Breeding, Coimbatore, India, pp19-28.

Varghese, M., Son, S., Ravi, N. and Lindgren, D. 2002b. Optimising selection in an open pollinated progeny trial of *Eucalyptus tereticornis*. Paper presented for International Conference on "Eucalypt productivity: a synthesis of physiological, environmental, genetic and silvicultural determinants of growth", Hobart, Tasmania, 10-15 November 2002.

Venkatesh, C.S. and Kedharnath, S. 1965. Genetic improvement of *Eucalyptus* in India. Silvae Genet. 14:141-176.

Wei, R.P. 1995. Predicting genetic diversity and optimizing selection in breeding populations. PhD Thesis, ISBN 91-576-4990-1, Swedish University of Agricultural Sciences, Umeå, 64p+12 chapters.

细叶桉实生种子园可育结实性变异及其对基因多样性的影响

MOHAN VARGHESE [1,2], N. RAVI [2], SEOG-GU SON [1,3], DAG LINDGREN [1]

[1] *Department of Forest Genetics and Plant Physiology*
Swedish University of Agricultural Sciences
SE 901 83, Umea, Sweden
[2] *Institute of Forest Genetics and Tree Breeding*
Coimbatore - 641002, India
[3] *Korea Forest Research Institute*
Cheongryangri 2-207, Seoul, Korea

本文对印度南部四年生细叶桉子代试验林的树木可育结实性变异进行了研究，旨在将其转变成实生种子园。对 200 株桉树的表型性状进行了虚拟选择，结果发现：根据树高性状选择，只有 18%的入选桉树具有可育结实能力，且在可育结实能力方面具有较大变异(近交系数为 A=17.4)，这将使得 12 个最具可育结实能力的个体具有过度的代表性，从而会导致较高的遗传漂流、近交衰退和较低的群体现状有效数量(N_s=8.5)。从每棵树采集同样数量的种子可以减少近交系数，使群体现状有效数量翻一倍。与 6 个高世代种子园相比，第一代种子园会有较多的遗传多样性流失，相对群体现状有效数量下降到原来的 65%。与选择出来的优树相比，额外保留 35%的父本可以 100%地提高第一代种子园的群体现状有效数量和有效方差群体的大小，但却使增益降低了 2%。两种策略(只保留父本和限制单株采种数量)在第二代都可以降低近交系数。两种策略的结合可以减少可育结实性变异对实生种子园的影响，这对制定热带桉树引种驯化的管理策略非常有用。

ADVANCES IN TISSUE CULTURE OF *EUCALYPTUS* IN CHINA

YAOJIAN XIE

China Eucalypt Research Center
Zhanjiang 524022, Guangdong, China

Tissue culture is a very important technology for the research and development of *Eucalyptus* plantation. Up to date, about 60 *Eucalyptus* species have been studied with tissue cultural methods including use of tissues, cells or organs from seed, shoot, floral bud, lignotuber, callus, root, anther, suspension, protoplast and somatic embryo. Plantlets have been obtained for most of the species studied.

1 Significance of Tissue Culture

Tissue culture is a technology to micro-propagate plants with small pieces of tissues (organs or cells) in culture medium under sterile conditions. In a broad sense, tissue culture includes culture of organ, tissue, embryo, seed, protoplast, floral bud, etc. In a narrow sense, plant tissue culture is an *in vitro* propagation technology, so-called micro-propagation or tube propagation (Pan and Dong 1995; Cao and Liu 1996). It has become an important part of *Eucalyptus* plantation management due to the following significances or characteristics:

- Fast mass-propagation of superior genetic materials. With tissue culture technology, newly bred or introduced varieties, individuals and the natural and man-made useful mutants can quickly be reproduced in a relatively short time, and widely cultivated in production within two years (Cao and Liu 1996).
- Sterilized culture and seedling propagation. Pathogens such as bacteria, fungi, and virus etc are killed, and the potential disease damage to the seedlings or plantations can be reduced. Pseudomonas sp. is the main disease considered in the cultivation of *Eucalyptus* in China (Cao and Liu 1996).
- Fast propagation of special breeding materials such as floral bud and multiploid. It is through asexual cloning that the special breeding materials can be mass propagated, and their superior performance can be obtained in plantations. For example, a multiploid is in general sterile and can only be reproduced by asexual cloning. While great progress has been achieved in poplars, the potential in eucalypts has yet to be explored (Cao and Liu 1996).
- Propagation of gene-engineered plants. The tissue culture of transgenic plant cell is the important step to obtain and mass-propagate engineered plants. There are two general approaches to obtain transgenic plants (Chen 1992): 1) use protoplast as the receptor, transfer the target gene into cell with plasmid as the carrier, and obtain the transgenic plant by means of cell

culture, and 2) use tissue or organ such as lamina, stem section, embryo, anther, ovary and embryonic suspension cell as the receptor, transfer the target gene into tissue or organ with Agrobacterium fumefaciens as the carrier or by using physical means, and obtain the transgenic plants by means of tissue culture.

- Propagation of in vitro stored germplasm. The sterile materials such as shoot and somatic embryo stored under ultra-low temperature conditions need to be propagated by using tissue culture technology before being mass propagated for other purposes.

2 The Advances in Tissue Culture of *Eucalyptus*

To date, about 60 species of *Eucalyptus* have successfully been propagated with tissue culture technology. Most of the species were first reported in the foreign countries, but *E. leichow* No.1 (a natural hybrid of *E. exserta* and *E. camaldulensis*), *E. urophylla* × *E.grandis* etc. in China (Ouyang et al. 1980; Ouyang et al. 1981; 9. Xie 1981; Zeng 1988). Micro-propagation of eucalypts with tissue culture technology in large scale is now very popular in many countries such as Australia, Brazil, South Africa, etc. In China, the institutes that have tissue culture facilities mass-producing eucalypt seedlings are: Chinese *Eucalyptus* Research Centre, Guangdong Academy of Forestry, Zhanjiang Forestry Institute, Leizhou Forestry Institute, Guangxi Academy of Forestry, Qinzhou Forestry Institute, Dongmen Forestry Center, Zhangzhou Forestry Tissue Culture Center, South China Tropical Agriculture University and Hainan Forestry Institute.

2.1 *Embryo Culture*

Seed embryo is used as the explants of culture. In cross breeding, the incompatibility of megagamete and androgamete may lead to the abnormal development of embryos. In this situation, we peel the immature embryo off, culture it in the medium, and obtain the normal plant. However, most embryo cultures use mature embryo as explants to induce the aseptic bud. The sterilized seeds provide aseptic materials for bud and callus culture. It may be said that the key to embryo culture is the sterilization of the seeds. The tolerance of seeds to sterilization depends on the size and thickness of the seed coat (McComb and Bennett 1986). *E. citriodora* and *E. diversicolor* seeds are small, and have a thin coat that can only be sterilized in 2% NaOCl for 20 min, whereas *E. marginata* seeds are large and have a thick seed coat can tolerate 5% NaOCl for 60-90 min (Bennett and McComb 1982; McComb and Bennett 1986). MacRae and van Staden (1990) found that eucalypt seeds could be well sterilized by using solution of half H_2O_2 (30%W/V) and half absolute ethanol for 10-20 min, followed by rinsing the seeds three times in sterile distilled water.

2.2 *Bud and Tender Shoot Culture*

Bud culture is most popular in plant tissue culture. Aseptic plantlets have successfully been obtained by using the sprout, bud, tender shoot, grafted bud, leader bud etc. of mature *Eucalyptus* trees (Furze and Cresswell 1985; Burger 1987). The best approach to micro-propagating *Eucalyptus* is to induce axillary and adventitious buds in large scale by using explants of different. These explants include: seedlings (Hartney and Kabay 1980), shoots of mature tree (Gupta et al. 1981: Rao and Venkateswara 1985), sprouts (Badia 1982; Hartney 1982a) or mature shoots successfully grafted on seedling rootstock (Franclet and Boulay 1982). With axillary and adventitious buds, roots are finally induced.

The length of used shoot is about 10-20 mm with 2-3 mm of diameter, left with 1-2 axillary buds, petiole and a little lamina. There are two kinds of buds on shoots: visible axillary buds (naked bud), and dormant buds that hide in the petiole tissue (subsidiary or sealing buds) (Pryor 1976). A sealing bud develops into secondary branch after the naked bud was burned or damaged by pest. Cluster buds derived from tissue culture may come from normal buds (naked or sealing buds) or adventitious buds, which become normal shoots through the further development of their axillary buds.

New shoots that do not stretch continuously sprout out from axilla of the shoot cultured, and finally become a bud cluster. Along with generation of new shoots, calli at the bottom of explants that contact the medium further develop. Increase of tumor-formed calli in size or number may sometimes results in more buds or shoots (Hartney 1982a).

A number of culture media that are often added with low concentration of growth hormone/phytokinin have been used to stimulate the proliferation of buds. Hartney (1982b) succeeded in the proliferation of buds with the medium of 1/2MS + 2%sucrose + NAA + BA. Liquid medium was used to stimulate the proliferation of buds in axillary culture, in which the explants were collected from the branch of 20 years old *E. citriodora* and the mature *E. tereticornis* (Mascarenhas et al. 1982). MacRae and van Staden (1990) found that gelatin medium performed better than agar medium in the culture and proliferation of the axillary buds of *E. grandis*.

Mass propagation of tissue culture depends on species, clone, and origin and age of explants. Hartney (1982b) concluded that 12 species of *Eucalyptus* had the potential to produce 100 million seedlings from one explant in the first year. With the use of family seedlings, micropropagation of *E. marginata* can further be accelerated.

Bud culture may be used for maintaining superior clones for long times. Hartney (1982a) stored the naked bud culture of *E. camaldulensis* and *E. grandis* on simple culture medium in a common family refrigerator for 8 months. Franclet and Boulay (1982) stored bud culture of *E. gunnii* and *E. dalrympleana* in a refrigerator for 6 months, and found that there was no negative effect on later growth. Muralidharam (1989) stored embryo culture of *E. citriodora* on B_5 medium at 10°C for 9 months without regeneration, but did not lose the characteristics of embryo.

Zhou (1989) successfully induced aseptic plantlets in the axillary bud culture of *E.* ABL12. Huang (1994) studied the main factors affecting the growth of axillary bud, proliferation and rooting in the shoot culture of *E. citriodora*.

2.3 *Lignotuber Culture*

Tumor-shape lignotubers usually generate from the base of the young *Eucalyptus* trees, which has a number of vegetative buds, vascular tissue and nutrient storage substance (Pryor 1976). Callus culture was successfully induced from lignotuber for: *E. bancroftii* (de Fossard 1974), *E. citriodora* (Aneja and Atal 1969), *E. oblipua, E. viminals* (Winton 1974) and *E. rubusta* (Goncalves 1975). However, regenerative plantlets were only obtained for *E. citriodora*. It is out of our expectation because a lignotuber usually contains lots of dormant buds that are often used in vegetative propagation.

Effective sterilization before collection of explants is probably the key to induce calli as well as plantlets. Bactericide, insecticide and antibiotic can be used for the purpose (Franclet and Boulay 1982; Bonga 1987). Boulay (1983) suggested that bactericide to be sprayed on the explants provider every two days a week before explants were collected. de Fossard (1981) recommended that the provider should be transferred into a clean greenhouse for a better sterilization and survival of explants.

2.4 *Root Culture*

Bachelard and Stowe (1963) maintained the culture of seedling roots in a liquid medium added with 14% coconut juice. The root tips may be sub-cultured in the same type of medium, although no bud was induced. Adam (1987) successfully cultured the isolated roots of *E. gunnii* on the MS solid medium without plant growth regulators.

2.5 *Floral Buds Culture*

Bennett and McCom (1982) reported that floral buds of *E. marginata* were soaked in 95% ethanol for 5 min, flamed, treated with NaOCl for 20 min, and then rinsed with sterile water 3 times. Bud coats were removed and stamens were carefully taken out for callus culture. This method resulted in a perfect sterilization.

2.6 *Callus Cultures*

Successful *in vitro* callus formation and culture have been reported in 30 *Eucalyptus* species. However, regeneration of plantlets has only been obtained for 12 species. A number of basic media and combinations with different concentrations of different plant hormones have been developed for the *in vitro* callus culture of *Eucalyptus*. Earlier studies included use of undetermined components of coconut juice, casein hydrolyzate and the extraction of yeast in the media (Aneja and Atal 1969; Ram et al. 1971). Although the additional ingredients for culture medium were well documented, the extraction of wheat buds was still used in the callus culture of *E. viminalis* and *E. coccifera*, and coconut juice was still used in the suspension culture of *E. gunnii* and the callus culture of *E. polybractea*.

Callus culture is usually carried out under dark or delicate light with the temperature at 25°C. The morphological characteristics of callus vary depending on origins of explants and culture media. While the callus produced from the cotyledon of *E. marginata* was thick and yellow, the callus from pistal filament culture was cloddy, with color changing from gray to yellow (Bennett and McComb 1982). The callus from cotyledon and hypocotyls of *E. citriodora* was thick and red (Lakshmi

1979); the callus from hypocotyls of *E. globules* was thick and cloddy with colors of red, yellow and light green (Oka et al. 1982).

It is difficult to regenerate bud from the callus of *Eucalyptus*. Aneja and Atal (1969) reported success in plant regeneration from the calli of lignotubers. Lee and de Fossard (1974) reported that shoot and lignotubers calli of *E. bancroftii* did not regenerate any organs under the conditions of growth hormone/kinetin=175 and 100, respectively. The cultures above were all carried on in the dark with temperature at 25°C. Yet, light is necessary in the process of bud regeneration from callus.

On White medium added with 1 mg/L IAA, Kitahara and Caldas (1975) induced regenerative hypocotyls calli, and roots and buds under fluorescence light at 12 hr/day for 60 days. The additional conditions necessary to regenerate buds from the cotyledon calli of *E. citriodora* were: 0.2 mg/L IAA plus 1 mg/L KT and 12 hr/day of fluorescence light (Lakshmi 1979). These buds could root in the medium only with growth hormone. Calli of many *Eucalyptus* species could root under the 12 hr/day of fluorescence light, but not in the dark (Goncalves et al. 1979). Regenerated buds from the pistal calli of *E. marginata* had relatively low rooting rate (5%), while regenerated buds from the cotyledon calli had 35% rooting rate (Bennett and McComb 1982).

From seeds and aseptic seedlings of *E. Leichow No.1*, Ouyang et al. (1980, 1981) generated calli, embryonic cell aggregate, and embryoid. Regenerative culture increased at a rate of 20-30 times with high viability; many complete plantlets differentiated, and the survival of transplanted seedlings was as high as above 95%.

2.7 *Anther Culture*

Anther culture was studied for *E. callaophla*, *E. citriodora*, *E. marginata*, *E. dalrympleama* and *E. grandis* (Goncalves 1975; Bennett and McComb 1982; Boulay 1987). Bennet and McComb (1982) obtained regenerative shoots from anther and induced roots. Boulay (1987) induced roots from anther calli in the media with different concentrations of plant hormones, but no buds were regenerated.

2.8 *Suspension Culture*

Sussex (1965) obtained friable calli from the hypocotyls of *E. camaldulensis* on White medium with 1 mg/L 2,4-D and 15% coconut juice. With the calli obtained, suspension cultures were established in the agitated liquid medium with the same growth hormone, and the single cells grew into nodular cell aggregates. When cultured in the agar medium, these cell aggregates resulted in friable calli again. Khanna and Staba (1968) tried to create suspension cultures by using seedling calli of *E. globules* on a modified MS medium, but the cell did not grow well. The concentrations of 2,4-D used in the callus and suspension cultures were 1 mg/L and 0.1 mg/L, respectively.

By using leaves of *E. gunnii* × *E. dalrympleana* hybrid and *E. gunnii*, friable calli were produced on MS medium with 2 mg/L 2,4-D (Teulieres et al. 1990). Suspension cultures were obtained through two steps by using MS or White medium. The first step involved the culture of the cell suspension for 2 months on MS medium with 0.1 mg/L BA and 3 mg/L 2,4-D or White liquid medium with 10% coconut juice and 3 mg/L 2,4-D. The second step was to maintain the cell suspension in the medium with 0.1 mg/L BA and 1.5 mg/L 2,4-D. The number of cells was doubled on the MS medium every 3 days, and the cell suspension may form calli when they were transferred onto solid media.

2.9 *Protoplast Culture*

The suspension cultures of *E. grandis, E. saligna* and *E. callaophla* have successfully been developed, which enable protoplast culture (e.g. Teulieres et al. 1990). Different from complete cells, protoplasts often differ in tolerance to low temperature, and their viabilities are correlated to the frost-resistance of the donor plant. Protoplast cells cultured on solid media may proliferate, and form calli. Teulieres et al. (1990) suggested use of this technique in selection and propagation of frost-resistance varieties.

Penchel and Kirby (1990) isolated protoplasts of *E. grandis* from hypocotyls, cotyledons, calli and cell suspensions derived from the internodes tissue of buds and *in vitro* tender leaves of mature trees. Protoplasts were produced from the cotyledon of *E. citriodora*, of which 80% were viable. They were cultured on the liquid-on-agarose medium at 27°C in the dark.

Protoplasts of *in vitro* seedlings of *E. saligna* cultured with *Hibiscus cannbinus* protoplasts could accelerate the division of cells in the early days, and stimulated the regeneration of plantlets (Ito et al. 1990). Protoplasts were also used in the transformation of DNA of *E. saligna* and *E. gunnii* (Ito et al. 1990; Teulieres 1990)

2.10 *Somatic Embryogenesis and Artificial Seeds*

Somatic embryogenesis may be obtained by using embryos of *E. citriodora* on semisolid B_5 medium with 3 mg/L BA and 50g/L glucose (Muralidharan and Mascarenhas 1987; Muralidharan et al. 1989). To initiate somatic embryos, it is necessary to soak seeds in sterile water with temperature at 29°C for 2 days before placement on medium. When transferred to Smith&McCown medium with 0.5mg/L BA and 10% coconut juice, somatic embryo can continuously differentiate. The isolated embryos can develop and grow in 1/2MS liquid medium without hormone. With embryonic cell aggregates, the differentiation process of somatic embryo can be accelerated. Normally, embryonic cell aggregates may be maintained over 36 months on the improved B_5 medium with 5 mg/L NAA in dark, without losing the viability. The improved B_5 medium contains 500 mg/L of hydrolase protein and glutamine respectively. However, the development and growth of embryos perform well on fresh B5 medium without adding hormone and needing light.

On MS medium with 0.1 mg/L NAA and 5 mg/L kinetin, somatic embryos can differentiate from calli derived from shoots of 4-year-old *E. grandis* trees. Somatic embryos can also be initiated from the friable calli cultured in liquid medium containing 1 mg/L of BA, kinetin, NAA and 2,4-D respectively (27°C, 16 hr/day or light). For example, somatic embryos can be induced and regenerative plantlets can be produced for *E. leichow No.1* (Ouyang et al. 1980, 1981). Somatic embryos can also be obtained from the calli culture of seedling hypocotyls and internode of *E.*

gunnii on two media with different hormone concentrations. By subculturing embryogenic calli, Boulay (1987) achieved secondary somatic embryogenesis.

Qin and Kirby (1990) induced embryo-like structures from seedling hypocotyls and cotyledon of *E. botryoides, E. dunnii, E. grandis, E. rudis* and young leaves of superior clones of adult *E. grandis*. The medium consisted of mineral element of MS, vitamin and amino acid of RV. Green protruberances developed slowly from cut surfaces of explants after cultured in the medium with 1.1 mg/L 2,4-D for 2 weeks. When transferred to the medium with 1.1 mg/L BA, the protruberances developed into adventitious shoots and embryo-like structure. If the calli induced from the explants are initially fast-growing and red or white, they could not further develop into shoots or embryo-like structures.

3 Conclusions

Tissue culture is of great significance in research and cultivation of *Eucalyptus*. To date, tissue culture techniques have been studied in about 60 *Eucalyptus* species. The techniques include cultures of embryo, shoot, tender stem, anther, suspension, protoplast and somatic embryogenesis. Most of them have generated complete plantlets. Further studies should probably focus on the cultures of *Eucalytpus* multiploid, somatogamy and transgenic cells. Undoubtedly, along with the progress and advance in other aspects of biological technologies, tissue culture becomes more and more important in the research and application of plantation management, seedling production, breeding and gene engineering of *Eucalyptus*.

References

Adam, S. 1987. Obtention de raciness transformees chez *Eucalyptus gunnii* H par Agrobacterium rhizogenes. In: "Annales de researches silvicoles", Assosiation Foret-Cellulose, pp7-21.

Aneja, S. and Atal, C. 1969. Plantlet formation in tissue cultures form lignotubers of *Eucalyptus citriodora* Hook. Curr. Sci 38:69-70.

Bachelard, E. P. and Stowe, B.B. 1963. Growth *in vitro* of roots of *Acer rubrum* L and *Eucalyptus camaldulensis* Dehnh. Physi. Plant. 16:20-23.

Badia N.K. 1982. Influence de la vitamine E sur la multiplication vegetative *in vitro* de l'*Eucalyptus rudis* Endl. De larix x *eurolepis* Henrg et de Quercus borealis Michx. Bulletin des Researches Agronomiques de Gembloux 17:219-226.

Bennett, I.J. and McComb, J.A. 1982. Propagation of jarrah (*Eucalyptus marginata*) by organ and tissue culture. Aust. For. Res. 12:121-127.

Bonga, J.M. 1987. Clonal propagation of mature trees: problems and possible solutions. In: "Cell tissue in forestry", General Technical Report P.S.W. 69, Vol.1, pp102-107.

Boulay, M. 1983. Micropropagation of frost resistant *Eucalyptus*. In: Proc. of Workshop on *Eucalyptus*, California, Sacramento, USDA Forest Service, General Technical Report, pp102-107.

Boulay, M. 1987. Recherches preliminaaires sur l'embryogenese comatique d'*Eucalyptus gunnii*. In: "Annales de researches silvicoles", Association Forét-Cellulose, pp23-37.

Burger, D.W. 1987. *In vitro* micropropagation of *Eucalyptus sideroxylon*. HortSci. 22:486-497.

Cao, Z. and Liu, G. 1996. Practical course on plant tissue culture. Gansu Science and Technology Publishing House, Lanzhou, China, pp1-8.

Chen, Z. 1992. Study on plant gene engineering. Beijing University Press, Beijing, China, pp1-7.

de Fossard, R.A. 1974. Tissue culture of *Eucalyptus*. Aust. For. 37:43-54.

de Fossard, R.A. 1981. Tissue culture propagation of *Eucalyptus ficifolia* F. Muell. In: Proc. of Symposium on "Plant tissue culture" (Peking), Pitman, Boston, pp425-438.

Furze, M.J. and Cresswell, C.F. 1985. Micropropagation of *Eucalyptus grandis* and *nitens* using tissue culture techniques. South African For. J. 135:20-23.

Goncalves, A.N. 1975. The growth and developmental physiology of *Eucalyptus* in cell and tissue culture systems. MSc Thesis. Ohio State University, Columbus.

Goncalves, A.N., Machado, M.A., Caldas, L.S., Sharp, W.R. and Mello, A.H. 1979. Tissue culture of *Eucalyptus*. In: "Plant tissue culture: principles and applications", W.R. Sharp, P.A. Larsen, E.P. Paddock and V. Raghavan (eds), Ohio State University Press, Columbus, pp509-526.

Gupta, P.K., Mehta, U.J. and Jagannathan, V. 1981. Tissue culture of forest trees: clonal propagation of mature trees of *Eucalyptus citriodora* and *Eucalyptus camaldulensis*. Plant Cell Rep. 2:296-299.

Hartney, V.J. 1982a. Tissue culture of *Eucalyptus*. In: Combined Proceedings of International Plant Propagator's Society, 32:98-109.

Hartney, V.J. 1982b. Vegetative propagation of *Eucalyptus* by *in vitro*. In: "Colloque international sur la culture *in vitro* des essences forestieres", IUFRO, Nangis, France, Association Forét -Cellulose, pp175-179.

Hartney, V.J. and Kabay, P.K. 1980. Vegetative propagation of *Eucalyptus* by tissue culture. In: Proc. of IUFRO Symposium and Workshop on "Genetic improvement and productive of fast growing tree species", Sao Paulo, Brazil, pp791-793.

Huang, S. 1994. Preliminary study on tissue culture of *Eucalyptus citriodora*. J. Guangxi Agri. Univ. 13:189-192.

Ito, K., Doi, K. Tatemichi, Y. and Shibata, M. 1990. Plant regeneration from protoplasts pf *Eucalyptus*. In: Proc. of VII International Congress on "Plant tissue and cell culture", Amsterdam, pp19.

Khanna, P. and Staba, E.J. 1968. Antimicrobials from plant tissue culture. Lloydia 31:180-189.

Kitahara, E.H. and Caldas, L.S. 1975. Shoot and root formation in hypocotyls callus of *Eucalyptus*. For. Res. 21:242-243.

Lakshmi, S.G. 1979. Morphogenesis and plant regeneration from cotyledonary culture of *Eucalyptus*. Plant Sci. Lett. 14:63-68.

Lee, E.C.M. and de Fossard, R.A. 1974. The effects of various auxins and cytokinins on the *in vitro* culture of stem and lignotuber tissue of *Eucalyptus bancroftii* Maiden. New Phytol. 73:707-717.
MacRae, S. and van Staden, J. 1990. *In vitro* culture of *Eucalyptus grandis*: effect of gelling agents on propagation. J. Plant Physi. 137:249-251.
Mascarenhas, A.F., Hazara, S., Potdar, U., Kulkarni, D.K. and Gupta, P.K. 1982. Rapid clonal multiplication of mature forest trees through tissue culture. In: Proc. of the fifth International Congress on "Plant tissue and cell culture", A. Fujiwara (ed), Tokyo, Japan, pp719-720.
McComb, J.A. and Bennett, I.J. 1986. *Eucalyptus* (*Eucalyptus spp*). In: "Biotechnology in agriculture and forestry", A.Y.P.S. Bajaj, Springer-Verlag, Berlin, Vol.1, pp340-362.
Muralidharan, E.M. and Mascarenhas, A.F. 1987. *In vitro* plantlet formation by organogenesis in *Eucalyptus camaldulensis* and by somatic embryogenesis in *Eucalyptus citriodora*. Plant Cell Rep. 6:256-259.
Muralidharan, E.M., Gupta, P.K. and Mascarenhas, A.F. 1989. Plantlet production through high frequency somatic embryogenesis in long term culture of *Eucalyptus citriodora*. Plant Cell Rep. 8:41-43.
Oka, S., Yeung, E.C. and Thorpe, T.A. 1982. Shoot formation in *Eucalyptus globulus* hypocotyls explants. N. Z. J. For. Sci. 12:501-509.
Ouyang, Q., Li, Q. and Pang, H. 1980. Preliminary report on embryogenesis in *Eucalyptus*. J. Plant Physi. 6: 429-432.
Ouyang, Q., Li, Q. and Pang, H. 1981. A study on embryoid generation from *Eucalyptus* callus. Sci. Silvae Sini. 17:1-7.
Pan, R. and Dong, Y. 1995. Plant physiology. The Advanced Education Press, Beijing, pp249-251.
Penchel, R.M. and Kirby, E.G. 1990. Isolation and culture of protoplasts from juvenile and adult *Eucalyptus grandis*. In: Proc. of VII International Congress on "Plant tissue and cell culture", Amsterdam, pp31.
Pryor, L.D. 1976. The biology of *Eucalyptus*. Edward Arnold, London, 82p.
Qin, C. and Kirby, E.G. 1990. Induction of shoots and embryo-like structures in cultures derived from juvenile and adult *Eucalyptus spp*. In: Proc. of VII International Congress on "Plant tissue and cell culture", Amsterdam, pp21.
Ram, H., Rao, C.H. and Mathure, J.M.S. 1971. Anthocyanin development *in vitro* tender internodes explants of Eucalytpus. Curr. Sci. 18:499-500.
Rao, K.S. and Venkateswara, R. 1985. Tissue culture of forest trees: clonal multiplication of *Eucalytpus grandis* L. Plant Sci. 40:51-55.
Sussex, I.M. 1965. The origin and morphogensis of *Eucalyptus* cell populations. In Proc. of International Conference on "Plant tissue culture", P.R. White and A.R. Grove (eds), McCuchan, Berkeley, pp383-391.
Teulieres, C., Teissie, J. and Boudet, A.M. 1990. Transient gene expression in protoplasts and intact cells of *Eucalyptus gunnii*: survie des cellules et de leurs protoplasts à basse temperature. In: "Annales de researches silvicoles", Association Forét-Cellulose, pp117-131.

Winton, L.L. 1974. Second addendum to the bibiography of tree callus cultures. In: "Genetics and physiology notes", The Institute of Paper Cemistry, Appleton, Wisconsin, No.19.

Xie, Q. 1981. Complete plant induction from tissue culture of *Eucalyptus leichow* No.1. In: "China cooperative science and technology of *Eucalyptus*: report", No.1, pp15-21.

Zeng, L. 1988. Study on tissue culture of *Eucalyptus urophylla* × *Eucalyptus grandis*. Qinzhou For. Sci. Tech. 1:1-8.

Zhou, Z. 1989. Study synopsis on tissue culture and seedling technology of *Eucalyptus* ABL.12. Guangdong For. Sci. Tech. 2:11-12.

中国桉树组织培养研究进展

谢耀坚

国家林业局桉树研究开发中心
524022 中国广东湛江

桉树组织培养在理论研究和实践应用中都具有非常重要的意义。迄今为止，全世界对约有 60 多种桉树进行过组织培养技术研究，包括胚培养、芽培养及嫩茎培养、花药培养、悬浮培养、原生质体培养、体细胞胚培养等技术，绝大部分成功获得完整植株。

A STUDY ON PHOTOSYNTHESIS OF THREE-MONTH OLD EUCALYPT CUTTINGS

ZHAOHUA LU, JIANMIN XU, RUXIANG CHEN, GUANGYOU LI, JIAYU BAI AND DEXIANG CHEN

Research Institute of Tropical Forestry, Chinese Academy of Forestry
Guangzhou 510520, Guangdong, China

The photosynthesis of six 3-month-old superior eucalypt clones that are widely extended for plantation purpose was tested. The light response saturation point of *E. urophylla* was between 1800-2000 μmol/m^2/s; the net photosynthetic rate (Pn) and irradiance had a quadratic correlation. Significant difference in photosynthesis of seedlings was found among eucalypt clones. Transpiration rate (E) and Pn of the eucalypt clones were higher than other local broad-leave tree species. Not only did some clones of *E. urophylla* × *E. grandis* and *E. urophylla* × *E. tereticornis* grow faster than clones of *E. urophylla* × *E. camalduensis* and *E. urophylla*, but also had higher water use efficiency (WUE) and photosynthesis ability. Daily variation of Pn, E and leaf conductance (Cond) of different clones are similar. Leaf intercellular CO_2 concentration (Ci) and Pn of different clones changed in opposite direction but showed the same pattern. Pn, Cond and WUE had positive correlation with the height, DBH and individual volume. Photosynthesis testing was suggested to assist early selection of superior eucalypt clones.

1 Introduction

Eucalyptus species have become one of the most important cultivated tree species in southern China due to its rapid early growth and multipurpose wood products, especially for wood pulp making since the systemic introduction from Australia and other countries (Qi 1989; Mckenney et al. 1991) in the 1980's. Development of high-yield eucalypt plantations with superior clones intensively bred and selected has brought in great economic benefits. Nowadays eucalypt clones can easily be mass propagated in large scale for high-input plantation management. Therefore, using superior eucalypt clones to develop plantation is an important way to increase productivity of eucalypt plantations for development of forestry and other relevant industries in China.

Photosynthesis is one of the most important bio-chemical reactions using the solar radiation on earth. Among the assimilative products synthesized during the net photosynthesis by plants, part is used to provide energy for the respiratory loss during life, part for respiratory loss of growth, and the rest for growth. Photosynthesis can influence the status of a plant's net productivity to a certain extent, and decide a plant's total productivity directly.

Significant differences in net photosynthesis and transpiration rate among eucalypt species were earlier reported (e.g. Chen et al. 1995). We choose six superior eucalypt clones that are widely planted in the southern China to determine the difference among clones, and to study the relationship between photosynthesis and growth and use of photosynthesis to assist early selection of eucalypt clones.

2 Materials and Methods

2.1 *Experimental Apparatus*

The LI-6400 Portable Photosynthesis System was used for this experiment.

2.2 *Experimental Materials*

All cutting seedlings for this test were selected from nursery (RITF/CAF in Guangzhou) at age of 3 months with similar height growth (mean of 3 cuttings, 25-30 cm). Some information about the six clones at a trial, and the growth at age of 2.5 years, were presented in Table 1. The third and fourth leaves of each seedling from the top were sampled to test their photosynthesis in April 2002.

Table 1. Basic information and 2.5-year growth of six eucalypt clones tested.

No.	Clone code	Species	Origin	Growth at 2.5-year old		
				Height (m)	DBH (m)	Volume (m^3)
1	TH9224	*E. urophylla* × *E.camalduensis*	RITF	10.69	7.41	0.01957
2	DH32-28	*E. urophylla* × *E. grandis*	DM	12.21	8.90	0.03224
3	DH19-6	*E. urophylla* × *E. grandis*	DM	12.16	8.88	0.03196
4	TH9113	*E. urophylla* × *E. tereticornis*	RITF	10.92	8.20	0.02448
5	TH9211	*E. urophylla* × *E. tereticornis*	RITF	11.34	7.85	0.02329
6	U1	*E. urophylla*	RITF	10.9	7.75	0.02182

Note: 2.5-year old growth was obtained from the field trial located at ChengMai, Hainan Island. The volume was calculated according to CSIRO, i.e. $V=H\times DBH^2/3$, (Li 2002).

2.3 *Light Response of Photosynthesis*

LI-6400 with the auto-testing program was used to measure the irradiance, increasing from 0 μmol/m^2/s to 3000 μmol/m^2/s, and the changes of photosynthesis characters at 3-minute interval. With the natural humidity, the CO_2 concentration was controlled at 360 μmol/m^2/s, and the temperature at 23-25°C.

2.4 *Physio-Ecological Characters of Photosynthesis*

Base on the curve of the light response, 1800 μmol/m^2/s of irradiance was further fixed in the following experiments. During daytime with sunshine, the CO_2 concentration was controlled at 360 μmol/m^2/s. The physio-ecological characters of clones were measured at 2-hour interval from 8:00am to 6:00pm. Each measurement was completed in half-hour to reduce the error.

2.5 *Photosynthesis Index*

Some of main photosynthetic indexes were calculated, such as net photosynthetic rate (Pn, μmol CO_2/m^2/s), transpiration rate (E, mol H_2O/μm^2/s), leaf intercellular CO_2 concentration (Ci, μmol/mol), leaf conductance (Cond, μmol CO_2 m/mol H_2O /m^2/s), and water use efficiency (WUE=Pn/E, μmol CO_2 m/mol H_2O) and WUE = Pn/E 。

2.6 *Statistic Analysis*

SAS version 6.12 was used to conduct the variation analysis, Duncan test and correlation analysis (SAS Institute 1988).

3 Results

3.1 *Light Response of Photosynthesis*

Mean net photosynthetic rate (Pn) of different clones against the photosynthetic photon flux density (PPDF) was plotted and regressed in Figure 1. Pn increased

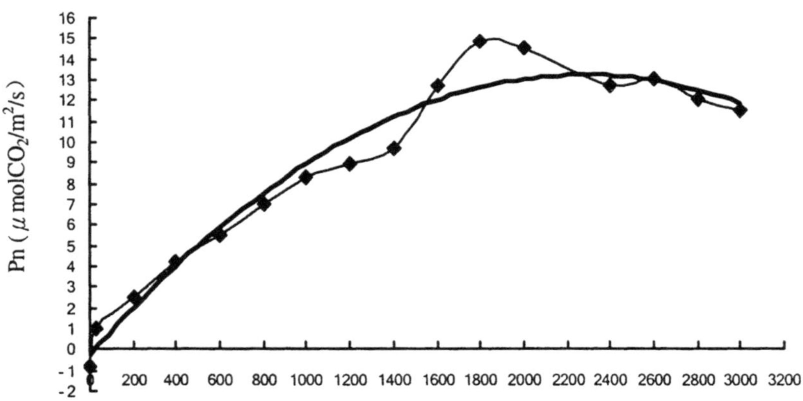

Figure 1. Light response curve of *E. urophylla*.

when the irradiance increased from 0 to 1800 μmol/m^2/s. The light response saturation point (LSP) occurred at 1800-2000 μmol/m^2/s, and from there, Pn slowed down and decreased a little with the irradiance, changing between 11.5-14.5 μmolCO_2/m^2/s. Photo-inhibition didn't occur evidently during the course of experiment. A significant quadratic relationship was found between net photosynthesis and irradiance:

$$Pn=3E\text{-}06PPDF^2+0.0118PPDF-0.2557,\ R^2 = 0.959.$$

3.2 *Comparison of Photosynthesis between Clones*

Measurements at 8:00, 10:00, 12:00, 14:00 and 16:00o'clock in a day were used for variation analysis and LSR test. Pn, E, Cond, and Ci were significantly different among clones (Tables 2 and 3). No.3 clone had the highest value of net photosynthesis and leaf conductance, and No.4 clone had the highest transpiration rate and concentration of leaf intercellular CO_2. The rankings of the clones tested in Pn, Cond, E and Ci were 3>5>2>4>6>1, 3>4>5>2>1>6, 4>3>5>2>6>1 and 4>2>6>1>3>5, respectively. The hybrid clones had higher photosynthesis than *E. urophylla* clone. No.3 clone (DH19-6, *E. urophylla* × *E. grandis*) had the highest photosynthesis, 50.56% and 50.65% higher in Pn and Cond, respectively, than No.1 clone (U1, *E. urophylla*). In general, hybrids of *E. urophylla* × *E. grandis and E. urophylla* × *E. tereticornis*, with high photosynthesis, grew faster than *E. urophylla* × *E. camalduensis* and *E. urophylla*.

Table 2. Variation analysis in photosynthesis of eucalypt clones.

Source of variation		Clones	Error	Total
DF		5	24	29
	SS	163.01	59.34	222.35
Pn	MS	32.6	2.47	
	F	13.19**		
	SS	21.06	15.52	36.59
E	MS	4.21	0.65	
	F	6.51**		
	SS	0.32	0.33	0.65
Cond	MS	0.064	0.01	
	F	4.69**		
	SS	2081.75	2789.5	
Ci	MS	416.35	116.23	
	F	3.58*		

Note: * and ** indicate significance at levels of 5% and 1%, respectively.

Table 3. Duncan test of difference among clones in different photosynthesis characters.

Pn	Clone No.		3	5	2	4	6	1
	Mean value		17.80	15.76	13.22	12.98	11.82	11.00
	Level	0.05	a	a	b	b	b	b
		0.01	a	ab	bc	bc	c	c
E	Clone No		4	3	5	2	6	1
	Mean value		6.986	6.984	6.528	6.374	6.140	5.012
	Level	0.05	a	a	a	a	a	b
		0.01	a	a	a	a	a	b
Cond	Clone No		3	4	5	2	1	6
	Mean value		0.815	0.795	0.699	0.687	0.565	0.541
	Level	0.05	a	a	ab	b	b	b
		0.01	a	a	ab	b	b	b
Ci	Clone No		4	2	6	1	3	5
	Mean value		293.06	292.85	288.91	286.07	274.67	273.57
	Level	0.05	a	a	ab	abc	bc	c
		0.01	a	ab	ab	ab	ab	b

The water use efficiency (WUE) of different clones was compared in Figure 2. In general, clones No.3, 5 and 2 had both high WUE (2.549, 2.415 and 2.074 μmol CO2·m/mol H2O, respectively) and growth (Table 1). In contrast, clones No.4, 1 and 6 had relatively both low WUE (1.858, 2.196 and 1.926 μmol CO2·m/mol H2O, respectively) and growth (Table 2). Obviously, clones No.3, 5 and 2 could sequester more CO2 at the same amount of water, or use water more efficiently than other clones.

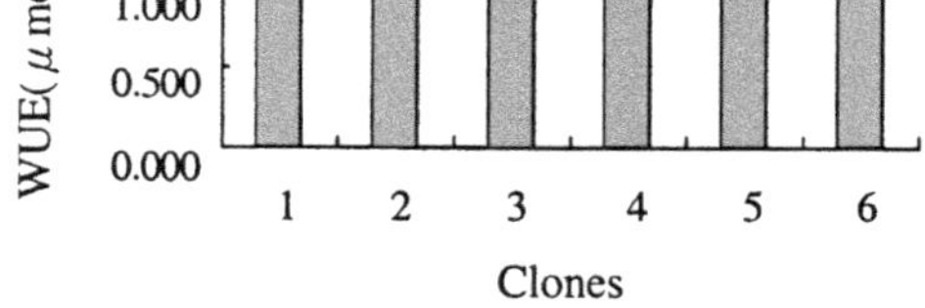

Figure 2. Comparison in WUE of clones.

3.3 *Diurnal Change of Pn, E, Cond and Ci*

In comparison with *Erythrina Crista-galli* Linn, *Saraca dives* Pierre., *Sterculia lanceolata* Cav., *pterocarpus indicus* Willd., *Castanaopsis hystrix* A.D.C., *Castanopsis patelliformis*, *Cryptocarya chinesis*, *Licuala spinisa* and *Lasianthus cyanocarpus* (Chen et al. 2001; Li 2002; Chen 2002), eucalypt clones studied in the present study had higher net photosynthesis and transpiration rate. The highest Pn and E reached 18.880 μmolCO_2/m^2/s and 7.88 mol H_2O /m^2/s, respectively. Pn and E did not change much from 8:00am to 4:00pm, but dropped sharply from 4:00pm to 6:00pm when the outside irradiance reduced evidently (Figure 3, A and B). The maximum Pn of the six clones occurred at different times: TH9211 and DH32-28 at 8:00am, TH9224 at 10:00am, DH19-6 and TH9113 at 12:00am, and U1 at 2:00pm, and the peak of Pn appeared at 10:00o'clock for most clones. It can be concluded from Figure 3 that photosynthesis of eucalypt clones tested was more active during the period of 10:00am-4:00pm than in the early morning or late afternoon. In order to avoid high temperature, the leaf temperature would reduce with water pervasion in the course of transpiration as a type of adaptation to high irradiance.

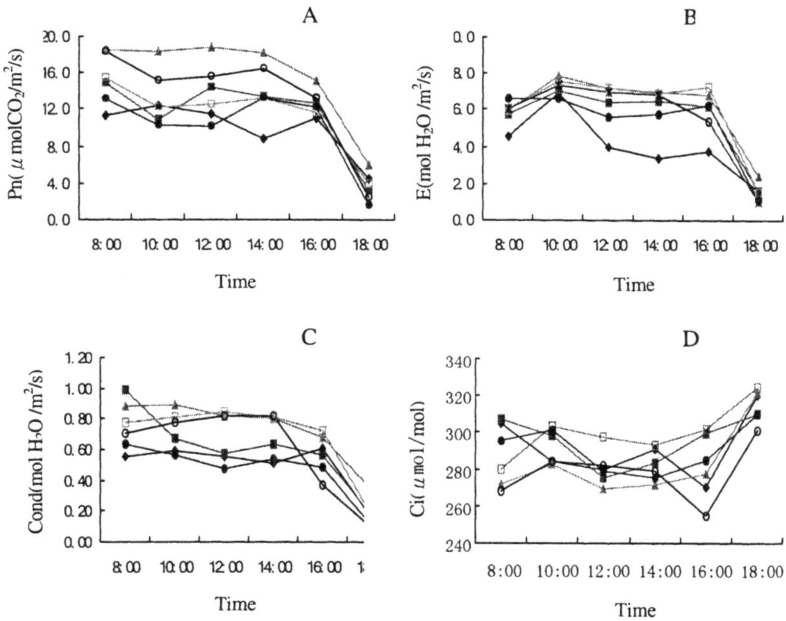

Legends: ♦ TH9224, ■ DH32-28, ▲ DH19-6, □ TH9113, ○ TH9211, ● U1

Figure 3. Diurnal changes of net photosynthesis rate (Pn, A), transpiration rate (E, B), leaf conductance (Cond, C) and Leaf intercellular CO_2 concentration (Ci, D).

Leaf conductance was relatively stable with a little change during the period of 8:00am-4:00pm, and dropped to the bottom at 6:00pm (Figure 3, C). However, the leaf intercellular CO_2 concentration (Ci) changed in the opposite direction of Pn during a day, although the rates differed (Figure 3, D and A).

3.4 *Correlation between Photosynthesis and Growth Characters*

Among the factors that mainly influence photosynthesis, irradiance, temperature and Cond. had the linear or non-linear relationships with net photosynthesis (Shen et al. 1998). Table 4 was the correlation matrix of the photosynthesis and 2.5-year growth characters. It was shown that Pn, E, and Cond influenced the growth of eucalypt cuttings at level of 0.1. Enhanced transpiration and conductance were also beneficial to the photosynthesis. Ci had significant negative correlation with Pn, and WUE. The results could be used to assist the early selection of superior eucalypt clones. Early selection of superior eucalypt clones would be more effective and accurate if photosynthesis would be considered along with other physiological characters and phenotype characters.

Table 4. The correlation matrix between the photosynthesis and characters of growth.

	Pn	E	Cond	Ci	WUE	H	D	SV
Pn	1							
E	0.6734*	1						
Cond	0.7432**	0.7784**	1					
Ci	-0.7698**	-0.1535	-0.2615	1				
WUE	0.7821**	0.0994	0.3251	-0.9191***	1			
H	0.6841*	0.5169	0.5396	-0.2555	0.4952	1		
D	0.6005	0.6774*	0.6997*	-0.0061	0.2313	0.9197***	1	
SV	0.6224*	0.6112	0.6465*	-0.0754	0.3204	0.9620***	0.9909***	1

Note: *, ** and *** indicate significance at levels of 10%, 5% and 1%, respectively.

4 Discussion and Conclusions

The net photosynthetic rate (Pn) of *E. urophylla* had a significant quadratic relationship with the irradiance, with the light response saturation point falling between 1800-2000 μmol/m^2/s. Pn, transpiration rate (E), leaf conductance (Cond) and leaf intercellular CO_2 concentration (Ci) significantly differed among the eucalypt clones. Pn and E of the six eucalypt clones tested were higher than other local broad-leave tree species. Photosynthesis of eucalypt hybrids with *E. urophylla* was higher than that of the pure *E. urophylla* clone. Clones such as *E. urophylla* × *E. grandis and E. urophylla* × *E. tereticornis* grew faster than others, e.g. *E.urophylla* × *E.camaldulensis* and *E. urophylla*, as the former photosynthesis was stronger than the later. Change over times during a day of Cond was similar to that of the net photosynthesis. However, Ci and Pn changed in the opposite direction but with the same pattern. Pn, E and Cond had weak but positive correlation with height, diameter at breast height and individual volume. To improve the selection efficiency, we suggest the use of photosynthesis data to assist early selection of eucalypt clones.

Acknowledgments

We thank Prof. Yide Li from RITF/CAF for advising the experiment design, Dr. Xiaoquan Zhang from CAF for reviewing the paper and valuable suggestions.

References

Chen, D. 2002. Research on photosynthesis of plants in tropical mountain rain forest under global change. Ms Thesis, Chinese Academy of Forestry.

Chen, T., Deng, D., Pan, Y., Huang, S. and Deng N. 1995. A preliminary study on physio-ecological characters of *Ecalyptus*. In: "Studies on the eucalypts ecosystem of short rotation in Leizhou, Guangdong Province", T. Zeng, Y. Liu and G. Fu (eds), China Forestry Publishing House, pp13-20.

Chen, Y., Gong, M., Chen, Y. and Wang, F. 2001. Effects of inoculation with 11 ectomycorrhizal fungal isolates on growth and photosynthesis of *Castanopsis hystrix* saplings. For. Res. 14:514-522.

Li, G. 2002. The photosynthesis and water use efficiency of eight garden tree species. For. Res. 15:291-296.

Mckenney, D.W., Davis, J.S., Turnbull, J.W. and Searle, S.D. 1991. The impact of Australian tree species research in China. Canberra: ACIAR Economic Assessment Series, 1991(12):6-7.

Qi, S. 1989. *Eucalyptus* in China. China Forestry Publishing House, Beijing, pp21-35.

SAS Institute 1988. SAS/STAT User's Guide for Personal Computers. Release 6.03 edition. Gary, NC, USA. SAS Institute.

Shen, Y., Shi, J. and Xu, D. 1998. Dynamic photosynthesis. Science Publishing House, Beijing, pp126-152.

桉树无性系苗期光合作用特性研究

陆钊华，徐建民，陈儒香，李光友，白嘉雨，陈德祥

中国林业科学研究院热带林业研究所
510520 中国广东广州

对 6 个现有桉树人工林高产无性系的苗期光合作用特性进行测定，结果表明：尾叶桉光响应的饱和点在 1800-2000 μmol/m²/s 之间，净光合速率 Pn 与 PPFD 呈显著的二次曲线相关关系。桉树无性系具较高的净光合速率与蒸腾速率，不同无性系各项光合生理生态指标差异显著，尾×巨桉、尾×细桉等生长量较好的无性系较尾×赤桉和尾叶桉无性系的水分利用效率更高，光合能力更强。净光合速率、蒸腾速率及气孔导度三者日变化进程较为相似，而胞间 CO_2 浓度的日变化方向则与净光合速率的则相反。净光合速率、气孔导度与蒸腾作用与树高、胸径、单株材积等生长指标存在较显著的正相关关系，可应用于优良无性系的早期选择研究。

III. Genetic Testing and Improvement

LOW-INPUT TREE BREEDING STRATEGIES

DAG LINDGREN

Department of Forest Genetics and Plant Physiology
Swedish University of Agricultural Sciences
SE-901 83, Umeå, Sweden

Sophisticated or futuristic tree improvement, which requires large resources, stable funding, competent and permanent technical staff and well-developed infrastructure, is well represented in the qualified literature, in scientific research, in available expertise and in formal training programmes. In contrast, methods particularly suitable for breeding in resource poor situations may be neglected or used inefficiently, as little thought is given to adequately develop, implement and optimise these methods. The use of low-input, simple, robust and cheap methods may actually be justified also in comparatively resource rich environments, but suppressed because they do not seem sophisticated and advanced enough.

A number of low-input techniques for seed production and long-term breeding are discussed. In particular: phenotypic selection, absence of testing, unidentified plants, no grafting, and forwarding the breeding population by wind pollination instead of controlled crosses. Sustainable and optimally used low-input strategies require advanced quantitative genetic considerations, in particular concerning the management of relatedness. Vegetative propagation may be a valuable tool both in poor and rich situations.

1 Introduction

The aim of this lecture is to stimulate implementation of, research about, and development of low-input techniques in forest tree breeding. The term "low-input" breeding is not well established, possible alternative terminology is e.g. "low input", "low-intensity" (Lindgren 2000), "low budget", "low cost", "cheap", "simple" breeding.

2 Why Do We Need Low Input Tree Improvement Methods?

Low-input breeding is breeding that is cheap, convenient, simple, robust, low budget, local, requires little record keeping and central control, and is possible to run on a small and uncertain budget. In contrast, high-input breeding is less constrained by the budget and uses techniques that are relatively more expensive, complex and demanding on competence and organisation, but are expected to return higher and more reliable gains. It fills no useful purpose to formulate exact definitions of the alternatives, as the difference is not strict; it is partly a semantic problem; what is low-input depends on the exact circumstances and is often somewhat controversial. There are certainly different levels of high input, and few programmes would classify themselves as high input programmes. Examples of common techniques, which may be considered high-input, are grafted seed orchards, controlled crosses with full pedigree (known pollen parent) and genetic testing. These measures are usually justified and affordable for conifer species of major economic importance, but debatable for low-input programmes. Low-input improvement techniques are probably relevant to many improvement activities in developing countries, and in environments where competence is poorly developed and organisation unstable (Namkoong et al. 1980). Low-input techniques may be relevant for minor programmes. Examples are planting of trees in urban and rural areas for parks and landscapes, where some low-intensity strategies have been discussed, e.g. by Lagerström and Eriksson (1996). A strong candidate for low-input breeding is such as exotics, which are used in a small or experimental scale and have an uncertain future, but there is still a will to create improved options for the future to utilise the genetic resource. Most countries have a long list of less commercially important species, which have been tested in provenance trials and are potentially useful, but the plant demand is low and the future market doubtful. For such a situation, a breeder may prefer a low-input solution rather than just doing nothing. The techniques involved can play a role as a form of safety device and back-up to high-input breeding programmes. Genetic management of the highest ranking genetic tiers of the breeding stock with genetically narrow high-input aggressive breeding for high gain may be complemented with low-input breeding of lower tiers for wide diversity and large breeding population (e.g. nucleus breeding, cf. Cotterill 1989, or the more extreme stratified breeding suggested by Ruotsalainen and Lindgren 2000). Even in rather resource rich programmes, there is an evident tendency to overestimate the potential and underestimate the resources needed for the high-input methods, and thus even in those situations, it may be beneficial to think more in the direction of low-input methods. For eucalypts, there are many species and they are cultivated in widely different circumstances, so the thoughts in this paper may certainly be frequently applicable.

Low-input breeding techniques can, if used wisely, be useful for gene conservation, assuring that threatened genetic resources will still be available in the future. Gene conservation should make possible the harvesting of seeds with reasonable gene diversity, genetic variability and adaptation for renewing the gene resource and should preserve options to support future tree planting and to initiate a more advanced breeding programme sometimes in the future without starting from scratch. In many countries, phenotypically superior stands were selected as *in situ* gene reserves, where an additional effort may improve their usefulness. A low-input breeding programme can meet these goals, with emphasis on high genetic diversity and future options besides improving what now is considered important economic traits. A high-input programme with components of gene conservation may be less efficient if it directs the same resources to fewer parents, and a smaller breeding population will be accompanied by a larger loss of gene diversity. Or resources may be too small to support anything else than a low-input programme. Then the alternative to low-input may be no-input and thus no intentional gene conservation.

There may be a more immediate demand for improved regeneration materials produced in a cost-efficient way, but the required quantities are limited and budget constraints do not justify a more expensive, high-input approach. For many species, elements of the biology may not be sufficiently known and that may cause constraints, e.g. the reproductive biology may be too poorly known to make successful crosses. There may be reasons to avoid methods that depend on a stable access to funding, technical expertise and technical equipment. A small breeding unit may be less able to handle high-input techniques than a bigger unit. Costs are scale-dependent, thus a method, which is cost-effective in a large programme, may be less cost-effective in a small programme.

3 Why Have Low-input Methods Received Little Attention?

The development of tree breeding as a science is mainly driven by the needs identified by the large high-input programmes. Consequently, there is a risk that methods suitable for high-input programmes are routinely implemented into low-input programmes also, which may be less suitable in the latter case. There is rather little qualified scientific thought given to low-input breeding strategies, and not much is found about it in the more established literature. In the few more known examples existing (e.g. Namkoong et al. 1980; Shelbourne 1992; Barnes 1995), the focus is still on rather fancy techniques, and little effort is made on optimising real low-input programmes. Tree breeders generally feel more motivated to focus attention on how well things can be done rather than how cheaply they can be done. It is also natural that intensive programmes employ more, better trained and more specialised breeders, who work in more affluent and stable environments, and it is these breeders and their supporting scientists who create the literature and common knowledge. It is easier to motivate and get funding for fancy, futuristic and visionary techniques (like variants of molecular breeding), rather than argue for somewhat sloppy, low-input methods. Even in less-developed environments, there is a strong will to use "modern" methods at the research frontier, rather than being cost-efficient. Many of the advantages of high-input over low-input methods are that the materials become better known, the activity more scientific and the actions more predictable; this is attractive, and may over-shadow cost-efficiency, in spite of that the latter ought to be a more important goal of tree improvement. The outcome of planned tree breeding operations are usually based on uncertain predictions, analogies and beliefs rather than proven facts. This leaves much room for arguing that fancy methods may be economically efficient even when they are not. A robust, low-input strategy is more able to survive neglect for some time in an environment with irregular and unreliable funding and less sensitive to errors and mistakes. That the local management system is irregular and unreliable and that some budget downgrading is acceptable are arguments, which breeders will avoid to make in public even when they are true. The demand on genetic calculations may be higher for low-input methods, but these calculations may be more difficult to make and explain than with more accurate methods, thus low-input-methods may be more difficult to get acceptance by decision makers. Thus, low-input strategies have few spokesmen, scientific investigators and promoters.

In the following, some techniques will be discussed, which can be implemented in low-input breeding strategies. Even if the text is written with low-budget breeders in mind, the techniques may have relevance also for high-input situations. It can also be noted that, depending on special circumstances, it may sometimes be wise to use sophisticated techniques in low-budget environments. The suitability of a particular technique depends on many circumstances, and therefore no sharp borderline between high-input and low-input can be drawn; some concepts described as low-input can be implemented in some high-input strategies and vice versa.

4 Genotype Testing May Not Be Efficient in Low Input Breeding

Conventional progeny testing or clonal testing can be seen as a way of boosting the heritability when selecting individuals. There is little place for intensive genotype testing in low-input programmes. Testing requires field identities, recorded pedigrees, documentation, long-term planning, long-term co-ordination of activities, organisational stability and, often, clonal archives. A strategy including testing means a large long-term investment, which seems possible to justify only in situations when it seems certain the results will be utilised and appreciated in the future. There are, however, studies indicating that it is not a general rule that testing genotypes is an efficient use of resources even in high input environments. For instance, a study by Ruotsalainen and Lindgren (1998) showed that, with few exceptions, if offspring was generated with pollen as good as the tested population, forward selection was generally superior to backward selection (i.e. selection based on testing), indicating that progeny-testing is not very efficient. The study by Danusevičius and Lindgren (2003) presented at this symposium does not support that selection based on progeny-testing is efficient compared to phenotypic selection.

More accurate breeding values can be estimated using information from relatives in procedures like combined index formation, e.g. combining family performance with individual performance or BLUP techniques. These techniques are able to maximise genetic gain when selecting a certain number of individuals, but at the cost of increased relatedness. This cost may be found acceptable. Wei and Borralho (2000) found that group coancestry after unrestricted BLUP based selection in three trials with open pollinated progenies of *Eucalyptus urophylla* in southeastern China as less than 1 per cent, thus status number above 50, which is by no way problematic. However, initially the material comprised of more than 300 open-pollinated families, and each open-pollinated family carries a status number of four. Thus, the status number of the progeny tests was above 1000 and the status number was reduced by a factor of more than 10 by the selection. Therefore, unrestricted BLUP may still lead to an unacceptable accumulation of relatedness in more constrained and in multi-generation situations. The effort to maximise genetic gain without greatly reducing gene diversity, requires large breeding populations, and this can be very costly.

5 Gene Diversity in the Breeding Population Boosts Genetic Gain in the Production Population

The aim of forest tree breeding is to supply forestry with genetically improved regeneration material on a sustainable basis. A high genetic gain in the long-term breeding population is just a tool to achieve that goal. To get regeneration material, the breeding population is creamed off to achieve as high an immediate gain as possible combined with an acceptable level of gene diversity. If the breeding population is more diverse, selections for the production population can be done more aggressively, sacrificing a higher share of the gene diversity in the breeding population to get that lowest acceptable level. Thus, a more diverse breeding population results in a higher gain in the step when the breeding population is creamed off, when selection for the seed orchard is done. This effect partly compensates for the faster advancement of gain in the long-term breeding population, which is possible by caring less about the accumulation of relatedness (Rosvall 1999).

6 Phenotypic (Mass) Selection

Phenotypic selection (mass selection) means selection based on the appearance of the individual tree itself, not considering the performance of its relatives. The phenotype of a tree gives an estimate of its breeding value (=phenotype×heritability).

In a study of the genetic gains obtainable from a range of alternatives, Shelbourne (1992) found that phenotypic selection in unpedigreed stands, although somewhat lower in genetic gain than more elaborate strategies, still seemed favourable when possibilities for faster generation turn-over and lower cost was taken into account. By considering the performance of the relatives, a greater gain can be achieved at the same selection intensity, but the additional gain is accompanied by a correspondingly higher relatedness among the chosen individuals. For instance, while selecting among the first generation of offspring, selected trees will be sibs to a higher extent, if sib performance is considered. If comparisons consider gene diversity as well as gain, the advantage for phenotypic selection is strong. When compared at the same gene diversity among the selections and the same selection intensity, phenotypic selection and combined-index selection (with restrictions on the highest number of offspring per parent) produce approximately the same genetic gain (Wei 1995, Andersson et al. 1998a, 1999). Phenotypic selection seems to be competitive for achieving genetic gain at a given level of gene diversity even when repeated for many generations in material with complex and unbalanced pedigrees (Spanos et al. 1996; Andersson 1999). When only the initial offspring is considered, it may be argued that combined-index selection or similar methods is able to produce a higher genetic gain than phenotypic selection, and foresters may not care that this is associated with a large reduction in gene diversity. This can be considered an argument against phenotypic selection in the short time perspective. Reduction in gene diversity will, however, lead to reduced gain in later generations, so after around five generations the *maximum* gain obtainable has been found to be about as high for phenotypic selection as for combined-index selection with restrictions (Andersson et al. 1998b). These findings concerned with forest tree breeding are supported by similar results by animal geneticists (e.g. Quinton et al. 1992).

Phenotypic selection, without knowing or caring about the pedigree of the trees, has in some studies been found to be about as efficient for long-term breeding as if the pedigree was known and used as an aid in selection (e.g. Wei 1995; Andersson 1999). For many model situations with selection among related trees with known pedigree phenotypic selection is the best selection method in the meaning that no other way of selection combines higher breeding value and at the same time higher gene diversity. Phenotypic selection was rather efficient, but it was just one among a whole array of efficient solutions, which differ in their relative emphasis on gain versus gene diversity. One of the options a breeder loses by using phenotypic selection is the option to make a deliberate choice between emphases on gain or gene diversity. Numerical simulations indicate that phenotypic selection often results in an intuitively appealing compromise and seldom results in a drastic increase in relatedness (e.g. Andersson 1999). Average relatedness can initially be set at a sufficiently low level just by selecting a sufficient number of trees (=wind-pollinated families). Later selections can be made by phenotypic selection in plantations of descendants from the initial selections. However, the number of selections transmitting genes to the following generations must be sufficiently high. What can be considered sufficient is mainly dependent on the average relatedness, which will build up as a result of the selections.

Phenotypic selection is the method used by *Nature* as the driving force for evolution, and has thus existed for millions of years. It may give some comfort to use a "natural" method; even if it seems unsatisfying that modern *Science* has difficulties to do better than *Nature*. The similarity to *Nature* is a strong argument that it is sustainable. That a method is similar to *Nature* can be regarded as a reason to use it in itself. It is very simple and it does not require sophisticated training. The selection situation is similar to forestry practice (e.g. similar mixture of genotypes). It minimises or eliminates the work of experimental layout and documentation. It does not use a complicated algorithm that works like a "black box", able to disguise severe mistakes. It is cheap. It is easy to keep track of data. Heritability in phenotypic selection is usually rather low (5-20%). That means - in contrast to selections based on genotype testing - that even the characters not regarded as desirable will remain among the trees selected to constitute the parents of the following generation, thus phenotypic selection has attractive properties for gene conservation. In many countries supporting expensive breeding programmes, much of the gain deployed to today's forests originates from phenotypic selection of plus-trees, thus it can be claimed to be the method with which forest tree breeding has the longest experience, and this experience is generally a positive one.

The accumulation and genetic structure following many generation phenotypic selections have to be predicted based on a set of assumptions concerning the relatedness structure. It is possible to do such calculations for general cases, which usually will be satisfactorily reliable. Specific circumstances may, however, cause considerable deviations from those predictions in specific cases; this is an advantage of working with known pedigrees.

Selection based on the phenotype needs not depend on actual measurements, i.e. a subjective evaluation of all characters of a tree (including its performance compared to its neighbors) can be made in the field. Such an evaluation may actually be more accurate than an objective measure, as all characters can be jointly considered including the local environment (thus the performance of the neighbors). It does not depend on a long chain of actions on different places and by different people where mistakes or delays may easily occur. In practical selection, there is usually a considerable amount of subjectivity at the final stage in a decision procedure; this cannot be circumvented by high-input techniques.

Phenotypic thinning for unpedigreed seed production areas and also plus tree selection have been discussed, advocated and applied e.g. by Harwood et al. (1996). As low-input breeding must usually rely on phenotypic selection, it must be pleasing for those applying it to know that it is now acknowledged by modern science as an efficient tool (Wei 1995; Andersson 1999). Thus, seed production areas can be established, and the resulting forests later creamed for the best unpedigreed trees, whose descendants are used for establishing new seed production areas, making low-input breeding sustainable over multiple generations, provided the tree numbers involved are sufficiently large.

7 Vegetative Propagation

Some species are as easy to propagate and handle as vegetative propagules as they are from seed, or vegetative propagation may actually be the only practical way to produce plants. There are many potential advantages with vegetative propagation (e.g. Ahuja and Libby 1993; Lindgren 2002). Often there are technical difficulties with the propagation method, and under such circumstances investment in development is not worthwhile for low-input breeding (clonal forestry is actually rather seldom practiced even for high-input breeding). If reasonably large, well-adapted and well-formed trees can be selected and vegetative propagated for practical use without added costs or complications, it seems recommendable to use that option. Vegetative propagation is an evolutionary dead end, thus actions must be considered to assure recombination even for cases where the main propagation methods are vegetative. This can be done by establishing "gene resource plantations" (see below). It must be assured that relatedness remains low and gene diversity high, but that can be done with calculations in combination with sufficient numbers.

A possible application is to use phenotypic selection in a "gene resource plantation" as a first step. In a second step, vegetative propagules of the best selections could be placed together in a stand. A stand with some type of selected clones would actually be equivalent to a grafted seed orchard. The pollen would come from good phenotypes. If clones were placed in row plots, the worst looking clones could be removed. Clonal identity and field maps would not be required for this operation; whether or not it is a good clone can be judged by evaluating the row plot, without knowing the identity of the clone. Seeds from the better clones could be used for the next cycle of gene resource plantations. Coancestry and inbreeding can be kept under control by calculations in the same way as for phenotypic selection, clonal selection can be considered equivalent with phenotypic selection with high heritability. Inbreeding may increase but experiences in connection to Finnish spruce seed orchards indicate that this increase may be small (Nikkanen 2002). To use clone plantations for seed collection has been suggested as a cheaper alternative to seed orchards of Norway spruce in Sweden (Lindgren and Karlsson 1993). Above it was argued that phenotypic selection often was a competitive alternative. Clonal testing instead of phenotypic testing can be seen as a way as increasing the heritability and thus the associated selection gain. A breeding strategy based on clonal testing is generally more powerful than a strategy built on phenotypic or progeny testing (Danusevičius and Lindgren 2002a and 2002b). This ought to be true even if the clones are unidentified, thus planting clonal rows and select the good rows ought to have potential.

8 Estimates of Accumulation of Relatedness Based on Fertility Variations

Low-input programmes must be concerned with inbreeding, relatedness and diversity. These factors must be predicted to manage gene resource plantations and to plan seed collection. In high-input programmes, known pedigrees, individual identification and selection algorithms that utilise this knowledge can keep control. In low-input programmes, the actual operational control may be relaxed, and thus it becomes more important to forecast what will happen by appropriate use of theoretical predictions and by reasonable estimates of key factors. What happens depends mainly on the gene pool of the population and variations in the contributions of individuals to the next generation. To predict what happens is an advanced operation. The likely consequences may be forecasted by simple heuristic rules, tables and instructions for less-advanced users, but as future low-input breeders will very likely have access to competence and computers, the need of predictions will seldom be bottlenecks.

The gene pool of the offspring is the same as the gene pool of the successful gametes of the parents; this connection links generations. It seems natural to link fertility to successful gametes, but it is not known what gametes will be successful in advance, and there will be stochastic variation, which is high if low-input measures are used. It is thus more useful to define fertility as a characteristic of the parental genotype. Fertility is defined as "a parent's *ability* to produce successful gametes". The *true* number of successful gametes per parent is both technically and principally difficult (or impossible) to estimate (e.g. there is no unequivocal definition of how long the zygote that results from a successful gamete must survive to be characterized as "successful"). Quantitative estimates of variation among trees in female or male reproductive structures can be made even for a low-input programme, and it seems likely that such counts will be sufficiently accurate for most situations. Anyway, the differences in fertility among considered objects or years are likely to be more important than the inaccuracies of the estimation method.

A quantification of fertility differences among a group of parents in probabilistic terms can be made as a basis for predictions and theoretical development. The sibling coefficient (Kang 2001) refers to the probability that two gametes, chosen randomly from the gene pool of gametes, originate from the same parent, compared to that in the gene pool of the parents. It is associated to the probability that individuals share the same parent, and thus are sibs. Mathematically, *sibling coefficient,* Ψ*,* can be defined as $N\Sigma p_i^2$, in which N is the number of individuals and p_i is the relative fertility of individual i. Relevant theory is developed in the PhD theses by Bila (2000) and Kang (2001) and papers in them. "The effective number of parents" can be expressed as N/Ψ, which can be viewed as identical to the classical "variance effective number". *The sibling coefficient* is a function of the coefficient of variation for fertility; Ψ=2 corresponds to a coefficient of variation of 100%. *Sibling coefficient* for a forest stand may typically be 2 (Bila and Lindgren 1998). Ψ=2 means that there will be twice as many sibs among the seeds as expected if all mating were equally frequent. Ψ=2 means that relatedness and later inbreeding will build up over generations twice as fast as if there were equal mating. Reasons to suggest "sibling coefficient" to be useful in predictions of generation shifts are that it is independent of the number of members in the population, that it focuses on probabilities and that it has transparent interpretations as mentioned above. An example of this calculation technique is demonstrated in Bila et al. (1999). The loss over generations is predicted when seeds are collected and a small sample of these seeds is used to replace the stand (a gene conservation stand). If a limited number of offspring is considered, the successful gametes can be seen as obtained by sampling from all gametes. An option to increase the effective number, thus to reduce group coancestry, loss of gene diversity and subsequent inbreeding, is to keep female fertility constant by mixing the same amount of seeds from all trees. This is a rather effective measure. The technique can also be used to trade off gain and gene diversity in the seed crop or gain and gene diversity in the stand itself (low-input breeding). The idea is that inbreeding and the associated phenomenon can be kept manageable and balanced against cost, gain and other desiderata by management technique and numbers, rather than by keeping exact pedigrees. Such calculations are needed for a low-input programme, but can be generalized in tables. There can be a tremendous variation in the magnitude of fertility variation. For example, Varghese et al. (2003) reported a sibling coefficient, Ψ=17.4 in a first generation progeny trial of *Eucalyptus tereticornis* in south India studied at four years of age with the intention to convert it to a seedling seed orchard. Only 18% trees were fertile out of 200 trees selected for phenotypic superiority. Most cases with high sibling coefficient are objects which are young or flower poorly (Kang 2001) or are not well adapted or it happens to be a special year, and probably not important overall, and even the seeds from the object just described would not have a dangerously low gene diversity.

9 Use More Offspring from the Best Parents

Increase in relatedness and breeding value are the major outcomes of selection, and there are optimal combinations of them in the sense that under given constraints and gene diversity, there exists an optimal strategy that maximises gain. An optimal strategy may be conservative or aggressive depending on the demand of gene diversity. It is a good breeding practice to allow the better trees to be over-represented in both breeding populations and production populations. Truncation selection is the common way of achieving this; it draws a strict border between pass and fail. A more gradual differentiation in treatment of materials relative to their goodness is more optimal. Sophisticated algorithms for identifying such strategies have been developed for deployment (linear deployment, e.g. Bondesson and Lindgren 1993; Wei and Lindgren 1995) and selection (group-merit selection, Lindgren and Mullin 1997;

Andersson 1999). While paternal fertility may be unknown, it is still possible to trade off against female fertility, and thus pick more seeds and plant more plants from the best female parents. Fernandez and Toro (2001) applied integer mathematical programming in a selection scheme on an open-pollinated population in *Eucalyptus globulus*, resulting in a large reduction in loss of gene diversity at a small cost of genetic gain compared to truncation selection. There seems to be a need to develop techniques that are simple to handle, but still close to optimal for low-input breeding. Many optimization techniques require plants to be identified by family within gene resource plantations; this is an argument for family identification in the field. These techniques will contribute a better guarantee that diversity is preserved and make the balance between gain and diversity more of a deliberate choice. I doubt it is efficient keeping identifications in a low-input programme for this reason only, but if this is done for other reasons, then techniques using this knowledge could be used. Another application could be to consider the family value for low heritability characters or characters such as survival, as a guide to how many to select from each family, but to select phenotypically within family (preferable for other characters). To harvest cones from fewer trees than are left as possible pollen parents can be seen as using more offspring from better trees, the best trees will be used both as seed parents and pollen parents, while more trees will be used at least as pollen parents.

10 Grafted Seed Orchards Are Not Low-input Options

Professional breeders spend much attention on situations where large investments are made to establish installations for seed production. The idea to use grafted superior individuals as parents is attractive. If superior trees are grafted, all their genes are utilised, while if just their open pollinated offspring is used, only half of the genes will be from the selected trees. The intensively managed grafted pine seed orchard has worked rather well for many of the most influential breeding programmes and has thus become the standard model. Subsequently, breeders acting in poor environments often also see the grafted orchard as the ideal because of influences by training and literature. However, the use of grafts demands organisation, planning and development of practical skills. Grafting demands skilful experienced workers and this is complicated if there is not some scale and continuity. Grafts are sensitive to damage, sensitive to competition from other vegetation, expensive, not easy to keep in backup reserve, and need continuous tending by experienced professionals. The intensive management at the start of the seed orchard means a large investment, which is justified only when the advantages are large and certain; one prerequisite for that is stability over time. They are not likely to remain operative over long periods if funding for their management is withdrawn. They constitute a type of land management, which is not typical and may therefore cause irritation and requires special competence. Often grafts are used on other land and places than the trees were adapted to, and that may contribute to problems with flowering. The large investment, the need for attention and perhaps also need for internal pollination may force grafted seed orchards to be concentrated in large expensive units, where they can be conveniently managed, and they may be less sutiable to programmes with small seed needs.

It seems that different variants of seedling seed orchards (Varghese et al. 2000) often offers a better alternative for low-input cases. Unpedigreed seedling seed orchards established with equal mix of seedlings from at least 50 superior unrelated trees ought to be sufficient to initiate a local small-scale breeding programme as well as to supply seeds after phenotypic thinning.

There may be a place for low-input seed production units even in high-input programmes mainly supported by grafted seed orchards. Lindgren and Karlsson (1993) suggested for Norway spruce in Sweden that the two projections of the need for seed are made: the lowest estimate could be met by high input grafted seed orchards, and low-input gene resource plantations using rooted cuttings could be used to meet the difference between the high and the low estimate. The economic loss by investments, which never get a return, will be low if the low estimate will be realised. If the actual seed need development hits the high estimate, the loss in genetic quality will be limited, in particular considering the possibilities to expand on the best of the grafts (e.g. by cutting propagation). If the seed need is low, the "gene resource plantations" can be used for wood production.

11 Gene Resource Plantation

"Gene resource plantations" are suggested as the solution to some of the problems discussed above. This concept is often similar to or may be interpreted as a low-input variant of "breeding seedling orchard (BPO)" (e.g. Barnes 1995) or "unpedigreed (or half-pedigreed) seed production areas" (Harwood et al. 1996). Often these may be established as a mixture of half-sib families obtained from open pollination on selected trees (from a previous cycle of gene resource plantations). Similar plantations constitute a part in many breeding programmes, but when documented, they still often appear without focus on minimising costs and maximising simplicity or optimising the procedures used. The gene resource plantations may be established in a similar fashion to regular plantations. If the gene resource plantation is similar to - and is managed in ways not too different from - an ordinary stand, it does not require much special competence. It can be managed with the existing competence and organisation of forest caretakers. There are fewer constraints on the location. These features may make it easier to find localities where flowering, seed production, seed quality, land access, representatively for plantations, difference for desirable characters and pollen isolation will be favourably combined. A gene resource plantation can be small and cheap, and the same concept can be used for many species and sub-populations. A gene resource plantation can sometimes be regarded as a seedling seed orchard, but seed production may not be the only aim. It is comparatively easy to arrange for multiple use so that it serves concurrently all or some of the functions of conservation of the species; conservation of the genetic diversity of the species; seed source for plantations; production of desirable commercial products; and as a local demonstration plot, while retaining options to initiate a more regular improvement programme. If there are subsides for commercial plantations, these can be applicable for the stand, making it an economically attractive alternative. The idea to harvest commercial seeds in objects that look like a forest or progeny test has been used, for example in improvement of coastal Douglas-fir in Oregon and Washington (Silen and Wheat 1979).

Plantation practices commonly employed in many poor circumstances results often in rather irregular plantations. Seedling quality is often low, planting quality poor, replanting common, survival low and stocking patchy; species are sometimes mixed and agroforestry is often practiced. Gene resource plantations are likely to function better than regular progeny tests or grafted seed orchards under these conditions, as they are robust and less sensitive to irregularities. However, the gain by phenotypic selection will be larger if selection is practiced in uniform plantations, and the risk for loss of whole families is larger if conditions are sloppy, thus improved plantation techniques probably combine well with many variants of low-input breeding.

If a gene resource stand is to function as a seed source, some investment should be made in thinning and managing the stand for that purpose. Seed collection may use methods that damage the production of the forest, e.g. the most economic way of harvesting the seeds may be to fell trees. These costs appear close to seed harvest, when it is known that the seeds are needed and the investment worthwhile. It is not like a grafted seed orchard, which depends on a large investment long time before the benefit appears, and the investment make it doubtful to let seed collection harm future seed production.

A plus tree in a plantation represents a good tree while a seedling from a plus tree just has a good seed parent. The idea of clonal seed orchards is that the pollen parents are also selected. However, this expectation of pollination with plus tree pollen parents has not been realised as expected in many seed orchards, we must often accept that up to half of the pollen parents are found outside the orchard (Lindgren 1991). Thus the advantage of the better control of fathers in clonal seed orchards compared to a gene resource plantation is often less than was expected some decades ago.

Authorities (like EU, OECD and national boards of forestry) feel an interest to keep genetic diversity under control, but the only factor that can be conveniently and objectively audited, is census numbers. Therefore, much legal attention and restrictions go to numbers, and these numbers tend to be high. This attitude by authorities will probably seriously constrain innovations and gain. Seedlings are genetically different and it is therefore likely that legal worries will be less severe for plants from seedling seed orchards (seedling stands). Seeds are much easier to handle than vegetative propagules. The gain from the initial step of selecting seed parents is only half of that of selecting the whole genotype, which is a severe draw back. However, half-sib families have four times the effective size (status number) as clones from the same plus tree (Lindgren et al. 1996). Thus, some of the gain lost in the first step can be recovered by using the excess diversity to make more radical selection later. Another way of phrasing it is that a plantation obtained from open pollination of 100 trees can harbour up to four times higher effective number as a seed orchard with 100 clones.

12 Genetic Management of Gene Resource Plantations

"Gene resource plantations" can initially be established by selecting desirable trees in the forest, harvesting seeds from them, use a number of seeds from each mother tree (a limited number from each mother) to produce plants, and establish one or (mostly) several plantations from that. Desires on the site may include accessibility, not very small area, and appropriate for the species (more factors are discussed above). There is a minor advantage if there is some isolation from possible pollen sources outside the stand. It can be advantageous to establish a gene resource plantation similar to a commercial plantation. That means that selection effects will make the genetic constitution of the materials more suitable for plantation forestry than natural seed sources, that management will fit better into forest operations and that it is more likely there is a commercial use for the biomass production besides its genetic functions.

The planting should probably use a rather close spacing to allow later thinning. Identities of trees need not necessarily be known (thus no records on the family identities or plantation plans are required), as general calculations can be used to assure that the number of trees and harvested mothers is large enough. To keep known identities require: 1) keeping family identities known at plantation; 2) mapping the area after plantation; 3) placing identification markers in the field; 4) maintaining identifications; and 5) keeping maps available for long time. Thus, known tree identities require considerable cost, fails sometimes, is more requiring on competence and means that there must be organisational stability.

The gene resource plantation should probably be subject to rather intense silviculture and thinning so that desirable, well-adapted phenotypes are favoured. At thinning less desirable phenotypes are removed. Sufficient spacing will probably stimulate more seeds lower down in the crown. Thinning can be problematic, too hard thinning can be a stress factor, thus for many species thinning should be mild but frequent and this may be a problem in a low-input programme. Establishment and management depends partly on the aims; if the goal is more to support a good seed source, then close spacing with subsequent frequent thinning is desirable, while if the aim is just conservation, close spacing is not desirable. Even if the purpose is gene conservation, it should be remembered that the healthy well-adapted trees remaining after thinning and selected for harvesting would probably also be favoured by natural selection. Thus, a gene conservation goal is probably better met by a larger number of trees and more sites than by little thinning. A close initial spacing and subsequent thinning increases the cost. This is more motivated the more certain is the need for improved seeds; if it seems too uneconomic it could be cancelled. There could be specific considerations that need to be made, e.g. if females and males are different trees, or if it is predominantly a selfing species, or if consideration must be given to pollen vectors. If seed production is an important aim, spacing could be rather wide compared with a stand where wood production is the main aim. If the seed requirement continues, it is probably a good idea to replace the gene resource stand by a more improved gene resource stand, rather than heading for using the initial stand as a seed source as long as physically possible. The gene resource plantation must be renewed at some interval, and it may be practical to combine cutting with seed harvest.

The effect of the different actions could be evaluated if certain steps were taken. If seeds are stored from different stages of the programme (seeds from the initial selections in the forest, seeds from the resulting improved stand, and seeds with which the next-generation stand will be established), the option to establish comparative trials exists. Some long-term storage of seeds is recommended when it is possible and simple. If this is not the case, it may not be worthwhile or not even practically possible to conserve at the different stages.

The system could be complemented with a more gene conservation approach. Stands could be managed by systematic thinning or other criteria than commercial.

13 Provenances in Gene Resource Plantations

The provenance question is always somewhat uncertain. Something may be known or guessed and a gene resource plantation, planned to serve as a local seed source, should not contain genetic material that is likely to be inferior. On the other hand, it should cover gene mass from some range and have diversity in characters. It may often be wise to arrange a gene resource plantation with known identities and a layout identical to a combined provenance - half sib - trial, like what was suggested by Nanson (1972). Small plots (single-tree plots (by far most efficient but also rather technical demanding) or five-tree row plots may be best) could be used on the provenance level, so as not to leave unproductive gaps if unsuitable provenances are thinned out. If there is interest in acquiring more scientific information, the trial could be measured before first thinning and the results evaluated. In practice, the major cost with provenance trials is probably not their establishment, but rather their measurement, evaluation and presentation. Their establishment could be justified as gene resource plantations, even if they are never measured. It is important, however, that thinning is not delayed. It may be noted that (provided tree identification remains) even a phenotypically thinned trial can provide provenance information. If there is a provenance variation in a gene resource stand, and the stand is phenotypically thinned and the best looking trees harvested for seeds, this is a way to utilise the well-adapted suitable provenances, as such provenances will be over-represented among trees selected for harvest. If much of the gene resource stand turns out to be comprised of inferior provenance material, the next cycle of gene resource stand could contain fresh collections, including more material from the areas where the good origins grow. While, if much of the provenance material seems acceptable and the range satisfactorily covered, the stand could serve as the seed source for the next cycle. Actually, this type of gene resource plantation could be more efficient than the currently most used procedure for non-priority species, that is to first identify the best provenances and then initiate seed collections based on that knowledge.

It may be a good idea to let a species adapt a generation before starting seed production. Part of this adaptation process can take place in test stands but hardly grafted seed orchards.

Acknowledgements

Many people have contributed to the discussion around this manuscript and thoughts during many years. Among those who have commented or contributed recently I would like to mention D. Danusevičius, S. Ruotsalainen and M. Varghese. Huogen Li is acknowledged for writing the Chinese abstract.

References

Ahuja, M.R. and Libby, W.J. (eds) 1993. Clonal forestry I. Genetics and biotechnology and II. Conservation and application. Springer-Verlag, Berlin, Heidelberg.

Andersson, E.W. 1999. Gain and diversity in multi-generation breeding programs. Acta Univ. Agri. Suec., Silvestria 95, 42p+4chapters.

Andersson, E.W., Spanos, K.A., Mullin, T.J. and Lindgren, D. 1998a. Phenotypic selection can be better than selection for breeding value. Scand. J. For. Res. 13:7-11.

Andersson, E.W., Spanos, K.A., Mullin, T.J. and Lindgren, D. 1998b. Phenotypic selection compared to restricted combined index selection for many generations. Silva Fenn. 32:111-120.

Barnes, R.D. 1995. The breeding seedling orchard in the multiple population breeding strategy. Silvae Genet. 44:81-88.

Bila, A.D. 2000. Fertility variation and its effects on gene diversity in forest tree populations. Acta Univ. Agri. Suec. Silvestria 166, 31p+4chapters.

Bila, A.D. and Lindgren, D. 1998. Fertility variation in *Milletia sthulimannii, Brachystegia speciformis, Brachystegia bohemi* and *Leucaena leucocephala* and its effects on relatedness in seeds. For. Genet. 5:119-129.

Bila, A.D., Lindgren, D. and Mullin, T.J. 1999. Fertility variation and its effect on diversity over generations in Teak plantation (*Tectona grandis* L.f.). Silvae Genet. 48:109-114.

Bondesson, F.L. and Lindgren, D. 1993. Optimal utilization of clones and genetic thinning of seed orchards. Silvae Genet. 42:157-163.

Cotterill, P.P. 1989. The nucleus breeding system. In: Proc. of the 20th Southern Forest Tree Improvement Conference, Charleston, South Carolina, pp36-42.

Danusevičius, D. and Lindgren, D. 2002. Efficiency of selection based on phenotype, clone and progeny testing in long-term breeding. Silvae Genet. 51:19-26.

Danusevičius, D. and Lindgren, D. 2003. Clonal testing may be the best approach to long-term breeding of *Eucalyptus*. In: "*Eucalyptus* plantations – research, management and development", R.-P. Wei and D. Xu (eds), Proc. of International Symposium on "*Eucalyptus* plantations", Guangzhou/Zhaoqing, China, 1-6 September 2002, World Scientific, Singapore, pp192-210.

Fernandez, J. and Toro, M.A. 2001. Controlling genetic variability by mathematical programming in a selection scheme on an open-pollinated population in *Eucalyptus globulus*. Theor. Appl. Genet. 102:1056-1064.

Harwood, C.E., Nikles, D.G., Pomroy, P. and Robson, K. 1996. Impact of thinning via phenotypic selection on the genetic base of planted seed production areas. In: "Tree improvement for sustainable tropical forestry", M.J. Dieter, A.C. Matheson, D.G. Nikles, C.E. Harwood and S.M. Walker (eds.), Proc. of QFRI-IUFRO Conference, Caloundra, Queensland, Australia, 27 October to 1 November 1996, pp148-153.

Kang, K.S. 2001. Genetic gain and gene diversity of seed orchard crops. Acta Univ. Agri. Suec. Silvestria 187, 75p+11chapters.

Lagerström, T. and Eriksson, G. 1996. Improvement of trees and shrubs by phenotypic selection for landscaping in urban and rural areas - A Swedish example. For. Landsc. Res. 1:349-366.

Lindgren, D. (ed) 1991. Pollen Contamination in Seed Orchards. In: Proc. of the Meeting of the Nordic Group for Tree Breeding 1991, Swedish University of Agricultural Sciences, Department of Forest Genetics and Plant Physiology, Report 10:1-120.

Lindgren D. 2000. Low-intensity tree breeding. In: "Rapid generation turnover in the breeding population and low-intensity breeding", K. Lundkvist (ed), Department of Forest Genetics, Uppsala, SLU, Sweden, Research Notes 55, pp37-48.

Lindgren, D. 2002. Advantages of clonal propagation. In: M. Welander and L.H. Zhu (eds), Proc. of Workshop on "High quality birch – clonal propagation and wood properties", Alnarp, 53, 27-28 August 2002, SLU, SE 230.

Lindgren, D. and Karlsson, B. 1993. Cheaper improved Norway spruce seeds for Sweden. In: "Norway spruce provenances and breeding", V. Rone (ed), Proc. of IUFRO S2.2-11 Symposium, Latvian Forestry Research Institute, Riga, Latvia, 14-18 September 1993, pp224- 230.

Lindgren, D. and Mullin,T.J. 1997. Balancing gain and relatedness in selection. Silvae Genet. 46:124-129.

Lindgren, D., Gea, L.D. and Jefferson, P.A. 1996. Loss of genetic diversity monitored by status number. Silvae Genet. 45:52-59.

Namkoong, G., Barnes, R.D. and Burley, J. 1980. A philosophy of breeding strategy for tropical forest trees. Dept of Forestry, Oxford, Tropical Forestry Papers 16, 67p.

Nanson, A. 1972. The provenance seedling seed orchard. Silvae Genet. 21:243-249.

Nikkanen, T. 2002. Functioning of a Norway spruce (*Picea abies* (L.) Karst.) seed orchard. PhD Thesis, Metsäntutkimuslaitoksen tiedonantoja 850, 58p.

Quinton, M., Smith, C. and Goddard, M.E. 1992. Comparison of selection methods at the same level of inbreeding. J. Anim. Sci. 70: 1060-1067.

Rosvall, O. 1999. Enhancing gain from long-term forest tree breeding while conserving genetic diversity. Acta Univ. Agri. Suec. Silvestria 109, 65p+4 chapters.

Ruotsalainen, S. and Lindgren, D. 1998. Predicting genetic gain of backward and forward selection in forest tree breeding. Silvae Genet. 47:42-50.

Ruotsalainen, S. and Lindgren, D. 2000. Stratified sublining: a new option for structuring breeding populations. Can. J. For. Res. 30:596-604.

Silen, R.R. and Wheat, J.G. 1979. Progressive tree improvement program in coastal Douglas-fir. J. For. 77:78-83.

Shelbourne, C.J.A. 1992. Genetic gains from different kinds of breeding population and seed or plant production population. Suid-Afri. Bosb. 160:49-65.

Spanos, K.A., Andersson, E.W. and Lindgren, D. 1996. Multi-generation comparison between restricted phenotypic and combined index selection for genetic gain and diversity. In: "Conservation of forest genetic resources", M. Kurm and Y. Tamm (eds), Proc. of Nordic Group for Forest Genetics and Tree Breeding Meeting, Estonian Agricultural University, Tartu, Estonia, 3-7 June 1996, For. Studies 28:86-100.

Varghese, M., Nicodemus, A., Nagarajan, B., Siddappa, K.R.S., Bennet. S.S.R. and Subramanian, K. 2000. Seedling seed orchards for breeding tropical trees. Scroll Press, Institute of Forest Genetics and Tree Breeding, Coimbatore, India, 126p.

Varghese, M., Ravi, N., Son, S.-G. and Lindgren, D. 2003 Variation in fertility and its impact on gene diversity in a seedling seed orchard of *Eucalyptus tereticornis*. In: "*Eucalyptus* plantations – research, management and development", R.-P. Wei and D. Xu (eds), Proc. of International Symposium on "*Eucalyptus* plantations", Guangzhou/Zhaoqing, China, 1-6 September 2002, World Scientific, Singapore, pp111-127.

Wei, R.-P. 1995. Predicting genetic diversity and optimizing gain in breeding programmes. PhD Thesis, ISBN 91-576-4990-1, Swedish University of Agricultural Sciences, Umeå, Sweden, 64p+12 chapters.

Wei, R.-P. and Lindgren, D. 1995. Optimal family contributions and a linear approximation. Theor. Pop. Biol. 48:318-332.

Wei, R.-P. and Lindgren, D. 1993. Phenotypic selection was more efficient than combined index selection when applied on full sibs of lodgepole and Scots pine. Swedish University of Agricultural Sciences, Department of Forest Genetics and Plant Physiology, Report 11:289-292.

Wei, X. and Borralho, N.M.G. 2000. Genetic gains and levels of relatedness from best linear unbiased prediction selection of *Eucalyptus urophylla* for pulp production in southeastern China. Can. J. For. Res. 30:1601–1607.

低投入林木育种策略

DAG LINDGREN

Department of Forest Genetics and Plant Physiology
Swedish University of Agricultural Sciences
S-901 83, Umeå, Sweden

未来精细的林木改良离不开丰富的遗传资源、稳定的财政资助、熟练而相对固定的技术人员，以及完备的基础设施，这在相应的文献资料、专家建议、科学研究及正规的培训计划中均有所体现。然而，对于未来的育种方法，尤其是适用于资源非常贫乏场合的育种方法却很少被关注。实际上开发和优化一种低投入的育种方法是非常有必要的，这种简单而廉价的方法通过适当的调整也可适用于资源丰富的情形，但往往被视为不够精细而遭受冷遇。当一个树种的资源增长时低投入技术更是被忽视。

本文对种子生产和长期育种中的一系列低投入技术和方法进行了讨论。特别地，着重讨论了表型选择、不进行遗传测定、育种材料无需进行系谱鉴定、不采用嫁接、用自由授粉取代控制授粉等低投入育种技术与方法。一个可持续的低投入育种策略需要考虑群体的数量遗传背景，尤其要考虑育种群体中亲缘关系的管理。无性繁殖技术无论是对于资金充足的场合还是资金缺乏的场合均是一个很重要的工具。

USE OF SITE-SPECIES MATCHING AND GENETIC GAIN TO MAXIMISE YIELD - A SOUTH AFRICAN EXAMPLE

TAMMY L. SWAIN AND ROBIN A.W. GARDNER

Institute for Commercial Forestry Research (ICFR)
Box 204, Sabie, 1260, South Africa

The sites available to the South African forestry industry for commercial afforestation of eucalypts for pulp and paper production are diverse, and often termed as low in productivity. To manage the various limiting factors in the temperate areas, such as severe frost, low temperatures, occasional snow and drought and high altitude pests, a combination of matching site and species, together with the use of genetically improved material, is recommended to maximise gains from the sites.

To this end, site-species trials were established over several sites to investigate alternative species for growth on high altitude sites. Species such as *E. nitens* and *E. macarthurii* grow very well on the sites which are more extreme for cold and frost, *E. dunnii* and *E. smithii* on the more intermediate sites, and the new species, *E. badjensis* and *E. benthamii* have shown great potential as possible alternatives to *E. nitens* and *E. macarthurii* respectively. Breeding programmes have been initiated to test provenance and progeny in these species.

These species will be discussed with reference to site suitability, and results from provenance/progeny trials in the various species will be presented, including yield, pest and disease and pulping information.

1 Introduction

South Africa is a diverse country with a wide range of sites established to commercial forestry. The forestry sites range from just above sea level to 1900 m in altitude, with mean annual temperatures of 14°C to over 20°C, and annual rainfall ranging from 650 mm to approximately 1100 mm per annum. Historically, *Eucalyptus* species have been grown on the more sub-tropical/tropical Zululand Coastal Plains for pulp and paper production, and on the more temperate, high altitude sites of Mpumalanga (MPU) for mining timber. *Eucalyptus grandis* was the main species grown in Zululand for pulp, and *E. nitens*, *E. macarthurii*, *E. fastigata* and *E. elata* were traditionally grown on the high altitude sites for mining timber, although the latter two were limited by varying degrees of intolerance to cold, frost and snow. However, the shift in focus from mining timber to pulp and paper production by the South African Forestry Industry in the early 1980s engendered a need to grow *Eucalyptus* species for pulp and paper production on the colder, higher altitude sites of MPU, and the highland areas of western KwaZulu-Natal (KZN) and the northern portion of the Eastern Cape (Figure 1). Many of these sites are termed "low productivity" in terms of their rainfall, temperature or soil depth, and only certain species will grow on these sites. Apart from low temperatures and related frosts, there is also the risk of heavy snow at least once in a rotation. Thus there was a need to investigate species which, firstly, could grow on these relatively low productivity sites and, secondly, still produce suitable pulp yields.

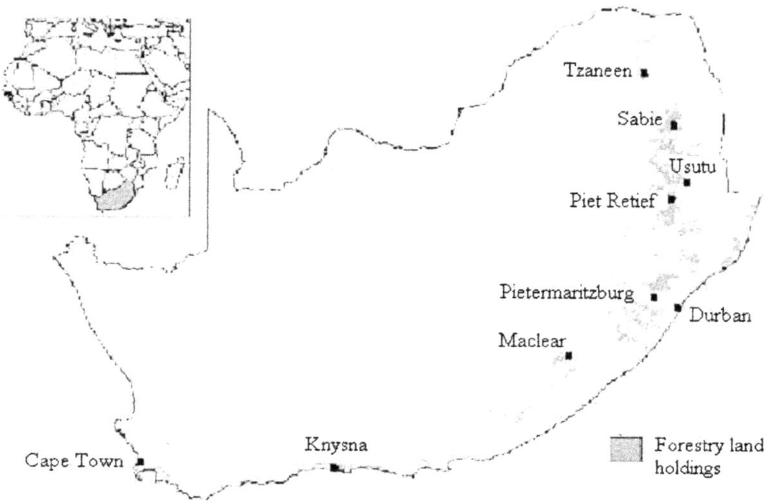

Figure 1. Map of South Africa and current forestry land holdings.

Results from an early Institute of Commercial Forestry Research (ICFR) site-species interaction trial series, established over a range of high altitude sites in Mpumalanga province (Schönau and Gardner 1991) showed *E. macarthurii* to be the only species suitable for planting on low quality sites experiencing regular heavy frosts during winter. *E. macarthurii* and *E. nitens* were known as the most frost tolerant and snow tolerant commercial eucalypt species in South Africa at the time, respectively, and, based on the information available then, both of these species could be grown on a nine year rotation for pulp and paper production.

A further series of site-species interaction trials was then established by the ICFR from 1990 to 1992 to investigate potential alternatives to *E. macarthurii* and *E. nitens* for colder, higher altitude sites, particularly in areas where site productivity is regarded as low. These trials identified several species over time, i.e. *E. badjensis* and *E. benthamii*, and confirmed that *E. smithii* and *E. dunnii* also had a role to play on certain sites (Darrow 1994, 1996; Gardner 1999). Breeding programmes have since been developed for all of these species. However, apart from identifying species with commercial potential, for which breeding programmes could be developed, the site-species interaction trials have played another important role by providing valuable information regarding the adaptability of the species to different site types and environmental macro-factors such as snow, drought, etc.

Although the shift in emphasis in South African forestry from tree growth for mining timber production to pulp and paper production began in the early 1980s, intensive research into pulp properties of different eucalypt species only began in

South Africa in the early 1990s. Up until that time, and into the mid/late 1990s, *E. grandis* at low altitude sites, and *E. nitens* and *E. macarthurii* at high altitude sites, were considered acceptable for pulping. At the same time, many species were being investigated for their potential to grow on the cold, high altitude sites, and breeding programmes being initiated, with very little, if any, knowledge being available on the pulping properties of these species. Then, however, Clarke (1995) reported on intensive research carried out into the wood properties of several CTE species, and these results have started to influence species choice. For example, species such as *E. smithii* have become more preferred in terms of pulp and paper production, and *E. macarthurii*, conversely, has become less favoured.

Information gathered from the site-species interaction trials, and the later provenance/progeny trials established in the respective breeding programmes, has made siting of species on low productivity sites a more informed process. This ultimately assists the forestry industry in achieving optimum growth performance on all sites, with the least risk of crop failure. However, the choice of CTE species, based on the information now available regarding wood properties, often conflicts with the species which should be chosen, based on site information.

Asking the following questions can, however, assist with successful commercial plantation establishment:

- Which species will grow best on the site?
- Is improved seed of this species available? If not, which provenances of the Australian material, for which seed is available, performs best under South African summer rainfall growing conditions?
- If improved seed is available, what level of improvement is available, and what sort of gains can be expected?
- What is known about other factors, such as pulp properties, which will influence the choice of material available?

The information necessary to make informed choices on the cold, high altitude sites, based on the site-species interaction trials and provenance/progeny trials, is summarised in this paper.

2 Species Choice

From the site-species interaction trials, provenance/progeny trials and general experience, a table has been developed (Table 1) whereby cold tolerant eucalypt (CTE) species can be identified, and compared with each other for site requirements and limitations, as well as other growth characteristics. Details of mean annual precipitation (MAP), mean annual temperature (MAT) and altitude ranges for each of the CTE species, as experienced by these authors, can be found in Table 1. Relative rankings of the species for characteristics such as cold, frost, snow, disease and insect tolerance are also included. Species such as *E. elata*, *E. fastigata* and *E. fraxinoides* have been included in Table 1, as although they are not discussed in this paper, they are well known CTE species, and are useful as a comparative measure.

Table 1. Summary of cold tolerant eucalypt species grown in South Africa, with relative rankings for growth characteristics.

Species	Optimum MAT range (°C)	Minimum MAP range (mm)	Altitudinal range (m)[1]	Minimum soil depth (cm)[2]	Comparative ranking[3]						
					Cold tolerance 1-very high 4-fair	Frost tolerance 1-very high 4-slight	Snow tolerance 1-high 4-sensitive	Drought tolerance 1-high 4-sensitive	Snout beetle resistance 1-resistant 4-highly susceptible	Phytophthora resistance 1-resistant 4-highly susceptible	Termite resistance 1-resistant 4-highly susceptible
E. badjensis	14.5-17.0	800-900	1100-1600	30	2	2	2	2	3	2	1
E. benthamii	14.5-17.5	800-900	1100-1600	45	1	1	4	3	2	1	1
E. dorrigoensis	14.5-17.5	780-950	1100-1500	45	2	1	4	3	2	1	1
E. dunnii	15.0-19.0	800-950	900-1350	35	4	4	4	1	4	1	1
E. elata	14.5-18.5	825-980	1000-1350	50	3	3	3	4	2	1	4
E. fastigata	14.0-17.0	800-950	1200-1600	45	1	3	2	2	1	4	3
E. fraxinoides	14.0-16.0	780-900	1200-1700	30	1	4	2	2	2	4	3
E. macarthurii	14.0-18.0	780-925	1150-1500	40	2	1	4	2	3	1	1
E. nitens	14.0-16.0	825-950	>1350	45	1	2	1	3	2	1	4
E. smithii	15.0-18.0	830-950	1100-1500	40	3	3	3	2	3	4	1

[1] The altitude tolerance of a species becomes higher with decreasing latitude, although the lower altitude thresholds generally apply to KZN, and higher thresholds to MPU.

[2] Soil depth is very circumspect, as several of the species will thrive in shallow soils at the higher altitude, higher rainfall ends of their ranges, on conditions that occasional mists supplement monthly rainfall throughout the year.

[3] The scores for the comparative rankings are generalised for each species, as within each species, provenance differences may exist for some of these characteristics. Tolerance of some of the new species may also improve with selection and breeding.

Other general information about each species, which cannot easily be included in table, but which is useful when deciding on a suitable species for a site, is summarised below.

2.1 *E. nitens*

E. nitens was traditionally grown for mining timber in South Africa, and is the most snow and cold tolerant of the eucalypts grown in South Africa (Darrow 1996; Gardner and Swain 1996). The species is frost tolerant, but not as hardy as *E. macarthurii* (Darrow 1994, 1996). *E. nitens* does not coppice easily, and any ability to coppice appears to decrease with age.

The species is very sensitive to fire, and the roots of *E. nitens* are susceptible to attack by wood eating termites during the first couple of years of growth (Herbert 1993). *E. nitens* is not generally susceptible to damage by the *Eucalyptus* snout beetle (*Gonipterus scutellatus*), however, it has been known for both the larvae and the beetle to browse on *E. nitens*, should the preferred host, *E. dunnii,* and even *E. macarthurii* and *E. smithii,* be unavailable. Although there is a biological control or *Gonipterus scutellatus* in South Africa, namely the wasp *Anaphes nitens*, which parasites the eggs of the snout beetle, this does not over winter successfully at high altitudes, which is why *Gonipterus scutellatus* is a problem at altitudes over 1300 m.

2.2 *E. macarthurii*

E. macarthurii was also grown for mining timber in South Africa, and is the most frost tolerant of the commercially grown CTEs in the country. Within its correct temperature range (Table 1), *E. macarthurii* is the most hardy of the eucalypt species, producing commercially viable yields with relatively low risk on low productivity sites. On many sites, *E. macarthurii* is often the only species that will grow. Although *E. macarthurii* is the most frost tolerant of the CTEs, it is not quite as cold tolerant as *E. nitens* and has proved to be far less able to withstand heavy snowfalls than was previously thought. Heavy snowfalls cause stem breakages anywhere along the length of the tree, and bending of the stems is common. The species was rated as only "slightly hardy" to snow, similar to *E. dunnii* in this respect, but hardier than *E. grandis* and *A. mearnsii*. (Gardner and Swain 1996). *E. macarthurii* coppices easily.

E. macarthurii is considered to be resistant to root damage from termites (Herbert 1993). The new foliage, however, is susceptible to defoliation by the larvae of the snout beetle, should the preferred food source of the larvae and beetles, i.e. *E. dunnii*, be unavailable.

2.3 *E. smithii*

E. smithii has traditionally been planted in the cooler areas of South Africa for its foliar cineole content, cineole being a leaf oil which has high economic value (Poynton 1979; Boland et al. 1992). However, research in the past decade has shown that it has very rapid early growth and high yields in short rotations, which makes it suitable for pulp and paper production.

E. smithii is moderately frost tolerant, but at the higher end of its altitudinal range should be planted on sloped sites where cold air can drain off. The species is not as frost tolerant as *E. macarthurii*, but is similar to *E. dunnii* in this respect (Darrow 1996). During the severe drought which occurred in South Africa from 1991 to 1993, *E. smithii* proved to be one of the most drought hardy species, as tolerant of drought as *E. dunnii* (Darrow 1994).

The species has been rated as "moderately hardy" to snow. It is at much higher risk than *E. nitens* and *E. fraxinoides* on high altitude sites, but does not sustain as much damage as *E. dunnii* and *E. macarthurii* at the same sites (Gardner and Swain 1996). *E. smithii* coppices well.

E. smithii is highly susceptible to *Phytophthora* root rot, and it is essential that the species be planted on sites that allow for good drainage of water from the soil profile. In areas where the MAT exceeds 19°C, the risk of *Phytophthora* infection is increased. Past experience indicates that the avoidance of a) rough handling/damaging of seedlings during the transportation and planting process and b) establishment on ex-eucalypt, hot-burn sites, lowers the risk of *Phytophthora* infection even further.

The species shows reasonable resistance to termites (Schönau and Gardner 1991), but can suffer some damage from snout beetle in the first few years of growth, particularly on the colder, high altitude sites (Darrow 1994), and if the preferred host, *E. dunnii,* is not available.

2.4 *E. dunnii*

In South Africa, *E. dunnii* has a growth rate, form and gum bark similar to that of *E. grandis*, although its timber quality, until recently, was considered locally to be somewhat inferior, especially with regard to splitting. However, growth has been good in short rotations in the cooler growing areas on sites too cold and/or too dry for *E. grandis*, and it is thus grown mainly in the cooler areas of the country, being classified as a cold tolerant eucalypt in South Africa. Of the cold tolerant eucalypt species grown in South Africa, *E. dunnii* has the best natural stem form, with very little taper.

E. dunnii grows better than *E. grandis* in cooler conditions (MAT<16.5) and is far more frost and snow tolerant than *E. grandis*. However, in comparison to other CTE species such as *E. macarthurii* and *E. nitens*, *E. dunnii* shows only moderate tolerance to frost, and thus frost pockets and sites exposed to chilling winds should be avoided. *E. dunnii* is considered to be even more drought tolerant than *E. macarthurii* in South Africa (Darrow 1994). The species has been classified as "slightly hardy to snow", and is as snow tolerant as *E. macarthurii*, but more so than *E. grandis*. Damage during heavy snows has been mainly broken stems and crowns (Gardner and Swain 1996). Although the species suffers stem or tip damage due to heavy frosts, and is susceptible to crown damage by strong winds and snows, it demonstrates a remarkable ability to recover from most forms of stem breakage or damage that occur in the first half of the rotation, resulting in an acceptably straight bole by the end of the rotation (Gardner and Swain 1996). *E. dunnii* coppices easily in most cases.

In cooler areas *E. dunnii* is susceptible to defoliation, often severe, by the snout beetle. The species again demonstrates its ability to recover from damage and grows through the depleted foliage, maintaining its good stem form. Although the yield lost due to defoliation has never been scientifically quantified, growth measured at 66 months in three ICFR provenance/progeny trials indicates that severe loss occurs. Growth at two sites, where no snout beetle damage occurred, was markedly better than the third site, which was severely defoliated by snout beetle (*Gonipterus scutellatus*) larvae almost a year after establishment, i.e. average tree height was 16% and 25 % better at the first two sites, respectively, than the third site, average diameter at breast height (dbh) was 18% and 14 % better, and average individual tree volume was 69% and 61% better, respectively (Swain et al. 2001). However, as correct weed control was also not maintained in the trial damaged by *Gonipterus* during the first two years after establishment, the combination of these two factors resulted in late canopy closure and slower growth than in the other trials, making a direct comparison due to *Gonipterus* damage difficult.

2.5 *E. badjensis*

This species is relatively new in South Africa, having only been introduced into site-species interaction trials in 1990 and 1991. However, trial results have indicated that *E. badjensis* may be an alternative pulping species to *E. macarthurii*, *E. smithii*, *E. dunnii* and/or *E. nitens* on cold, high altitude sites (Darrow 1996; Gardner 1999; Gardner 2001). The expected niche for this species is on sites of moderately high altitude (1200 m to 1500 m), which are too cold or prone to frost for acceptable growth of species such as *E. grandis, E. dunnii* and *A. mearnsii*, but where exceptionally heavy winter snows are unlikely.

In South Africa, *E. badjensis* is only a little less frost hardy than *E. macarthurii* and *E. benthamii*, is as hardy as *E. nitens*, and is more so than *E. smithii*. It has an optimum MAT range between 14.5°C and 18°C, and, as such, is as cold tolerant as *E. nitens*, the most cold tolerant of the eucalypts grown in South Africa. The species appears to be almost as drought tolerant as *E. dunnii*. *E. badjensis* does not suffer top breakage or damage due to winds, and has been classified as "moderately hardy" to snow. Early indications are that the species is slightly more snow tolerant than *E. smithii*, far more so than *E. dunnii* and *E. macarthurii*, but not as tolerant as *E. nitens* (Gardner and Swain 1996). The trees begin with a rapid early growth, both in diameter and height, and this may account for some of the cold and frost tolerance in the species. *E. badjensis* coppices easily.

E. badjensis is susceptible to defoliation by larvae of the snout beetle (Darrow 1994), particularly if the preferred species, *E. dunnii*, is not available to the beetle. However, trees in ICFR site-species trials and later provenance trials, indicate that, although shoots are attractive to the snout beetle, the recovery of the canopies is rapid, as in *E. dunnii,* and thus this may not be a major shortcoming in the species.

Until recently, *E. badjensis* appeared to be resistant to *Phytophthora*. However, new information from ICFR site-species interaction trials indicates that, at altitudes below 1100 m, the species is susceptible to one strain of the root-rot fungus, this strain being relatively new to South Africa (Maseko, pers. comm.).

2.6 *E. benthamii and E. dorrigoensis*

(*E. dorrigoensis* was formerly a provenance of *E. benthamii* (Dorrigo Plateau), which was reclassified in the late 1990s.)

E. benthamii and *E. dorrigoensis* are extremely frost tolerant, having demonstrated similar frost hardiness to *E. macarthurii*, and as such, may be alternative species options to the latter on frost prone, high altitude sites. The species can be planted in areas with a MAT as low as 14.5°C and a maximum of 17.5°C, and again, are as cold tolerant as *E. macarthurii*, although not quite as tolerant as *E. nitens* and *E. badjensis*. *E. benthamii* is almost as drought hardy as *E. dunnii* at the cold end of its MAT range (14.5°C), but requires a far higher MAP at warmer sites (>16.5°C). During the severe drought in South Africa from 1992 to 1994, it was noted that *E. benthamii* does not tolerate drought well compared to *E. badjensis* or *E. smithii*, and appeared slightly more sensitive to drought than *E. macarthurii* on warm sites (Darrow 1996; Gardner and Swain 1996; Gardner 1999). *E. benthamii* has performed relatively poorly in terms of snow tolerance - early indications are that the species is only slightly more hardy than *E. dunnii*, and the species has been classified as "slightly tolerant" to snow (Gardner and Swain 1996). The species has also experienced some top breakage in heavy winds, and in this regard, appears to be similar to *E. dunnii* (Gardner 1999). In establishing plantations with the species, it is therefore important to avoid exposed ridges and hilltops which are exposed to strong winds and/or heavy snow falls. *E. benthamii* coppices extremely well, and very little damage due to snout beetle has been found.

Regarding *E. dorrigoensis*, it is expected that this species will not differ too markedly from *E. benthamii* in terms of drought, wind and snow tolerance, but may be better with respect to stem form (Gardner 2000). Early results from ICFR provenance/progeny trials indicate that *E. dorrigoensis* is extremely frost tolerant (Swain 1999).

3 Australian Material - Provenance Choice

The South African Forestry Research Institute (SAFRI) established trials investigating provenance and family differences in *E. nitens* from 1982 to 1988. Further *E. nitens* trials were established by the ICFR in 1992, as were trials of *E. macarthurii, E. smithii, E. dunnii, E. badjensis* and *E. benthamii* from 1984 to 2000. These provenance/progeny trials have indicated that provenance differences exist for growth, and other characteristics, in these species when grown in South Africa. The provenance differences are summarised below, indicating which provenances performed best under South African summer rainfall conditions. Table 2 provides details of the origins of the provenances for each species.

3.1 *E. nitens*

ICFR trials testing nine Australian at six sites in the summer rainfall region of South Africa indicated that the more northern New South Wales (NSW) provenances of Ebor, Barren Mountain and Barrington Tops (Table 2) are distinctly better suited to growth in the South African summer rainfall region than the southern NSW provenances of Bendoc, Glenbog and Tallaganda. However, one of the more southern provenances, Badja, was the top-performing provenance at two of the sites.

Table 2. Details of provenance origins of the CTE species in Australia.

Provenances	Latitude (S)	Longitude (E)	Altitude (m)	Number of mother trees
E. nitens				
Ebor/Barren Mountain	30°29'	152°24'	1460-1560	27
Barrington Tops	31°55'	151°30'	1450	20
Tallaganda	35°49'	149°30'	1180-1450	46
Badja	36°01'	149°34'	880-1300	28
Glenbog	36°38'	149°24'	900-1200	31
Bendoc	37°12'	145°52'	790-1040	5
Penny Saddle	37°47'	146°16'	900	2
E. macarthurii				
Paddy's River	34°38'	150°07'	600	20
Penrose	34°39'	150°14'	620	13
Bowral	34°32'	150°21'	660-700	7
Canyonleigh	34°36'	150°06'	600	11
Hanging Rock	34°40'	150°10'	590	5
Boyd River	33°58'	150°03'	1200	17
Thiurat Rivulet	33°59'	150°05'	1080	14
E. smithii				
Wombeyan Road	34°21'	150°12'	700	10
Wingello SF	34°42'	150°10'	650	10
Robertson to Kiama	34°28'	150°21'	500	5
Nerriga	35°05'	150°07'	680	10
Tallaganda	35°23'	149°37'	950	10
Larry's Mountain	35°51'	150°01'	260	10
Belimba Fire Trail	36°04'	149°41'	1040	10
Narooma	36°12'	150°07'	80	10
Mt Dromedary	36°18'	150°00'	450	11
Bega	36°35'	149°52'	360	11
Towamba	37°05'	149°47'	220	10
Mine's Rd/Nadgee SF	37°27'	149°51'	200	7
Mt. Buck	37°36'	148°30'	300	9

Table 2 (*continued*). Details of provenance origins of the CTE species in Australia.

Provenances	Latitude (S)	Longitude (E)	Altitude (m)	Number of mother trees
E. dunnii				
1st trial series: Dead Horse Track	28°25'	152°20'	600-700	35
Unumgar	28°25'	152°42'	300	2
Oaky Creek	28°26'	152°22'	650-750	15
South Yabra SF	28°36'	152°29'	540	14
Boomi Creek	28°28'	152°40'	600	11
Acacia Creek	28°23'	152°20'	600-750	8
2nd trial series: Boomi Creek	28°25'	152°41'	300	26
Wallaby Creek	28°28'	152°29'	340	34
Moleton NSW	30°09'	152°53'	450-550	36
Koreelah SF	28°16'	152°32'	600-650	73
Bald Knob SF	28°21'	151°33'	510	8
Beaury SF	28°30'	151°26'	440	12
South Yabra SF	28°36'	152°29'	540	22
E. badjensis				
Badja SF	36°02'	149°34'	1000	40
Brown Mountain, via Nimmitabel	36°29'	149°19'	1100	20
E. benthamii				
Kedumba Valley A	33°49'	150°23'	140	30
Kedumba Valley B	33°49'	150°23'	140	18
Bent's Basin A (18857)	33°49'	150°22'	0	1
Bent's Basin B (19374)	33°52'	150°38'	40	4
Nepean River, Camden A	34°02'	150°41'	80	4
Nepean River, Camden B	34°02'	150°41'	80	3
Nepean River, Wallacia	33°52'	150°38'	40	1

The Victorian provenances did not perform well and it is likely that the juvenile susceptibility of these provenances, particularly Penny Saddle, to leafspot (various *Mycosphaerella)*, set the growth of the trees back, and the trees never recovered after this early setback. The more northern NSW provenances show greater resistance to *Mycosphaerella* through their earlier development of intermediate and adult foliage (Swain et al. 1998). Ebor/Barren Mountain appears to be the most cold tolerant of the *E. nitens* provenances tested, and Tallaganda provenance appears to be the most frost tolerant of the provenances (Swain et al. 1998). It is suspected that there are provenance differences with regard to snow tolerance, but this should be investigated in further site-species interaction trials.

3.2 *E. macarthurii*

Results of ICFR *E. macarthurii* provenance trials comparing material from Australia with material selected from local plantations, have shown that local selections perform significantly better than most Australian provenances in terms of both yield and stem form (Stanger 1993; Swain et al. 1999). It is therefore recommended that seed from South African *E. macarthurii* seed orchards be used rather than Australian seed. However, of the Australian provenances tested in South Africa, the material from Penrose and Paddy's River performed better than the others (Table 2) (Swain 1995).

3.3 *E. smithii*

ICFR provenance/progeny trials at six different sites have shown that the Wombeyan Road provenance from NSW in Australia performed well overall. Other provenances which can be used for the establishment of commercial plantations in South Africa are Tallaganda, Mount Dromedary and Larry's Mountain (Table 2). Generally, it appears that the more northern provenances, except Mount Dromedary, are suited to these sites and that, with the exception of Larry's Mountain, the provenances which have done well in South Africa are from relatively higher altitudes. Mount Buck and Mine's Road/ Nadgee are not recommended (Swain et al. 2000).

3.4 *E. dunnii*

Results from measurements done in ICFR provenance/progeny trials show that *E. dunnii* is one species where there are not marked differences between provenances for growth in South Africa. The first trial series indicated only that Dead Horse Track provenance performs better than Acacia Creek and Unumgar (Table 2) (Swain and Gardner 2000). Sixty-six month measurements in the second trial series showed that Moleton and Beaury SF provenances perform significantly better than Wallaby Creek and Boomi Creek (Table 2) for total volume, height and diameter at breast height (Swain et al. 2001). There were no other significant provenance differences in either trial series.

3.5 *E. badjensis*

Results from measurements done at 66 months in two ICFR provenance/progeny trials show that seedlots from the Badja provenance in Australia perform significantly better than those from the Brown Mountain provenance (Table 2). There were no significant differences between the two *E. badjensis* provenances for stem form (Swain and Gardner 2001).

3.6 *E. benthamii*

Results from measurements at 66 months in three *E. benthamii* provenance/progeny trials have shown that the Kedumba Valley provenances are best suited to the South African summer rainfall growing conditions. At all three sites the two Kedumba

Valley provenances were significantly superior to the other provenances, with the exception of Nepean River (Camden A), and Bent's Basin A at some sites. The Nepean River (Wallacia) provenance had significantly poorer growth than all other material at two of the sites, and was only better than Dorrigo Plateau and Nepean River (Camden A) at one site (Table 2). There were no significant differences between the provenances for stem form.

Overall growth of *E. benthamii* was best at the coldest, highest site, indicating that the species can withstand the low temperatures associated with the site (Swain and Gardner 2001).

4 Improved Material and Genetic Gain

Ideally, genetic gain is calculated from genetic gain/seed source trials, where seed from top families is compared to unimproved and commercial material. However, with the exception of *E. macarthurii* and *E. nitens*, formal genetic gain trials have not yet been established for the CTE species, and, in the case of *E. nitens*, these trials are still very young. Thus, for the purpose of this paper, gain has been calculated by taking the average growth of the top 15 families in a trial over either the commercial control, or if no such control was included, over the trial mean. After much discussion between tree breeders in the South African Forestry Industry, this figure of the top 10 to 15 families has been decided upon to make up superior commercial bulks of improved seed. These top families are selected on the basis of yield/growth, disease and insect tolerance (should such differences exist at a family level) and wood properties. Seed from remaining families is then bulked to form an improved, but not superior, grade of commercial seed. This means that, in some cases, the most superior genetic material possible is being deployed commercially in South African forests, short of using clonal material. As rooting of large numbers of CTE cuttings is very difficult to achieve, production of clonal material for commercial deployment is not a viable option at present.

Table 3 presents the percentage increase in final dbh of the top 15 families, over either the trial mean or the commercial control, for all the CTE species discussed in this paper, with the exception of *E. macarthurii*.

In cases where improved seed is available from ICFR seed orchards, it is recommended over the Australian seedlots.

The second generation of *E. macarthurii* was established in 1993 and 1994 (Swain 2000), and tremendous gains have been made in this breeding programme at the ICFR, with increases of between 10 to 15% being realised for dbh. Table 4 summarises some of the gains that were realised for second generation material over the SAMTMA control, the latter being the South African Mining Timber Manufacturer's material that was available commercially in the early 1990's. Survival of the improved material is far better than the unimproved material, which markedly increases the volume produced in the second generation (Swain 2000).

The stem form of the second generation material shows dramatic improvement over both the parent material (Swain et al. 1999) and the unimproved *E. macarthurii* which is still seen in some commercial plantations. The improved stem form will have a direct impact on final timber volume obtained from these trees, as it appears that bark strips far easier from a straight tree than from the corkscrews and bends associated with unimproved *E. macarthurii*, and logs will be able to be stacked and transported far more efficiently than in the past.

Improved second generation seed is readily available in South Africa and should be used for all commercial plantings of this species.

Table 3. Percentage increases in gain of the top 15 families of the different CTE species over either the trial mean or the commercial control.

Species	Trial site	% increase in dbh of top 15 families over the trial mean	% increase in dbh of top 15 families over the commercial control	Availability of improved seed
E. nitens	Amsterdam, MPU (101 mths)	8	-	✓
	Jessievale, MPU (101 mths)	7	-	
	Daspoort, MPU (110 mths)	13	-	
	Helvetia, MPU (110 mths)	10	-	
	Woodstock, MPU (86 mths)	15	-	
	Babanango, KZN (110 mths)	12	-	
E. smithii	Babanango, KZN (110 mths)	12	-	×
	Frischgewaagd I, KZN (87 mths)	10	36	
	Frischgewaagd II, KZN (87mths)	12	21	
	Hlelo, MPU (100 mths)	12	-	
E. dunnii E90/01 trial series	Kia Ora, KZN (112 mths)	9	-	✓
	Wetherby, KZN (112 mths)	7	-	
	Frischgewaagd, KZN (112 mths)	12	-	
E. dunnii E90/02 trial series	Maybole, KZN (66 mths)	10	16	✓
	Babanango, KZN (66 mths)	11	15	
	Hlelo, MPU (66 mths)	8	5	
E. badjensis	Lothair, MPU (66 mths)	11	-	×
	Paulpietersburg, KZN (66 mths)	12	-	
E. benthamii	Mossbank, KZN (66 mths)	13	-	×
	Utrecht, KZN (66 mths)	12	-	
	Panbult, MPU (66 mths)	11	-	

Table 4. Percentage increases in gain of improved *E. macarthurii* over the commercial control.

	Site	% increase of top 15 families over SAMTMA commercial control
P1 second generation trial series	Pinewoods, KZN	10% (for dbh)
	Vlakkloof, MPU	12% (for dbh)
P2 second generation trial series	Pinewoods, KZN	13% (for dbh) 9% (for height)
	Vlakkloof, MPU	23% (for dbh) 14% (for height)

However, a major limitation of some of the CTE species is that production of improved seed is difficult, as the species are shy, or reticent, flowerers under South African growing conditions, i.e. they take many years to flower, without the application of growth retardants such as Paclobutrazol, and once they do so, are often not prolific flowerers. *E. nitens* is such a species and is reportedly slow in reaching reproductive age even under natural growing conditions in Australia (Eldridge et al. 1993). A further limitation is that, due to the flowering window for *E. nitens* in South Africa being from April to September, there are very few natural pollinators such as bees or insects active during these winter months, resulting in low cross pollination once the trees have reached reproductive maturity (Jones 2002). *E. smithii* is also a shy flowering species, and although part of the flowering window falls over autumn and spring, allowing for open pollination in seed orchards, this species retains its developing flowering buds for 15 to 18 months before they open, again delaying seed production. *E. dunnii* flowers at a relatively early age, in comparison to *E. nitens* and *E. smithii*, but also holds its developing buds for 15 to 18 months, followed by another 12 to 15 months after seed capsule development, for seed maturation. Little is known about the flowering of *E. badjensis*, but it too, appears to be shy flowering in South Africa, not having flowered by eight years of age in a seed orchard stand that has had Paclobutrazol applied. *E. macarthurii*, *E. benthamii* and *E. dorrigoensis*, however, flower naturally from about four years of age in a seed orchard, and once flowering is established, do so prolifically.

Thus, the percentages presented in Table 3 only provide an estimate of genetic gain, or improvement, as not all of those top families may flower and produce seed. However, as research into flowering of the CTE species progresses, and more information becomes available on how flowering can be manipulated and seed production improved, it is hoped that these gains will be realised.

5 Pulp Properties

Following the initial break-through work on pulping properties of nine eucalypt species by Clarke in 1995, pulp sampling of CTE species has been a fairly regular occurrence. Results of pulp sampling of different species and provenances, at different ages, grown on various sites in South Africa have been summarised in Table 5. It is now established that site conditions play a dramatic role in influencing the ultimate wood properties of a species or clone (Gardner et al. 2001), and Table 5 is an attempt to show the ranges of pulp yields obtained over a range of sites. Species should thus be compared with other species at the same site (Site number). For example, although *E. smithii* is the species of choice based on pulp properties, and can produce pulp yields of up to 55% on certain sites, this can drop to below 50% at a poor site, which is then in the range of species such as *E. badjensis* at the same site. The majority of the results summarised below were obtained using the dissolving pulp process (Clarke 2000a). Kraft pulping (Swain and Gardner 2000) is also commonly used in South Africa, but it should be noted that the dissolving pulp process generally produces pulp yields two to three percent lower than kraft pulp yields.

Another factor to be considered is that different processors pulp to different kappa numbers, which often makes it difficult to compare results from different mills/processors.

The fast growth, high density and exceptional pulp properties of *E. smithii* make this species a commercially desirable species, as long as it is planted on sites which are well drained and which do not encourage *Phytophthora* root rot. However, unimproved *E. badjensis* was second to *E. smithii* for mean pulp yield on a shallow-soil site, producing 45.3% to *E. smithii*'s 47.4% (Gardner 2001). Early indications are that *E. badjensis* pulps at least as well as *E. grandis* and *E. nitens*, with a range of variation existing to enable selective improvement for these characteristics (Gardner 2001; Swain and Gardner 2001).

E. nitens has good kraft pulping properties (Clarke 2000b), and dissolving pulp yields ranging from 43.5% to 51.5% (Clarke 1995; Gardner 2001). Differences also exist between *E. nitens* provenances for pulping properties, with Ebor having the best pulp yield following kraft pulping, compared to Brown Mountain, Barrington Tops and Tallaganda, the latter having the lowest pulp yield (Clarke 2000b).

The splitting of *E. dunnii* has made it unsuitable for mining timber in the past, and it was not initially highly sought after for pulp production. However, results of several pulping studies (Clarke 1995; Arbuthnot 1999; Swain and Gardner 2000) have shown that it has comparable, and often better, pulping properties than *E. grandis,* depending on site. Results are still awaited from *E. benthamii* pulping studies.

6 Conclusions

The site-species interaction trials have proved to be extremely important for the South African Forestry Industry in South Africa. They have identified species with commercial potential for growth on high altitude forestry sites, and for which breeding programmes could be developed. They have also provided information on the adaptability of various cold tolerant eucalypt species to different site types and environmental macro-factors.

Table 5. Summary of pulp properties of CTE species at several sites.

Species	Reference	Site No.	Pulping process	Age of material	Basic density (kg/m³)	Density range (kg/m³)	Avg. pulp yield (%)	Pulp yield range (%)
E. grandis	Clarke, 1995	1a	Dissolving	6 yrs	410	-	48.7	44.2-52.1
	Clarke, 1995	1b	Dissolving	6 yrs	450	-	47.0	42.1-51.0
E. nitens	Clarke, 1995	1a	Dissolving	6 yrs	490	-	48.9	45.6-51.2
	Clarke, 1995	1b	Dissolving	6 yrs	560	-	48.5	45.5-51.0
	Gardner, 2001	2	Dissolving	9 yrs	554	487.5-677.2	43.5	41.7-44.3
E. macarthurii	Clarke, 1995	1a	Dissolving	6 yrs	509	-	47.1	43.4-50.4
	Clarke, 1995	1b	Dissolving	6 yrs	570	-	45.0	42.6-48.6
	Gardner, 2001	2	Dissolving	9 yrs	569	466.8-678.4	43.2	41.3-47.5
E. smithii	Clarke, 1995	1a	Dissolving	6 yrs	521	-	51.0	47.0-55.5
	Clarke, 1995	1b	Dissolving	6 yrs	570	-	51.2	48.2-54.0
	Gardner, 2001	2	Dissolving	9 yrs	586	527.0-617.6	47.4	45.5-48.6
E. dunnii	Clarke, 1995	1a	Dissolving	6 yrs	497	-	48.4	46.8-51.5
	Clarke, 1995	1b	Dissolving	6 yrs	530	-	46.1	43.0-49.0
	Arbuthnot, 1999	3	Kraft*	10 yrs	509	444.0-561.0	50.3	45.9-51.9
	Swain and Gardner, 2000	3	Dissolving	10 yrs	560	490.0-660.0	45.9	42.8-49.0
E. badjensis	Gardner, 2001	2	Dissolving	9 yrs	540	436.3-645.8	45.3	43.1-46.9
	Swain and Gardner, 2001	4	Dissolving	3.5 yrs	528	326.7-734.6	45.1	41.8-47.2

* Kraft pulping normally process pulp yields two to three percent higher than dissolving pulp yields.

This research led to the initiation of breeding programmes for *E. badjensis, E. benthamii* and *E. dorrigoensis* in the 1990s, adding to the programmes already existing for *E. nitens, E. macarthurii, E. smithii* and *E. dunnii*. Further species information was gathered from the provenance/progeny trials in the breeding programmes, as well as information on provenance differences within the species. Apart from information on the yield, or growth, of species, provenances and individual families, characteristics such as disease and pest resistance, stem form, frost, cold, snow and drought hardiness have also been evaluated for each of the above. All of this information has been extremely valuable in siting the correct species on the high altitude, often low productivity, forestry sites in South Africa. As more knowledge is gained about the pulping properties of the CTE species, and as selections are made for these properties within the breeding programmes, the value of the material being deployed commercially will increase, with species which have been improved for wood properties, being established on the these sites.

Seed orchards in the different breeding programmes have started producing improved seed of some the CTE species, and information is available as to how much gain can be expected by deploying this material commercially.

Optimum production is obtained firstly by choosing the correct species, and if possible, provenance of that species, for the site. The use of improved material, should it be available, for that species, significantly increases the yields that can be expected off that site.

Thus, information gathered from the site-species interaction trials, and the later provenance/ progeny trials established in the respective breeding programmes, has made siting of species on low productivity sites a more informed process. This, together with the use of improved material for plantation establishment, ultimately assists the forestry industry in achieving optimum growth performance on all sites, with the least risk of crop failure.

References

Arbuthnot, A. 1999. Personal communication. Mondi Richard's Bay Mill.

Boland, D.J., Brooker, M.I.H, Chippendale, G.M., Hall, N., Hyland, B.P.M., Johnston, R.D., Kleinig, D.A. and Turner, J.D. 1992. Forest trees of Australia. CSIRO, Australia.

Clarke, C.R.E. 1995. Variation in growth, wood, pulp and paper properties of nine eucalypt species with commercial potential in South Africa. PhD Thesis, University College of North Wales, Bangor.

Clarke, C.R.E. 2000a. Are *Eucalyptus* clones advantageous for the pulp mill? South African For. J. 190:61-65.

Clarke, C.R.E. 2000b. Wood and pulp properties of four New South Wales provenances of *Eucalyptus nitens* grown on a warm and a cold sited in South Africa. Appita J. 53:231-236.

Darrow, W.K. 1994. Species trials of cold-tolerant eucalypts in the summer rainfall zone of South Africa. ICFR, Pietermaritzburg, South Africa, ICFR Bulletin 10/94.

Darrow, W.K. 1996. Species trials of cold-tolerant eucalypts in the summer rainfall zone of South Africa: results at six years of age. ICFR, Pietermaritzburg, South Africa, ICFR Bulletin 9/96.

Eldridge, K.G., Davidson, J., Harwood, C. and Van Wyk, G. 1993. Eucalypt domestication and breeding. Oxford University Press, Inc, New York.

Gardner, R.A.W. 1999. Site-species interaction. ICFR, Pietermaritzburg, South Africa, ICFR Annual Research Review 1999.

Gardner, R.A.W. 2000. Report on 1997 ICFR/CSIRO collaborative trip in northern New South Wales, Australia, and subsequent deployment of the seedlots in South Africa. ICFR, Pietermaritzburg, South Africa, ICFR Bulletin 15/00.

Gardner, R.A.W. 2001. Site-species interaction studies with cold tolerant eucalypts at high altitudes in South Africa. In: Proc. of IUFRO Working Group 2.08.03 Conference on "Developing the eucalypt of the future", Valdivia, Chile, 10-15 September 2001.

Gardner, R.A.W. and Swain, T.L. 1996. Snow damage to timber plantations in KwaZulu-Natal during July 1996 and subsequent species recommendations. ICFR, Pietermaritzburg, South Africa, ICFR Bulletin 8/96.

Gardner, R.A.W., Oscroft, D.O. and Ntombela, K.A. 2001. Final report on the 1992 Zululand site-species interaction trial series SGE.32 to SGE.34. ICFR, Pietermaritzburg, South Africa, ICFR Bulletin 17/2001.

Herbert, M.A. 1993. Site requirements of exotic hardwood species. ICFR, Pietermaritzburg, South Africa, ICFR Bulletin 3/93.

Jones, W.R. 2002. Breeding systems of some cold tolerant *Eucalyptus* species. MSc Thesis, University of Natal, Pietermaritzburg. South Africa.

Masheko, B. 2002. Personal communication. FABI, University of Pretoria.

Poynton, R.J. 1979. Tree planting in southern Africa. Vol.2: The eucalypts. Govt Printer, Pretoria.

Schönau, A.P.G., and Gardner, R.A.W. 1991. Eucalypts for colder areas in Southern Africa. In: Proc. of the IUFRO Symposium – "Intensive forestry: the role of eucalypts", Durban, 2-6 September 1991, ICFR, Pietermaritzburg, Vol.1, pp467-479.

Stanger, T.K. 1993. ICFR Annual Research Report 1993. ICFR, Pietermaritzburg, South Africa.

Swain, T.L. 1995. ICFR Annual Research Report 1995. ICFR, Pietermaritzburg, South Africa.

Swain, T.L. 1999. ICFR Annual Research Review 1999. ICFR, Pietermaritzburg, South Africa.

Swain, T.L. 2000. ICFR Annual Research Review 2000. ICFR, Pietermaritzburg, South Africa.

Swain, T.L. and Gardner, R.A.W. 2000. *Eucalyptus dunnii* provenance/progeny trials in KwaZulu-Natal, South Africa - final measurements and pulping properties. In: Proc. of Forest Genetics for the Next Millennium, IUFRO Working Party 2.08.01 Conference, 8-13 October 2000, Durban, South Africa.

Swain, T.L. and Gardner, R.A.W. 2001. New cold tolerant eucalypt species in South Africa - an update on provenance/progeny trials on *E. badjensis* and *E. benthamii*. In: Proc. of IUFRO Working Group 2.08.03 Conference on "Developing the eucalypt of the future", Valdivia, Chile, 10-15 September 2001.

Swain, T.L., Chiappero, C.C. and Gardner, R.A.W. 1998. Final measurements of six ICFR *E. nitens* provenance/progeny trials in the summer rainfall region of South Africa. ICFR, Pietermaritzburg, South Africa, ICFR Bulletin 5/98.

Swain T.L., Gardner, R.A.W. and Chiappero, C.C. 1999. 1984-1998: Fifteen years of South African *E. macarthurii* tree improvement trials at the ICFR. ICFR, Pietermaritzburg, South Africa, ICFR Bulletin 3/99.

Swain T.L., Gardner, R.A.W. and Chiappero, C.C. 2000. Final report on ICFR *Eucalyptus smithii* trials in the summer rainfall region of South Africa. ICFR, Pietermaritzburg, South Africa, ICFR Bulletin 14/00.

Swain T.L., Gardner, R.A.W., Chiappero, C.C. and Komakech, C. 2001. Mid-rotation results of a second *E. dunnii* provenance/progeny trial series. ICFR, Pietermaritzburg, South Africa, ICFR Bulletin 19/01.

利用适地适种和遗传增益提高产量 — 南非的一个例子

TAMMY L. SWAIN，ROBIN A.W. GARDNER

Institute for Commercial Forestry Research (ICFR)
Box 204, Sabie, 1260, South Africa

南非有许多立地条件可以营建桉树商业性纸浆林，但纸浆林的生产力很低。在这些立地条件下有太多因素影响人工林生产力，如严重的霜冻、低温、偶然的降雪、干旱和高海拔虫害等。我们认为，可以通过有效控制这些不良因素，综合利用适地适树和遗传改良等手段，来提高这些立地条件下桉树人工林的产量。

为达到此目的，我们在不同立地条件下建立了适地适树试验林，以研究适生于高海拔立地条件下的替代种。如亮果桉(*Eucalyptus nitens*)和毛皮桉(*E. macarthurii*)等树种在极度寒冷和霜冻严重的立地条件下长势很好，邓恩桉(*E. dunnii*)和谷桉(*E. smithii*)等树种适生于中等寒冷的立地条件；新种溪谷桉(*E. badjensis*)和本沁桉(*E. benthamii*)分别在替代亮果桉和毛皮桉这两个树种方面表现出巨大的潜力。这些树种的种源和子代测定等育种项目也已启动。

本文将讨论这些树种对不同立地条件的适应性，报道不同树种的种源/子代试验结果，包括林分产量、病虫害和制浆造纸等方面的信息。

GENETIC GAINS IN GROWTH OF *EUCALYPTUS UROPHYLLA* FROM SEED ORCHARDS IN WEST GUANGDONG

SHAOWEI HUANG [1], PULIN HE [2], BINGQUAN CHEN [1], KAI HUANG [3], YUAN CHEN [2], SHANWEN WANG [2], ZHENSHAO LUO [2] AND YILING ZHOU [4]

[1]*Forestry College, South China Agricultural University*
Guangzhou 510642, China
[2]*Leizhou State Forestry Bureau, Suixi County 524348*
Guangdong, China
[3]*Suixi Forestry Experiment Farm, Suixi County 524300*
Guangdong, China
[4]*Forestry Bureau of Suixi County, Suixi 524300*
Guangdon, China

Thirty-two open-pollinated families of *Eucalyptus urophylla* from the superior trees selected from a seedling seed orchard located in Suixi Forest Experiment Farm, Guangdong Province were included in a progeny test at Leizhou State Forestry Bureau to examine the genetic performance of the orchard. Compared to seed stand and imported commercial seed-lots, genetic gains in individual volume for the unrogued seedling seed orchard were 5.2% and 16.1%, respectively. For the advanced clonal seed orchard established with superior trees selected, the estimated gains would be 10.4% and 31.3%, respectively. Seeds from these orchards were recommended for developing plantations in the areas of West Guangdong.

1 Introduction

Eucalyptus urophylla S. T. Blake is becoming more and more important for developing commercial plantations in South China. Demand of seedlings and cuttings of this species has been rapidly increasing in the past years. Although vegetative cutting from superior individuals is now a common operation, use of seeds from seed orchards established with genetic improved materials generally results in plantations with high genetic diversity and high coppicing ability and growing potential of the coppices after harvesting. High genetic diversity may play an important role in reducing risks associated with epidemic diseases and pests. On the other hand, low-cost coppicing would be a main regeneration system for the next one or two rotations for many *Eucalyptus* planters in South China. Therefore, there should still be a great demand for improved seed orchard seeds while clonal plantations become popular.

In order to produce genetic improved seed, the first-generation seedling seed orchard of *E. urophylla* was established on the basis of provenance/family trial at Suixi Forestry Experiment Farm, West of Guangdong Province, in 1990 (Huang et al. 1999). The seed orchard was inspected and approved by the Committee of Improved Forest Tree Variety Examination of Guangdong Province in 1996. In 2002, a clonal seed orchard was established in the farm with 80 clones of superior trees selected from the first-generation seedling seed orchard. It is expected to produce seeds with much higher quality in short time. A progeny test established with seeds from part (randomly sampled) of the superior trees in 1997 is used to test the whole set of selections, which could be used to analyze the genetic performance of the first-generation and new clonal seed orchards. The present paper summarizes the results of the test and estimates the genetic gains of the orchards.

2 Materials and Methods

2.1 *Tested Families*

Thirty-two open-pollinated families of *Eucalyptus urophylla* from the seedling seed orchard in Suixi Forestry Experiment Farm, Guangdong Province were included in the progeny test. This seed orchard was converted from a provenance/family trial, which included 190 half-sib families within 8 provenances, with non-continuous single-tree plot design. Inferior families and individuals were thinned from the orchard. Before roguing, a total of 123 superior trees were selected from 84 top families, with no more than 2 trees from each family. Among the 123 selections, 32 trees were randomly sampled, and seeds were collected from these sampled trees. It should be mentioned that many non-superior trees were still kept in the converted orchard as pollen providers. Seeds for plantations were collected from those superior trees selected even after the seed orchard was rogued. The rouging may have minor effect on the pollen cloud, but was neglected in the present study. Therefore, the performances of the progenies in the test represented the selection consequence of the mother trees, or the genetic quality of the seeds collected for plantations from the first-generation orchard. Two mixed seed-lots, one from seed stand in Dongmen Forest Farm, Guangxi (CK1), and another (CK1) directly imported from Indonesia for commercial purpose were used as controls.

2.2 *Experimental Design and Measurement*

Seeds were sown in the spring of 1997, and seedlings were out-planted in the field in May of the year. Randomized complete block was employed, which had 10 replications with 4-tree non-contiguous plot for each family. In January of 2001, height (*H*) and diameter breast-height (*D*) were measured for all trees involved. Individual volume (*V*) was calculated with the following formula:

$$V = 3.14 \times H \times D^2 . \qquad [1]$$

2.3 *Statistical Analysis*

SAS/STAT® was used to conduct the statistical analysis. Analysis of variance was conducted by using GLM procedure while variance components were estimated by using VARCOMP procedure with REML (Restricted Estimation of Maximum

Likelihood) method (SAS Institute Inc. 1990). The statistic model (mixed model) used for estimation of variance component was

$$Y_{ijk} = \mu + B_i + F_j + BF_{ij} + \varepsilon_{ijk} , \qquad [2]$$

where Y_{ijk} was the observation of the kth individual tree in the jth family (plot) of the ith replication, μ was the overall mean of the test stand, B_i was the fixed effect of the ith replication, F_j was the random effect of the jth family, BF_{ij} was the random interaction effect of the ith replication and the jth family, and ε_{ijk} was the random experimental error associated with the ijk^{th} individual tree.

According to the variance components obtained, the genetic variance components and heritabilities were estimated. Genetic gain relative to the control for the first generation seed orchard was estimated with the following formula (Namkoong et al. 1966):

$$\Delta G = h_f^2 S_f , \qquad [3]$$

where S_f was the differential between the mean values of the families tested and either of the control seed-lots, and h_f^2 was the family mean heritability of the trait studied. Estimated genetic gain for newly established clonal seed orchard is the double of that estimated for the first generation seedling seed orchard (Hodge and White 1992).

3 Results and Discussion

3.1 *Differences among Families*

The results of analysis of variance were presented in Table 1. It was shown that differences were extremely significant among families. It was also shown that block and block x family interaction had significant effects on height growth, but had no effects on DBH and individual volume. Duncan's multiple-range test was conducted to express the difference among families (Table 2). Most families involved in the trial performed better than the controls. Twenty-two families, accounting for 68.8% of the families tested, were better than CK1 that was used as good seed sources for plantation purpose. However, there were still five families, accounting for 15.6%, which were inferior to the imported commercial seeds (CK2). The difference among families indicates that there is a great room to increase the gain of the seed orchard by further selecting families.

3.2 *Heritabilities and Estimated Genetic Gain*

With the variance components estimated, family heritabilities for height, DBH and individual volume were calculated according to Wright (1976). The heritabilities were 0.467, 0.687 and 0.664 for height, DBH and individual volume, respectively. Clearly, both high family heritabilities and differentials (Tables 1 and 2) favor family selection for increasing genetic gain.

Table 1. Analysis of variance for individual height (*H*), DBH (*D*) and individual volume (*V*).

Variance source	Mean square			F value			Pr>F		
	H	*D*	*V*	*H*	*D*	*V*	*H*	*D*	*V*
Block	12.1945	2.3543	0.0001534	4.08	0.43	0.82	<0.0001	0.9182	0.5962
Family	6.8183	16.9071	0.0005704	2.28	3.09	3.06	0.0001	<0.0001	<0.0001
Block×family	3.6366	5.1491	0.0001937	1.22	0.94	1.04	0.0484	0.6924	0.3754
Error	2.9888	5.4681	0.0001866						

Table 2. Duncan grouping of the families tested by using the mean individual volume.

Families	Duncan groups*	(line 2)	(line 3)	(line 4)	(line 5)	(line 6)	(line 7)
	a	b	c	d	e	f	g
4	x						
26	x	x					
5	x	x	x				
22	x	x	x	x			
8		x	x	x			
37		x	x	x			
38		x	x	x			
33		x	x	x			
34		x	x	x	x		
20		x	x	x	x		
18		x	x	x	x		
36		x	x	x	x		
7		x	x	x	x		
9		x	x	x	x		
10		x	x	x	x		
41		x	x	x	x		
12		x	x	x	x		
11		x	x	x	x		
27		x	x	x	x		
39		x	x	x	x		
40		x	x	x	x		
13		x	x	x	x		
CK1		x	x	x	x		
2		x	x	x	x	x	
32		x	x	x	x	x	
6			x	x	x	x	
3				x	x	x	x
28				x	x	x	x
CK2				x	x	x	x
16				x	x	x	x
25					x	x	x
1					x	x	x
30						x	x
29							x

* Families with the same line were not significantly different.

Genetic gain estimated for the first generation seed orchard resulted from one parent (mother) selection while genetic gain estimated for the clonal seed orchards was obtained from two parents selection. For the first case, S_f in the above formula [3] was the difference between the average of all the families tested minus the average of the control trees (CK1 or CK2). Genetic gain was estimated by using [3]. For the second case, the genetic gain was just the double of that for one parent selection (Hodge and White 1992). The average individual volume of all families tested was 0.024 m^3. Compared with the two controls, the selection differentials and genetic gains in individual volume estimated for two types of seed orchards were presented in Table 3. For the seedling seed orchard, gains relative to CK1 and CK2 were 5.21% and 16.13%, respectively. For the new clonal seed orchard, gains were 10.42% and 32.26%, respectively. Nevertheless, genetic gain in Table 3 for the clonal seed orchard might be over-estimated, due to the low selection intensity, the error of sampling the mother trees of families tested, the quality of the pollen cloud etc.

Table 3. Genetic gains in individual volume for two types of seed orchards.

	Seedling seed orchard		Clonal seed orchard	
	CK1	CK2	CK1	CK2
Mean of control (m^3)	0.0222	0.0193	0.0222	0.0193
Selection differential (m^3)	0.0017	0.0047	0.0017	0.0047
Genetic gain (%)	5.21	16.13	10.42	32.26

4 Conclusions

Genetic quality of the seeds collected from the superior trees of the seedling seed orchard at Suixi Forest Experiment Farm is better than that of seed-stand seeds as well as the imported commercial seeds. Further improvement of genetic gain would be obtained for the clonal seed orchards established with the selected superior individuals. It is strongly recommended that seeds produced from these seed orchards be used to replace the seed-stand and imported commercial seeds for development of short-rotation plantations in West Guangdong. Use of seeds from these orchards for other areas needs to be proved with proper tests.

References

Hodge, G.R. and White, T.L. 1992. Concepts of selection and gain prediction. In: "Handbook of quantitative forest genetics", L. Fins, S.T. Friedman and J.V. Brotschol (eds), Kluwer Academic Publishers, Dordrecht, pp140-195.

Huang, S., Zhong, W. and Huang, H. 1999. Study on genetic variation and early selection of the provenances and open-pollinated families of *Eucalyptus urophylla*. For. Res. 12:428-432.

Namkoong, G., Snyder, E.B. and Stonecypher, R.W. 1966. Heritability and gain concepts for evaluating breeding systems such as seedling orchards. Silvae Genet. 15:76-84.
SAS Institute Inc. 1990. SAS/STAT® User's Guide, version 6, fourth edition, Vol.2. SAS Institute Inc, Cary, NC, pp891-996 and pp1661-1673.
Wright, J.W. 1976. Introduction to forest genetics. Academic Press Inc, New York, pp239-252.

粤西尾叶桉种子园生长性状的遗传增益

黄少伟[1]，何普林[2]，陈炳铨[1]，黄凯[3]
陈源[2]，王善文[2]，骆振绍[2]，周一灵[4]

[1]华南农业大学林学院，中国广州 510642
[2]国营雷州林业局，中国广东雷州 524348
[3]遂溪县林业试验场，中国广东 524300
[4]遂溪县林业局，中国广东 524300

从广东省遂溪县林业试验场尾叶桉实生种子园收集 32 个自由授粉家系，在国营雷州林业局建立子代测定林，以检验种子园的遗传水平。结果表明，未经疏伐的实生种子园种子，以母树林种子和进口商品种子为对照时，单株材积的遗传增益分别 5.21%和 16.13%。若以该种子园的优树建立无性系种子园，则相应的遗传增益分别达到 10.42%和 31.26%。这说明，遂溪县林业试验场的尾叶桉种子园种子，其遗传品质较高，可以在粤西地区营造桉树人工林时采用。

CLONAL TESTING MAY BE THE BEST APPROACH TO LONG-TERM BREEDING OF *EUCALYPTUS*

DARIUS DANUSEVIČIUS [1,2] AND DAG LINDGREN [1]

[1] *Department of Forest Genetics and Plant Physiology*
Swedish University of Agricultural Sciences
S-901 83, Umeå, Sweden
[2] *Lithuanian Forest Research Institute, Girionys*
LT-4312, Kaunas Region, Lithuania

A computer simulator called breeding cycler has been developed to evaluate a complete cycle in balanced long-term breeding from parents to the next generation parents through testing selection candidates obtained by mating in the breeding population. The efficiency of a breeding program was defined as progress per year at a given annual cost (*GMG/Y*). Progress was measured as increase in a weighted average of breeding value and gene diversity, thus the loss of gene diversity is considered. Breeding cycler considers three testing strategies for evaluating candidates for the breeding population: *Phenotype, Clone* and *Progeny*. Breeding cycler requires a set of inputs, like annual budget, cost components, time components and genetic parameters, which were specified in a main scenario and each parameter was studied over a range with the aim to make the study relevant for much *Eucalyptus* breeding. Optimal test time and entry size was decided for each testing strategy and the testing strategies were compared. For the main scenario values, the maximum *GMG/Y* for the *Phenotype*, *Clone* and *Progeny* strategies was 0.42%, 0.636% and 0.331 %, respectively. The *Clone* strategy was superior and the *Phenotype* strategy was the second best under all input values studied. However, in the case of low heritability, cheap testing and long rotation the differences between *Phenotype* and *Progeny* strategies were small. Heritability, additive variance at mature age, rotation age, plant-dependent cost and time needed to produce the test plants had the strongest effect on *GMG/Y*. As increase of genotype-depended costs had a small effect on *GMG/Y*, investments in shortening the time until the females reach fertile age or in improving the success of cloning may pay off. In conclusion, to achieve the greatest benefit in long-term breeding, the breeding strategy shall primarily be based on clonal testing if this is practically possible, except for high heritability, when *Phenotype* testing can be the first choice. If cloning is not available, *Phenotype* strategy could be preferable at short rotations and high testing costs. The situation where *Progeny* testing is most favourable is when the heritability is very low, rotation is greater than 20 years, test plants are cheap to obtain, annual budget is large and flowering is early.

1 Introduction

Global growth of human population and increased concern on conservation of forest ecosystems lead to a reduction in area of commercial forest plantations per capita. However, namely the commercial forest is the basic source to meet the increasing demand for paper and wood products. For the world to be supplied with the wood on a long-term sustainable basis, it needs to invest much more in the development of high-yielding, short-rotation plantation forests (Fenning and Gershenzon 2002), and *Eucalyptus* seems to be a suitable tree species for such investment (Eldridge et al.

1993). Considering this growing pressure to produce more wood per unit area, the investments must return the maximum benefit per unit of time on a sustainable basis and be cost efficient (Carlisle and Teich 1978; Lindgren et al. 1989; Fernandez and Toro 2001).

In terms of sustainability or long-term gain, the benefit is an optimum balance between the progress in genetic gain and the associated loss of gene diversity in the breeding population (Lindgren and Mullin 1997). The long-term benefit may be defined as annual progress in group merit, calculated as a weighted average of genetic gain and gene diversity (Wei and Lindgren 2001). Breeding strategy based on recurrent cycles of balanced within family selection allows obtaining near-optimum gain per unit of gene diversity lost in long-term selection (Dempfle 1975; Rosvall 1999).

Long-term breeding programs of commercial *Eucalyptus* species are usually based on recurrent selection among and within full-sib or polycross families by using combined index selection, while maintaining discrete or overlapping generations (Eldridge 1985; Baudouin et al. 1997; Borralho and Dutkowski 1998; McRae et al. 2001). Combined index selection optimizes genetic gain by combining the information of an individual phenotype and its relatives without considering the associated loss of gene diversity (increasing relatedness) at the cost of which the additional gain may be obtained. Though, this loss of gene diversity may not always be considered important (Wei and Borralho 2000), it has a potential to be serious and this progress in gain obtained from combined index selection can be consumed by the loss of gene diversity in the future. Thus, if compared at the same diversity level, combined index selection may not be that favorable (Wei 1995; Andersson et al. 1998). This is actually the reason why within family selection is used in some forest tree breeding programs (e.g. conifers in Sweden, Danell 1993).

There is evidence of relative advantages of clonal testing in breeding of *Eucalyptus* (e.g. Shelbourne 1991; Araujo et al. 1996). However, knowledge on relative benefit of the cycling strategies based on phenotype, clone or progeny testing by simultaneously considering genetic gain, diversity, time and costs is limited. Optimization of these strategies regarding the number of genetic entries, test plants and testing time would also be of interest. In our earlier study, we did the corresponding comparison between the cycling strategies with the scenarios for northerly coniferous species (Danusevicius and Lindgren 2002).

Objective of this study was to compare annual progress in group merit gain between breeding strategies based on testing of *Phenotypes*, *Clones* and *Progeny* for selection of parents for the next breeding cycle assuming balanced within family selection and by considering the genetic parameters, cost and time components. The input values were chosen to suite the most likely scenarios for breeding of eucalypts.

2 Materials and Methods

2.1 *Long-Term Breeding Programme and Cycling Strategies*

The long-term breeding programme under which the alternative cycling strategies were compared is based on the Multiple Population Breeding System (MPBS) (Eldridge1985; Namkoong et al. 1988). The idea of MPBS is to maintain a metapopulation split into a number of unrelated breeding populations, one for each of the adaptive environments. Within each breeding population, the breeding is carried out by the means of continuous cycles of double-pair mating among 50 breeding population members and balanced within family selection of one individual per full-sib family as a parent for the following breeding cycle.

Our task was to compare the following cycling strategies within one breeding population to generate material for ranking of the candidates in a target trait or an index and the subsequent selection of one individual per full-sib family: (1) *Phenotype* strategy: full-sibs are planted in a test, the best phenotype is selected according to its phenotypic performance, (2) *Clone* strategy: full-sibs are cloned and ramets of each clone are planted in a test. The best ortet is selected based on performance of the cloned copies, and (3) *Progeny* strategy: open-pollinated or polycross progeny of the full-sibs are planted in a test and the parent to the best progeny is selected (Figure 1).

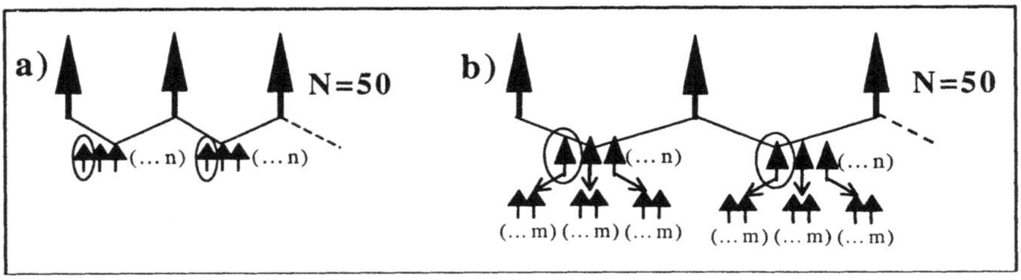

Figure 1. The principle for the long-term breeding considered. Double-pair mating is performed among 50 breeding population members (*N*=50, that it is 50 individuals only for the specification of diversity loss). That results in a candidate population. At the end of the breeding cycle the presumed best individual in each of the 50 families will be selected among the candidates and form the next generation of the breeding population. Three alternative testing strategies are evaluated (a) *Phenotype* testing: where the phenotype of the candidates is the test criterion, (b) *Clone* strategy the candidates are cloned and the clonal copies are tested, (c) *Progeny* strategy: OP or polycross seed are collected from the candidates and tested.

The testing was assumed to be carried out in a single constant environment (no G x E interaction). No C-effects (non-genetic causes of variation, e.g. maternal or cloning effects) or epistatic variance were considered. Breeding value of selected founders was set to zero (used as reference for genetic gain).

2.2 *Simulation Model*

The infinitesimal genetic model was assumed (infinite number of unlinked loci each with a small effect). The simulations were based on the main and the alternative scenarios (Table 1). While testing an alternative value of a parameter, all the other parameters were kept at the values for the main scenario. A deterministic computer simulator BREEDING CYCLE ANALYZER based on MS Excel workbook was used.[a]

[a] This program is available on the WWW page at: http:www.genfys.slu.se/staff/dagl.

Table 1. Parameters for the main and alternative scenarios. When an alternative value was tested, all other values were kept at the main scenario. BP means breeding population. If at a given scenario different values were given for different breeding strategies, the breeding strategy is indicated in the parentheses. All costs are expressed per breeding population member.

Parameters	Main scenario	Alternative scenarios
Additive variance (σ_A)	1	-
Dominance variance, % of the additive variance in BP (σ_D)	30	0; 100
Narrow-sense heritability in BP (h2) (obtained by changing σ_E)	0.2	0.05; 0.5
Additive standard deviation at mature age (σ_{Am}), %	10	5; 20
Diversity loss per cycle, %	0.5	0.25; 1
Rotation age, years	10	4; 20
Time before establishment of the selection test (T_{Before}), years	1 (*Phenotype*) 1 (*Clone*) 4 (*Progeny*)	2 (*Phenotype*) 2 (*Clone*) 2; 6 (*Progeny*)
Recombination cost (C_{Recomb}), $	30	15; 50
Cost per genotype (C_G), $	0.1 (*Clone*) 1 (*Progeny*)	1; 5 (*Clone*) 0.1; 5 (*Progeny*)
Cost per plant (C_P), $	1	0.5; 3
Cost per year and BP member (budget constraint)	10	5; 20
Group merit gain per year (*GMG/Y*)	To be maximized	

Group merit gain per year (*GMG/Y*) (Wei and Lindgren 2001) was the parameter to be maximised when searching for the best cycling strategy at a given total cost of one complete breeding cycle:

$$GMG = G - c\Theta ,$$

where *GMG* is the group merit gain obtainable from selection, *G* is the estimated additive genetic gain at rotation age (average breeding value of the individuals selected), *c* the is a weighting factor between loss of genetic diversity and genetic gain and also converting gain and diversity to the same scale, and Θ is the raise in group coancestry per breeding cycle, which, assuming that each parent contributes two offspring to be used as the parents in the next breeding cycle, was estimated as:

$$\Theta = 0.25 / n ,$$

where n is the number of the individuals selected. Note, the raise in group coancestry per breeding cycle is dependent on breeding population size only. However, cycling time may vary depending on the breeding strategy and, thus, diversity loss per unit of time may be variable.

The genetic gain at rotation age from within family selection following each breeding strategy was predicted according to the following formulas (Lindgren and Werner 1989):

2.2.1 *Phenotype* Strategy (selection based on phenotype),

$$G = \frac{\sigma_{Am} r_{J-M} i \sigma_A}{\sqrt{\sigma_A^2 + \sigma_D^2 + \sigma_E^2}},$$

2.2.2 *Clone* Strategy (selection based on clonal test), and

$$G = \frac{\sigma_{Am} r_{J-M} i \sigma_A}{\sqrt{\sigma_A^2 + \sigma_D^2 + \sigma_E^2 / n}},$$

2.2.3 *Progeny* Strategy (selection based on half-sib progeny test),

$$G = \frac{0.5 \sigma_{Am} r_{J-M} i \sigma_A}{\sqrt{0.25 \sigma_A^2 + (0.75 \sigma_A^2 + \sigma_D^2 + \sigma_E^2) / n}},$$

where G is the additive genetic gain in percentage, σ_A^2 is the additive variance, σ_D^2 is the dominance variance, σ_E^2 is the environmental variance, n is the number of plants per family, σ_{Am} is the standard deviation in breeding value of the selected individuals for a target trait at rotation age and is given as percentage of the average breeding value of the unimproved individuals for this trait (one standard deviation is equal to 10 %), i is the selection intensity estimated in the units of standard deviation of the mean of the selected individuals from the family mean by the aid of an approximation by Burrows (1975), and $r_{J\text{-}M}$ is the juvenile-mature (*J-M*) genetic correlation estimated according to the formula by Lambeth (1980) with an adjustment for the ratio of selection age to rotation age (Q) being close to 0 or 1:

if $0<Q<0.1$, then $r_{J\text{-}M}=3.108Q$;

if $0.1 \le Q \le 0.9$, then $r_{J\text{-}M}=1.02+0.308\text{Log}(Q)$; and,

if $0.9<Q<1$, then $r_{J\text{-}M}=0.988+0.12(Q-0.9)$.

Note that the scenarios at short rotation age may be interpreted as simulation of high *J-M* genetic correlation. Lambeth's (1980) approximation of age-age correlations may give a reasonable prediction of true age-age genetic relationships in *Eucalyptus* (Greaves et al. 1997).

The variable parameters to optimise the strategies for each of the scenarios in Table 1 were the following: age of selection in the selection test, family size for testing (*Phenotype* strategy), family size and number of ramets for clonal testing of each family member (*Clone* strategy), family size and number of half-sib progeny for testing of each family member (*Progeny* strategy).

2.3 *The Input Values*

The input values were chosen to be suitable for breeding of *Eucalyptus*. A most likely value was identified as the main scenario value. Then an upper and a lower reasonable bound for each value were identified.

Additive variance at mature stage (σ_{Am}) was set to 10 % for the main scenario and to 5% and 20% for the alternative scenarios. If it is at the lower value, genetic gain is given less importance versus gene diversity and, if it is at the higher value, vice versa.

For the main scenario, the weighting factor for loss of diversity (*c*) was set to 100 to make it compatible with genetic gain. This means that a 1% change in genetic gain is regarded as equally important as 1% change in gene diversity. This results in 0.5% loss of diversity per cycle. For the alternative scenarios, the weighting factor was set to result in a lower and to a greater penalty for loss of diversity (i.e. emphasizing the importance of gene diversity), leading to a loss of diversity of 0.25% and 1%, respectively.

Genetic parameters for the main scenario were chosen to represent height growth traits of *Eucalyptus*. The initial additive variance within family was set constant to 1 and the initial dominance and environmental variances were expressed as ratios of the additive variance. Additive variance within family is only dependent on the inbreeding of the parents (Foulley and Chevalet 1981). As inbreeding is low for all-important scenarios, the additive variance within family can be considered a constant. Presence of epistatic variance was ignored. For the main commercial *Eucalyptus* species, a mid-rotation narrow-sense heritability usually varies at about 0.4 - 0.5 for adaptive traits, rooting ability, wood quality traits and between 0.1 and 0.2 for height growth traits (Volker et al. 1990; Borralho and Wilsson 1994; Lemos et al. 1997; Kube et al. 2001; Sanhueza et al. 2002). Wood volume, basic density and pulp yield are the target parameters to be improved for commercial eucalypts in southeastern China (Wei and Borralho 1998, 1999).

Costs were expressed in "$" as "cost-units". The principle for setting the costs for different operations within a breeding cycle was the following: cost per test plant (plant-dependent cost) was set to 1$ and all the other costs were expressed as the ratios from 1$. Additional cost associated with age of the selection test was not considered. The total cost per cycle and breeding population member was calculated as:

$$C_{Per\ cycle} = C_{\mathrm{Re}comb} + C_{Init} + n(C_G + mC_P),$$

where C_{Recomb} is the cost for recombination among breeding population members, C_{Init} is the cost for initiation of the test (planning and localization of the test and the associated traveling costs), C_G is the cost per genotype, i.e. cost dependent on the type of reproductive material (genotype-dependent cost), C_P is the cost per test plant (plant-dependent cost), n is the number of genotypes (ortets for clonal test of female parents for *Progeny* test) and m is the number of plants (number of ramets per clone in clonal test or number of seedlings per family in progeny test).

Genotype-dependent costs include production of ortets in nursery or, alternatively, treatment to get tissue culture and growth of culture (the *Clone* strategy) and production of female parents, polycross or open-pollination, seed collection, and seed extraction (the *Progeny* strategy). Plant-dependent costs include plant production, establishment and maintenance of the selection test. The cost settings for the main scenario were the following: (1) cost for recombination was 30$, (2) cost for initiation of the selection test was 0 for all the three cycling strategies, (3) genotype-dependent cost for clone strategy was assumed to make 1/10 of the cost per plant, i.e. was set to 0.1$ per one ortet; genotype-dependent cost for the *Progeny* strategy was assumed to be 10 times higher than that for the clone strategy and was set to 1$ per one female parent, and (4) plant-dependent cost was 1$ per test plant for all the strategies; there may be some difference in plant-dependent cost between the strategies as regards type of the test plants (seedlings or ramets), however, we consider it as insignificant, as establishment, maintenance and assessment of the selection test make up the major part of plant-dependent cost.

The simulations were run with the annual budget constraint of 10$ per breeding population member for the main scenario and 5$ and 20$ per breeding population member for the alternative scenarios.

The time of one breeding cycle was subdivided into the following components:

$$T_{Cycle} = T_{Recomb} + T_{Before} + T_{Test} + T_{After} ,$$

where T_{Recomb} is the time needed for recombination (crossing and seed production), T_{Before} is the time needed to produce plants for the selection test (i.e. time from seeding to planting in the test), T_{Test} is the time needed for testing and selection of individuals as the parents for the next breeding cycle, and T_{After} is the time from selection of the new parents to harvest of their seeds for the next breeding cycle.

For the main scenario, the following timing of certain operations was chosen: the time for recombination was set to 1 year for all the alternatives; selection test may be established with 1-year-old seedlings (for the *Phenotype* strategy, T_{Before} was set to 1 year); 1 year is enough to produce some 15 cuttings from each full-sib candidate (for the *Clonal* strategy, T_{Before} was set to 1 year); the full sib candidates reach the reproductive maturity at the age of 2 and it takes 2 years to get open-pollinated or polycross seeds and produce the test plants (for the *Progeny* strategy, T_{Before} was set to 4 years). T_{After} was set to 1 year for all the strategies (assuming that a crossing archive is established at the same time as seeding of full-sib families in the nursery).

2.4 *The Optimization*

The optimization of a cycling strategy was carried by adjusting the selection age for each test size in a stage-wise manner to fit into a budget restriction (large test size is associated with high costs, which need to be shared among years) a combination of test size and selection age providing the maximum *GMG/Y* may be found.

3 Results

For the main scenario values, the maximum *GMG/Y* from the *Phenotype, Clone* and *Progeny* strategies was 0.420%, 0.636% and 0.331%, respectively (Table 2). This ranking remained stable for all the scenarios investigated (Figures 2 and 3). In case of high narrow-sense heritability, the *Phenotype* strategy was almost as good as the *Clone* strategy (Figure 3b). Except for the scenario with high heritability, the superiority of the *Phenotype* over the *Progeny* strategy was not large and in some cases minor (e.g. Figures 2b, 2c, 2e and 3c).

Increase of dominance variance slightly reduced *GMG/Y* obtained from the *Clone* strategy and had a little effect on the other strategies (Figure 2a). Increase of narrow-sense heritability markedly increased *GMG* for all breeding strategies (Figure 2b). For the *Phenotype* strategy it basically means higher genetic gain with more or less the same resources, as the selection age and family size did not change much (Table 2). For the *Clone* and *Progeny* strategies, it means that more candidates can be tested with less clonal copies or progeny and the selection can be made earlier (Table 2). When genetic gain was given a relatively higher importance (a higher additive standard deviation at mature age), the response by the *Clonal* strategy was the strongest, as it provides a higher precision in predicting the genetic values of the candidates (Figure 2c). Effect of diversity loss per cycle was comparably small (Figure 3d). As expected, annual budget was an important consideration for all strategies. However, there may be a threshold value where allocation of more resources may not markedly increase the benefit (Figure 2e). The increase of the total budget seemed to be an important consideration when comparing the phenotype and progeny strategies. Short rotations favoured the *Clone* and *Phenotype* strategies, the later gaining a reasonably large advantage over the *Progeny* strategy at rotation of 4 years (Figure 2f). Being a fixed value, recombination cost had an equal and small effect on all the strategies, simply affecting the sharing of resources among the operations within a breeding cycle (Figure 3a). Genotype-dependent cost was less important than plant-dependent cost (Figures 3b and 3c). Decrease of plant-dependent cost was especially beneficial for the *Clone* and *Progeny* strategies, whereby at a cheap testing scenario the *Progeny* strategy yielded similar gains as the *Phenotype* strategy (Figure 3c). On the other hand, when the testing was expensive, the *Phenotype* strategy appeared to be superior over the *Progeny* strategy and yielded not markedly lower *GMG/Y* than the *Clone* strategy (Figure 3c). Time before establishment of selection test was an important parameter when ranking the *Phenotype* and *Progeny* strategies. At a shorter time before establishment of the selection test, the *Progeny* strategy yielded similar *GMG/Y* as the *Phenotype* strategy (Figure 3d).

Effect of the input parameters on test size and testing time was the following: if heritability is high, more genetic entries with less plants from each entry or less *Phenotypes* need to be tested and, in the case of the *Clone* strategy, testing time was relatively shorter; shortening the rotation age resulted in a marked decrease of test size for all the strategies; high budget or reduced costs allowed to test more genetic entries or more plants to achieve a greater benefit over a shorter time, but the effect of genotype-dependent cost was relatively weaker; weights on genetic gain (σ_{Am}) or diversity, time before establishment of selection test, recombination cost seemed to have a little effect on test size and time (Figures 2 and 3, Table 2).

Table 2. Group merit gain per year (*GMG/Y* in %) for main and alternative scenarios for each testing strategy. Optimum number of test plants and optimum selection age are given. For an alternative value of a parameter, all others were kept the main scenario.

Parameter	Value	*Phenotype* strategy			*Clone* strategy				*Progeny* strategy			
		GMG/Y	Age to select	Family size	*GMG/Y*	Age to select	Ortet No.	Ramet No. per ortet	*GMG/Y*	Age to select	Female parent No.	*Progeny* No. per parent
σ_D^2, % of σ_A^2	0	0.416	5	64	0.652	7	9	9	0.329	9	6	20
	30*	0.420	5	64	0.636	7	10	8	0.331	9	6	20
	100	0.432	5	64	0.611	8	12	7	0.339	9	6	20
h^2	0.05	0.278	6	68	0.479	10	8	13	0.238	9	5	25
	0.1*	0.420	5	64	0.636	7	10	8	0.331	9	6	20
	0.5	1.219	5	59	1.243	5	30	2	0.681	8	12	9
σ_{Am}, %	5	0.185	6	73	0.297	9	11	9	0.151	9	6	21
	10*	0.420	5	64	0.636	7	10	8	0.331	9	6	20
	20	0.895	5	60	1.318	7	10	8	0.695	9	6	21
Diversity loss, %	0.25	0.447	5	60	0.659	7	10	8	0.347	9	6	20
	0.5*	0.420	5	64	0.636	7	10	8	0.331	9	6	20
	1	0.369	6	73	0.595	9	11	9	0.299	9	6	20
Rotation age, years	4	0.592	4	46	0.812	4	7	7	0.389	4	5	15
	10*	0.420	5	64	0.636	7	10	8	0.331	9	6	20
	20	0.306	7	83	0.489	10	11	10	0.268	15	7	26

Table 2 (*continued*). Group merit gain per year (*GMG/Y* in %) for main and alternative scenarios for each testing strategy. Optimum number of test plants and optimum selection age are given. For an alternative value of a parameter, all others were kept the main scenario.

Parameter	Value	*Phenotype* strategy			*Clone* strategy				*Progeny* strategy			
		GMG/Y	Age to select	Family size	*GMG/Y*	Age to select	Ortet No.	Ramet No. per ortet	*GMG/Y*	Age to select	Female parent No.	*Progeny* No. per parent
Time before establishment of selection test, years	1Ph Cl	0.420	5	64	0.636	7	10	8				
	2	0.390	6	77	0.608	8	11	9	0.352	8	6	17
	4Pr								0.331	9	6	20
	6								0.314	9	7	20
Recombination cost, $	15	0.437	5	74	0.675	7	10	9	0.344	9	6	23
	30*	0.420	5	64	0.636	7	10	8	0.331	9	6	20
	50	0.394	7	55	0.586	9	10	8	0.308	9	6	18
Cost per test genotype, $	0.1Cl				0.636	7	10	8	0.338	9	7	18
	1Pr				0.615	8	9	9	0.331	9	6	20
	5				0.544	9	6	11	0.306	9	5	21
Cost per test plant, $	0.5	0.472	5	120	0.793	7	13	12	0.439	9	7	35
	1*	0.420	5	64	0.636	7	10	8	0.331	9	6	20
	3	0.332	7	25	0.417	10	7	5	0.196	9	6	7
Total budget, $	5	0.328	8	31	0.428	9	7	5	0.207	10	5	10
	10*	0.420	5	64	0.636	7	10	8	0.331	9	6	20
	20	0.489	5	143	0.841	6	14	12	0.464	9	8	35

Note: The symbol in the superscript of the parameter values indicates the main scenario value for: (*) all strategies, (Ph) *Phenotype* strategy, (Cl) *Clone* strategy, (Pr) *Progeny* strategy.

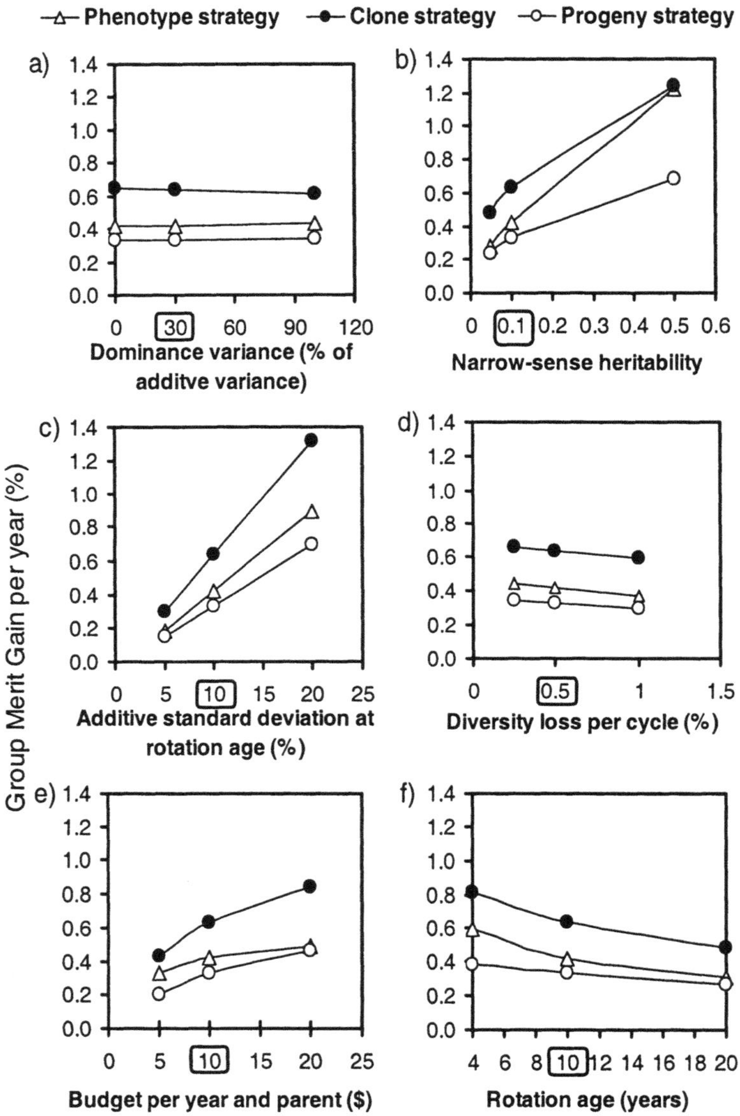

Figure 2. Comparison among the testing strategies for Group merit gain per year at the main and alternative scenario values for genetic parameters (plots a, b, c, d), total budget as the constraint (e) and rotation age (f). Scenarios at short rotation age may be interpreted as simulation for high juvenile-mature genetic correlation. The main scenario value is outlined on the X-axis.

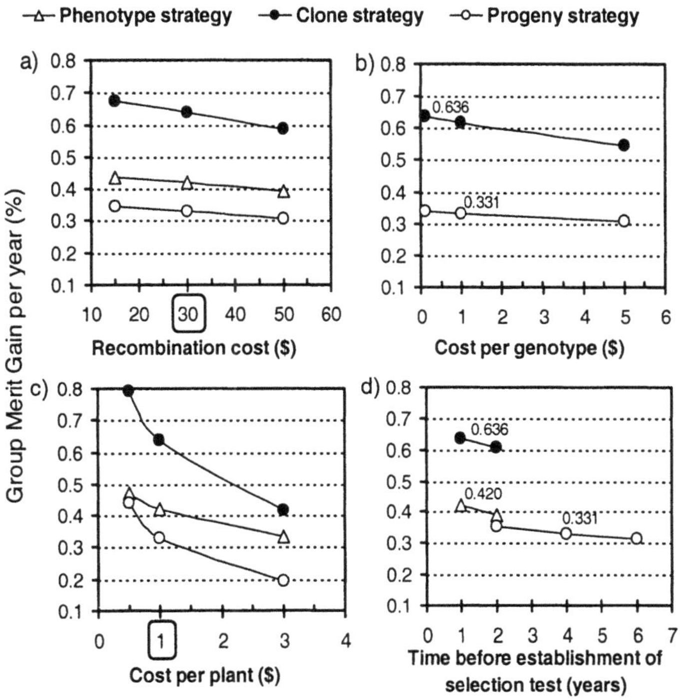

Figure 3. Comparison among the testing strategies for group merit gain per year at the main and alternative scenario values for cost components (plots a, b, c) and time before establishment of the selection test (i.e. time needed to produce seeds and plants for testing) (plot d). In the plots a) and c), the main scenario values are the same for all the test alternatives and are outlined on the X-axis. In the plots b) and d), the main scenario values are shown on the plot at the marker for a corresponding test alternative.

4 Discussion

Clonal testing was acknowledged as a valuable tool in breeding of *Eucalyptus* (e.g. Shelbourne 1991; Borralho et al. 1992a; Lambeth et al. 1994). However, it is associated with relatively higher costs and, in some cases, with investments into expensive propagation techniques. Our study considered the relative costs of the alternative cycling strategies and showed that, clonal testing still is the best approach to maximise the annual progress in group merit gain for all reasonable values of the input parameters, except for the scenario with high heritability (Figures 2 and 3). If heritability is high, there is no need to test copies of an individual, as its phenotype may provide as reliable information as its copies. Therefore, if narrow-since heritability for a target trait is higher than 0.4-0.5, phenotype-based selection may be a good alternative and the choices here may depend on factors other than the ones considered in this study (Figure 2b). Even if the cloning technique is cost-demanding (in our case, cost per genotype is high), clonal testing still is the best approach and the reduction in *GMG/Y* with increasing of genotype-dependent cost was comparably small (Figure 3b). Thus, investment into in vitro techniques or methods to increase the rooting ability for certain *Eucalyptus* species would be profitable. Substantial progress of in-vitro propagation techniques of *Eucalyptus* was reported (e.g. Reddy and Rockwood 1989; Bisht et al. 1999; Ruaud et al. 1999). Breeding towards a better rooting ability of *Eucalyptus* cuttings may also give a good result (Borralho and Wilson 1994; Lemos et al. 1997). However, if a cheap cloning technique is already in use, there would not be much benefit to develop new costly methods to shorten the time before establishment of selection test (i.e. to faster obtain the clonal copies), as it is already quite short in *Eucalyptus* and the response by *GMG/Y* would not be large (Figure 3d). However, development of cheap clonal testing methods may be considered, as it would markedly improve the efficiency of the clonal strategy (Figure 3c). High dominance variance seemed to have a minor effect on the benefit from the clonal strategy (Figure 2a). As the additive variance was kept constant, dominance variance increased at the cost of environmental variance, reduction of which has compensated the negative effect of increase in dominance variance. If genetic gain is given much less importance versus gene diversity, there is no marked difference on which strategy to use (Figure 2b). However, in reality, the situation would probably be the opposite, where genetic gain may obtain relatively higher weights. Then, clonal testing strategy is by far the best. The scenario at short rotation may be interpreted as selection for a trait with high *J-M* genetic correlation at a normal rotation. In the later case, owing its improved accuracy in predicting the breeding values, the clone strategy shall again be the first choice (Figure 2f).

In most cases, the *Phenotype* strategy was the second best, though; its superiority over progeny strategy was not large (Figures 1 and 2, Table 2). The main advantages of the *Phenotype* strategy are cheap costs and short time before establishment of selection test, which can even make it as the best choice for the target traits with high narrow-sense heritability (Figure 2b). In *Eucalyptus*, a relatively higher narrow-sense heritability was reported for the traits of an adaptive value, like frost or drought tolerance, wood basic density and rooting ability (Whiteman et al. 1996; Lemos et al. 1997). Thus, for long-term improvement of these traits, phenotype-based selection would be a good option. However, if the heritability is low, it may introduce an unacceptable error in prediction of the breeding values and shall better be avoided. The other situations where the *Phenotype* strategy can be favoured over the progeny strategy are short rotations and high testing costs (Figures 2f and 3c). Being relatively cheaper, the *Phenotype* strategy allows maintaining high selection intensity and yields a higher genetic gain than the *Progeny* strategy. It should be noted that, a shorter cycling means higher loss of diversity. This basically concerns the *Phenotype* strategy with the shortest cycles and, in case of small breeding population (diversity loss per cycle of some 1.5% in Figure 2d), it may become inferior to the *Progeny* strategy (Figure 2d).

For the *Progeny* strategy, shortening the time before establishment of selection test would be an important improvement, making the *Progeny* strategy as beneficial as the *Phenotype* strategy (Figure 3d), and this could be done at a higher cost, as the genotype-dependent cost had a little effect on *GMG/Y* from the *Progeny* strategy (Figure 3b). The investment here would be to speed up the reproductive maturity of the candidates and possibly use tissue culture techniques and indoor facilities (Hasan and Reid 1995; Häggman et al. 1996). A simple selection for early flowering may also give a good result in *Eucalyptus globulus* (Chambers et al. 1997). The other situations, in which the *Progeny* strategy may be preferred over the *Phenotype* strategy, are low heritability (Figure 2b), high budget (Figure 2e), long rotation (Figure 2f) and cheap testing (Figure 3c). The later being as important a measure to improve the *Progeny* strategy as speeding up the reproductive maturity. High budget situations may not be so realistic at present day funding tendencies. In *Eucalyptus*, heritability may increase with age (Bouvet and Vigneron 1995) and the accuracy of parental selection based on their progeny may be improved with age.

From the inputs considered, narrow-sense heritability, additive variance at mature age (weight for genetic gain), rotation age (*J-M* genetic correlation), time before establishment of selection tests and the testing costs had the strongest effect on annual progress in group merit gain (Figures 2 and 3). Therefore, these parameters shall be considered first when developing a long-term cycling strategy. As the effects of recombination cost and genotype-dependent cost were small (Table 2, Figure 3b), a relatively more costly technique may be used only if it would markedly improve another important parameter. In the case of genotype-dependent costs, this parameter is the time before establishment of selection test, e.g. time for production of female parents to be progeny tested. Development of cheap testing methods would be an efficient improvement of the clone and progeny cycling strategies (Figure 3c). Short rotation (high *J-M* genetic correlation) was least favourable for the *Progeny* strategy (Figure 2f). A reason for this may be its higher costs, which forced a longer testing time resulting in a lower progress in gain per unit time with increasing *J-M* correlation. The alternative scenario with high loss of gene diversity is adequate to simulation of a smaller breeding population (Figure 2d). As the increase in diversity loss per cycle had a minor influence on annual group merit gain, an idea may be to subline the breeding population according to the breeding values of the members (Ruotsalainen and Lindgren 2000). Another idea may also be to perform a two-stage selection by phenotypically selecting a higher heritability traits (adaptendess or for good rooting ability) at the first stage followed by selection based on progeny or clonal test at the second stage (e.g. Borralho et al. 1992b).

To summarise, the *Clone* cycling strategy is the best approach for all reasonable parameter values for *Eucalyptus*, except for high narrow-sense heritability. If cycling based on clonal testes is chosen, then its improvement efforts may consider development of cheap testing techniques. In the case of

Eucalyptus species for which cloning still is a difficult challenge, it would be worthwhile to invest into methods to improve the rooting ability or into in vitro techniques, as the increase in the genotype-depended costs (expensive cloning methods) seem to have a little influence on annual progress in group merit gain. On the other hand, if a cheap cloning technique is available, there is no need to use a more sophisticated and expensive one, as it will not add much benefit. The *Phenotype* cycling strategy can be the first choice at high heritability and, if cloning is not available, it could be preferable at short rotations and high testing costs. Otherwise, more thorough considerations between the *Phenotype* and *Progeny* strategies shall be made. The *Progeny* strategy can gain a priority over the *Phenotype* strategy when heritability is low, testing is cheap, and rotation is long and size of breeding population much smaller than 50 members. To improve the benefit from the progeny strategy, the most efficient investment is shortening the time before establishment the selection test (speeding up the first flowering) and this could be done at a higher cost.

Acknowledgements

We gratefully acknowledge "Carl Tryggers Stiftelse" for financial support. *Eucalyptus* breeding and parameters have been discussed with among others Daoqun Zang and Mohan Varghese. We are grateful to the symposium organizers for the opportunity to contribute with this presentation to the discussion on breeding efficiency in eucalypts.

References

Andersson, E.W., Spanos, K.A., Mullin, T.J. and Lindgren, D. 1998. Phenotypic selection can be better than selection for breeding value. Scand. J. For. Res. 13:7-11.

Araujo, J.A., Sousa, R., Lemos, L. and Borralho, N.M.G. 1996. Estimates of genetic parameters and prediction of breeding values for growth in *Eucalyptus globulus* combining clonal and full-sib progeny information. Silvae Genet. 45:223-226

Baudouin, L., Baril, C. and Clement Demange, A. 1997. Recurrent selection of tropical tree crops. Euphytica 96:101-114.

Bisht, P., Sharma, V.K., Joshi, I. and Kapoor, M.L. 1999. Micropropagation of newly produced F1 hybrid of *Eucalyptus* (*E. tereticornis* Sm. X *E. camaldulensis* Dehn. Soutern form). Silvae Genet. 48:104-108.

Borralho, N.M.G. and Wilson, P.J. 1994. Inheritance of initial survival and rooting ability in *Eucalyptus globulus* stem cuttings. Silvae Genet. 43:238-242.

Borralho, N.M.G. and Dutkowski, W. 1998. Comparison of rolling front and discrete generation breeding strategies for trees. Can. J. For. Res. 28:987-993.

Borralho, N.M.G., Almeida, I.M. and Cotterill, P.P. 1992a. Genetic control of growth of young *Eucalyptus globulus* clones in Portugal. Silvae Genet. 41:100-105.

Borralho, N.M.G., Cotterill, P.P. and Kanowski, P.J. 1992b. Genetic control of growth of *Eucalyptus globulus* in Portugal. 2. Efficiencies of early selection. Silvae Genet. 41:70-77.

Bouvet, J.M. and Vigneron, P. 1995. Age trends in variances and heritabilities in *Eucalyptus* factorial mating designs. Silvae Genet. 44:206-216.

Burrows, P. 1975. Variance of selection differentials in normal samples. Biometrics 31:125-133.

Carlisle, A. and Teich, A. H. 1978. Analyzing benefits and costs of tree-breeding programmes. In: "An International Review of Forestry and Forest Products", FAO, Unasylva 30.

Chambers, P.G.S., Potts, B.M. and Tilyard, P.A. 1997. The genetic control of flowering precocity in *Eucalyptus globulus* ssp. *globulus*. Silvae Genet. 46:207-214.

Danell, Ö. 1993. Breeding programmes in Sweden. 1. In: "Progeny Testing and Breeding Strategies", S.J. Lee (ed), Proc. of the Nordic Group of Tree Breeding General Approach, Edinburgh, October 1993, Forestry Commission, 128 (i-v).

Danusevičius, D. and Lindgren, D. 2002. Efficiency of selection based on phenotype, clone and progeny testing in long-term breeding. Silvae Genet. 51:19-26.

Dempfle, L. 1975. A note on increasing the limit of selection through selection within families. Genet. Res. 24:127-135.

Eldridge, K.G. 1985. Principles of forest genetics. In: Proc. of a Workshop on "Taxonomy and seed handling of Australian tree species in Eastern and Southern Africa, Harare, Zimbabwe", 8-12 July 1985.

Eldridge, K.G, Davidson, J., Harwood, C. and van Wyk, G. 1993. Eucalypt domestication and breeding. Oxford, Clarendon press, 288p.

Fenning, T.M. and Gershenzon, J. 2002. Where will the wood come from? Plantation forests and the role of biotechnology. Trends in Biotechn. 20:291-296.

Fernandez, J. and Toro, M.A. 2001. Controlling genetic variability by mathematical programming in a selection scheme on an open-pollinated population in *Eucalyptus globulus*. Theor. Appl. Genet. 102:1056-1064.

Foulley, J.-L. and Chevalet, C. 1981. Méthode de prise en compte de la consanguinité dans un modèle simple de simulation de performances. Ann. Génét. Sél. Anim. 13:189-196.

Greaves, B.L., Borralho, N.M.G., Raymond, C.A., Evans, R. and Whiteman, P. 1997. Age-age correlations in, and relationships between basic density and growth in *Eucalyptus nitens*. Silvae Genet. 46:264-270.

Hasan, O. and Reid, J. B. 1995. Reduction of generation time in *Eucalyptus globulus*. Plant Growth Reg. 17:53-60.

Häggman, H.M., Aronen, T.S. and Stomp, A.-M. 1996. Early flowering Scots pines through tissue culture for accelerating tree breeding. Theor. Appl. Genet. 93:840-848.

Kube, P.D, Raymond, C.A. and Banham, P.W. 2001. Genetic parameters for diameter basic density, cellulose content and fiber properties for *Eucalyptus nitensis*. For. Genet. 8:285-294.

Lambeth, C.C. 1980. Juvenile-mature correlation in Pinaceae and implications for early selection. For. Sci. 26:571-580.

Lambeth, C.C., Endo, M. and Wright, J. 1994. Genetic analysis of 16 clonal trials of *Eucalyptus grandis* and comparisons with seedling checks. For. Sci. 40:397-411.

Lemos, L., Carvalho, A., Araujo, J.A. and Borralho, N.M.G. 1997. Importance of additive genetic and specific combining ability effects for rooting ability of stem cuttings in *Eucalyptus globulus*. Silvae Genet. 46:307-308.

Lindgren, D. and Werner, M. 1989. Gain generating efficiency of different Norway spruce seed orchard designs. In: "Provenances, breeding and genetic conservation", L.-G. Stener and M. Werner (eds), Proc. of IUFRO WP S2.02-11 Meeting Norway spruce, Uppsala, Sweden, 1989, pp189-207.

Lindgren, D. and Mullin, T.J. 1997. Balancing gain and relatedness in selection. Silvae Genet. 46:124-129.

Lindgren, D., Libby, W.S. and Bondesson, F.L. 1989. Deployment to plantations of numbers and proportions of clones with special emphasis on maximizing gain at a constant diversity. Theor. Appl. Genet. 77:825-831.

McRae, T.A., Pilbeam, D.J., Gore, P.L., Dutkowski, G.W. and Apiolaza, L.A. 2001. Australian cooperative tree improvement strategy for *Eucalyptus globulus*. In: Proc. of IUFRO Conference Working Group 2.08.03 on "Developing the eucalypt of the future", Valdivia, Chile, 10-15 September 2001.

Namkoong, G., Kang, H.C. and Brouard, J.S. 1988. Tree Breeding: Principles and Strategies. Theor. Appl. Genet. 11 (Monographs), Springer-Verlag, 177p.

Reddy, K.V. and Rockwood, D.L. 1989. Breeding strategies for coppice production in a *Eucalyptus grandis* base population with four generations of selection. Silvae Genet. 38:148-151.

Rosvall, O. 1999. Enhancing gain from long-term forest tree breeding while conserving genetic diversity. PhD Thesis, Swedish University of Agricultural Sciences, Umeå, Sweden, Sylvestria 109.

Ruaud, J.N., Lawrence, N., Pepper, S., Potts, B.M. and Borralho, N.M.G. 1999. Genetic variation of in vitro rooting ability with time in *Eucalyptus globulus*. Silvae Genet. 48:4-7.

Ruotsalainen, S. and Lindgren, D. 2000. Stratified sublining: a new option for structuring breeding population. Can. J. For. Res. 30:596-604.

Sanhueza, R.P., White, T.L., Huber, D.A. and Griffin, A.R. 2002. Genetic parameters estimates, selection indices and predicted genetic gains from selection of *Eucalyptus globulus* in Chile. For. Genet. 9:19-29.

Shelbourne, C.J.A. 1991. Genetic gains from different kinds of breeding population and seed or plant production population. South African For. J. 160:49-66.

Volker, P.W., Dean, C.A., Tibbits, W.N. and Ravenwood, I.C. 1990. Genetic parameters and gains expected from selection in *Eucalyptus globulus* in Tasmania. Silvae Genet. 39:18-21.

Wei, R.-P. 1995. Predicting genetic diversity and optimizing selection in breeding programmes. PhD Thesis, ISBN 91-576-4990-1, Swedish University of Agricultural Sciences, Umeå, Sweden, 64p+12 chapters.

Wei, R.-P. and Lindgren, D. 2001. Optimum breeding generation interval considering build up of relatedness. Can. J. For. Res. 31:722-729.

Wei, X. and Borralho, N.M.G. 1998. Genetic control of growth traits of *Eucalyptus urophylla* S.T. Blake in southeast China. Silvae Genet. 47:158-165.

Wei, X., and Borralho, N.M.G. 1999. Breeding objectives and se-lection criteria for pulp production of *Eucalyptus urophylla* plantations in South East China. For. Genet. 6:173–192.

Wei, X. and Borralho, N.M.G. 2000. Genetic gains and levels of relatedness from best linear unbiased prediction selection of *Eucalyptus urophylla* for pulp production in southeastern China. Can. J. For. Res. 30:1601-1607.

Whiteman, P.H., Cameron, J.N. and Farrington, A. 1996. Breeding trees for improved pulp and paper production - A review. Appita J. 49:50-53.

桉树长期育种中的遗传测定效率

DARIUS DANUSEVIČIUS [1,2]，DAG LINDGREN [1]

[1] *Department of Forest Genetics and Plant Physiology*
Swedish University of Agricultural Sciences
S-901 83, Umeå, Sweden
[2] *Lithuanian Forest Research Institute, Girionys*
LT-4312, Kaunas Region, Lithuania

林木长期育种(或称多世代育种)中一个重要的环节就是对育种群体的亲本进行遗传测定。通常遗传测定的策略主要有三种，即表型测定、子代测定和无性系测定。采用何种策略主要根据测定效率(或称育种效率)。本研究采用年组合优势增益(*GMG/Y*)来衡量育种效率的大小，它是在给定的年度预算下综合考虑遗传增益和基因多样性损失两个因素，是两者的加权平均。基于以上思路，我们开发了一个计算机模拟程序，称为育种周期模拟程序(Breeding Cycler)，运行时要求输入一系列基本参数如年度预算、各阶段的具体花费和所需时间，以及相关的遗传参数等。本研究中所采用的参数及其变化范围主要是与大多数桉树(*Eucalyptus*)育种有关，我们据此优化各种测定方案的测定时间和规模，比较各种测定方案的优劣。

在主要方案(main scenario)下，表型测定、子代测定和无性系测定的年组合优势增益(*GMG/Y*)分别为 0.42%、0.331%和 0.636%。一般地说，无性系测定的效率最高，表型测定次之。然而，在遗传力低、测定花费少、轮伐期长的情况下，表型测定对子代测定的优势不大。遗传力、成林时的加性方差、轮伐期、测定材料费用、试验材料获得所需要的时间等因子对年组合优势增益(*GMG/Y*)影响很大。由于增加遗传型费用对年组合优势增益的影响较小，提早开花结实及促进无性繁殖将可获得较理想的效果。总之，无性系测定应是首选，表型测定在无性繁殖困难、遗传力高、轮伐期短、测定费用高时优势明显，而当遗传力非常低、轮伐期在 20 年以上、开花早、预算高、测定费用低等情形下，子代测定是最有利的选择。

STUDY ON COMBINED SELECTION OF PROVENANCES/FAMILIES OF *EUCALYPTUS TERETICORNIS*

JIANMIN XU, ZHAOHUA LU, GUANGYOU LI AND JIAYU BAI

Research Institute of Tropical Forestry, Chinese Academy of Forestry
Guangzhou 510520, Guangdong, China

The results of *Eucalyptus tereticornis* provenance/families trial involving 100 families within 14 provenances showed that there were highly significant differences in growth and other characteristics among families and provenances. Growth characteristics were closely correlated with stem form (*SF*) and under branch height (*CBH*) except branch size (*BR*). Individual and family heritabilities were relatively high. At an age of 7 years old, the mean values of height (*H*), *DBH* and individual volume (*V*) of the best provenance (No.13659) were 1.27, 1.32 and 2.14 times, respectively, of those of the worst provenance (No.13308). Mean individual volume of the best family (No.17) (0.06134 m^3) was 3.61 and 1.74 times of those of the worst family (No.33) and the whole population (all families), respectively. By using multi-trait index as the criteria, 22 superior families were selected, and the estimated gains were 7.31%, 6.39%, 18.01%, 13.77%, 8.22% and 5.36% for *H, DBH, V, CBH, SF* and *BR*, respectively.

1 Introduction

Eucalyptus tereticornis is one of the most widely introduced and cultivated eucalypts in the world due to its multipurpose wood products, rapid early growth, high resistant ability to both wind and drought, and low nutrient demand. This species is naturally distributed in Australia and Papua New Guinea with the most extensive latitudinal distribution in the genus *Eucalyptus,* extending from the southern Papua New Guinea to the southeastern Victoria of Australia (5°20′S-38°08′S). In the southern part of the natural distribution it generally occurs in the coastal areas; but in the areas with high rainfall it extends into the Great Dividing Range and tablelands of the northern Queensland. The altitude of the distribution ranges from near sea level to about 1150m in the northern Queensland, and up to 2000 m in Papua New Guinea (Jacobs 1981). In general, this species fairly prefers alluvial soils, sandy or gravelly loams with moisture but not waterlogged (Boland et al. 1984). In China, systemic studies on the variation of the introduced provenances and families of *E. tereticornis* have been conducted for more than 15 years, mainly aiming at selection of superior provenances and families to increase productivity of the plantations (Xu et al. 1993; Xu et al. 1995; Zhou and Liang 1994; Liang and Zhou 1995). This paper studies the genetic variation, correlation between traits and multi-trait selection of families of *E. tereticornis* based on an early provenance/family trial.

2 Materials and Methods

The experimental site was located at Zhinong town, Yangxi, Guangdong (22°30′N, 111°16′E and 36 m altitude). The lateritic red soil was developed from granite, which had the following properties: 1.65 g/kg organic matter, 0.81 g/kg hydrolytic nitrogen, 0.71 mg/kg available P and 3.9 pH value. The climate belongs to the tropical monsoon region with mean annual temperature of 23°C, mean annual rainfall of 2178 mm, and average 4-5 numbers of typhoons each year. The vegetation mainly consisted of *Echinochloa hainanensis* Chia., *Baeckea frutescens* L., *Melastoma candidum* D.Don, *Rhodomyrtus tomentosa* Hassk., *Dicranopteris dichotoma* Burm.f., and other thorn shrubs.

Seeds for this trial were supplied by the CSIRO Australian Tree Seed Center. There were 14 provenances comprised 100 families (Table 1). After clear-cut, the stumps were removed and the site was thoroughly cleared and ploughed. The planting spacing was 2 m × 3 m. Planting holes (40 cm × 40 cm × 40 cm) were dug and each was applied with 500 g barnyard manure, 250 g superphosphate and 100 g compound fertilizer as basal fertilizer, and covered with dirt. A randomized complete block design was used, with 30 replications each having single-tree plot (per family). All experimental trees were planted in May 1991. The height and diameter at breast height (*DBH*) of all the trees were measured every year. Stem form (*SF*) and branch (*BR*) were assessed after 32 months old, and under-branch-height (*CBH*) was measured at the age of 4 years. Stem form was scored as follows: 4 - straight, no knot and dead branch scar on stem; 3 - straight, few knots and scars; 2 - slight bent at one position, few knots and scars; and 1 – bent at more than two positions, few knots and scars. Branches were classified into 3 groups: 3 - small and uniform; 2 - medium and uniform; and 1 - large and forked. Both stem form and branch scores were transformed using arc-sine function before further analyzed.

Individual volume, *V*, was calculated with the following formula (Mckenney et al. 1991):

$$V = H \times DBH^2 / 3$$

Statistical analysis of variance, correlation and multi-trait index were conducted for height, *DBH*, *V*, *CBH*, *SF* and *BR* at age of 7 years by using SAS software (SAS Institute 1988) and SPQG *ver* 1.10 (Ye 1991). The following equations were used to calculate family (h_f^2) and individual (h_s^2) heritabilities:

$$h_f^2 = \sigma_f^2 / (\sigma_f^2 + \sigma_e^2 / r), \text{ and } h_s^2 = 4\sigma_f^2 / (\sigma_f^2 + \sigma_e^2),$$

where σ_f^2 was the family variance, σ_e^2 the error variance and *r* the replication. A multi-trait index (*I*) considering *n* traits was calculated for each family with the following function,

$$I = \sum_{i=1}^{n} b_i x_i = B'x, \text{ and } B = P_2^{-1} G_{21} A,$$

where b_i and x_i were the economic weight and the genetic performance, respectively, for *i*th trait, *B* and *x* the corresponding vectors of the economic weight and the genetic performance, respectively, P_2 the phenotypic covariance matrix, G_{21} the genetic covariance matrix, and A the vector of relative economy weights for all the traits considered.

Table 1. Seed sources of *Eucalyptus tereticornis* used in the trial.

Family No.	Seedlot No.	Location	Lat.	Long.	Alt. (m)	Viability (Grain/10g)
01-05	13308	Wallaroo State For, NSW	32°39′S	151°51′E	10	3560
06-10	13319	North of woolgoolga, NSW	29°55′S	153°12′E	30	5000
11-23	13398	East of kupiano, PNG	10°04′S	148°15′E	25	2600
24-29	13399	Oro Bay to Emo, PNG	08°57′S	148°28′E	200	5400
30-40	13418	Sirinumu Sogeri, PNG	09°30′S	147°26′E	580	6100
41-43	13446	North of Cardwell,QLD	18°16′S	146°00′E	40	6900
44-49	13544	40km N of Gladstone, QLD	23°44′S	151°01′E	10	9600
50-51	13659	1km North of laura, QLD	15°33′S	144°27′E	100	-
52-67	13661	Mt Molly, QLD	16°41′S	145°15′E	366	-
68-82	14212	5-12km S Helenvale, QLD	15°45′S	145°15′E	500	-
83-93	14424	Ravenshoe, QLD	17°39′S	145°21′E	700	-
94-96	15826	Ruth & Quartz CKs, QLD	15°43′S	144°37′E	120	-
97	16547	West Normanby River, QLD	15°46′S	144°58′E	140	-
98-100	16558	Oaky CK, Springmount, QLD	17°11′S	145°20′E	540	-

The economic weights of different traits were decided according to Cotterill and Jackson (1985).

Genetic gain for each trait was,

$$\Delta G = h_f^2 S \overline{X}^{-1}$$

where S was the selection differential between the mean of families selected and the population mean ($\overline{X}$).

3 Results

3.1 *Variance Analysis*

There were highly significant differences in all traits studied between provenances or families, and the differences between provenances were larger than those between families (Table 2). This indicated that in improving genetic gain, provenances plus families selection had great potential, and provenance component would contribute more than family component. At 7-year old, height, *DBH* and volume of the best provenance 13659 from Laura, QLD were 1.27, 1.32 and 2.14 times, respectively, of those of the worst provenance 13308 from Wallaroo State For, NSW (Table 3).

Table 2. Significant test (F value) of different effects on performances of *E. tereticornis* at different ages.

Source	DF	3-year				
		H	*DBH*	*V*	*SF*	*BR*
Block	29	2.4**	2.5**	3.9**	2.8**	2.2**
Provenance	13	15.1**	11.6**	15.1**	18.1**	9.5**
Family	99	4.0**	4.8**	4.5**	5.0**	2.6**
Source	DF	4-year				
		H	*DBH*	*V*	*CBH*	
Block	29	2.9**	2.0**	2.9**	2.6**	
Provenance	13	22.5**	13.4**	16.7**	45.4**	
Family	99	6.0**	5.2**	5.4**	8.4**	
Source	DF	5-year				
		H	*H*	*H*		
Block	29	3.5**	2.0**	2.1**		
Provenance	13	20.1**	13.6**	15.6**		
Family	99	5.9**	5.6**	5.5**		
Source	DF	6-year				
		H	*DBH*	*V*		
Block	29	4.1**	1.9**	2.5**		
Provenance	13	23.6**	16.3**	18.3**		
Family	99	6.4**	5.7**	5.8**		
Source	DF	7-year				
		H	*DBH*	*V*		
Block	29	3.2**	1.8**	2.6**		
Provenance	13	23.7**	17.0**	19.5**		
Family	99	6.4**	5.4**	5.8**		

There were 48 families exceeding the average volume of all families tested. The means of different traits and ranking in volume at 7-year old were given for the 20 best families (Table 4). The best family No.17 of provenance 13398 from East of Kupiano, PNG averaged 0.0613 m^3 individual volume, 3.61 times higher than the worst family No.33 of provenance 13418 from Sirinumu Sogeri, PNG, and 1.74 times higher than the population mean. The second best family (No.60) of provenance 13661 from QLD averaged 0.0571 m^3 individual volume, 3.36 times and 1.63 times higher than the worst family No.33 and the population mean, respectively.

Table 3. Duncan test of *E. tereticornis* performance among provenances (0.01).

Seedlot No.	*V*	*H*	*DBH*	*CBH*	*SF*	*BR*
13659 (QLD)	0.04874 a	11.63 a	10.40 a	4.96bc	3.037 ab	2.012b
13398 (PNG)	0.04421 ab	11.02 a	10.15 a	4.66 c	2.749 abc	1.821 b
16547 (QLD)	0.04243 ab	11.10 a	9.77 ab	5.41 ab	3.467 a	2.467 a
13418 (PNG)	0.04040 abc	10.56 bcd	9.94 ab	3.96 e	2.647 abc	1.752 b
16558 (QLD)	0.04028 abc	11.23 ab	9.75 ab	5.39 ab	2.752 abc	2.077 b
13544 (QLD)	0.03874 dbc	11.21 ab	9.50 ab	5.76 a	2.714abc	1.942 b
15826 (QLD)	0.03747 dbc	11.11 ab	9.63 ab	4.84 c	2.825 ab	2.028 b
14424 (QLD)	0.03623dbce	10.65 bc	9.49 ab	4.92 c	2.688 abc	1.961 b
13661 (QLD)	0.03310 dfce	10.13bcde	9.20 cb	4.51 dc	2.938 ab	2.016 b
13446 (QLD)	0.03051 dfge	9.99 cde	8.99 cd	4.06 ed	2.520 abc	1.866 b
14212 (QLD)	0.03016 dfge	9.63 f	9.00 cd	3.90 e	2.786 ab	1.817 b
13399 (PNG)	0.02820 fge	9.84 fde	8.51cde	4.15 ed	2.447 bc	1.774 b
13319 (NSW)	0.02721 fg	9.95 cde	8.25 de	4.73 c	2.171 c	1.904 b
13308 (NSW)	0.02279 g	9.16 f	7.89 e	3.93 e	2.130 c	1.806 b

Table 4. Means of different traits and ranking in volume of the 20 best families.

Family	*H*	*DBH*	*V*	*H/DBH*	*CBH*	*SF*	*BR*	Ranking
17	12.26	11.66	0.06134	1.05	4.59	2.585	1.667	1
60	11.58	11.57	0.05714	1.00	4.92	3.133	2.103	2
37	11.21	11.42	0.05485	0.98	3.95	2.683	1.807	3
21	11.96	11.29	0.05455	1.06	4.98	2.671	1.747	4
46	12.10	10.67	0.05437	1.13	5.86	3.033	1.933	5
51	12.24	10.97	0.05376	1.12	5.62	3.167	2.133	6
14	11.85	10.93	0.05149	1.08	4.58	2.945	1.790	7
19	11.61	10.51	0.04764	1.10	4.85	2.546	1.852	8
30	11.52	10.52	0.04674	1.09	3.91	2.577	1.619	9
18	11.03	10.22	0.04662	1.08	4.65	2.833	1.831	10
12	11.09	10.40	0.04623	1.07	4.71	2.896	1.929	11
23	11.08	10.24	0.04559	1.08	4.60	2.813	1.778	12
93	11.56	10.27	0.04544	1.13	4.99	2.834	1.969	13
83	11.38	10.41	0.04492	1.09	4.92	2.433	1.800	14
96	11.61	10.37	0.04477	1.12	4.66	3.073	2.217	15
38	11.14	10.10	0.04387	1.10	4.13	2.669	1.707	16
50	10.99	9.82	0.04371	1.12	4.32	2.907	1.891	17
98	11.16	10.13	0.04360	1.10	4.92	2.589	1.963	18
63	1129	10.41	0.04353	1.08	4.60	3.067	1.900	19
31	10.50	10.58	0.04322	0.99	3.52	2.776	1.740	20
Pop.	10.36	9.32	0.03516	1.11	4.50	2.70	1.890	
Mini.	8.29	7.14	0.01698	0.97	2.91	2.70	1.60	
Max.	2.26	11.66	0.06134	1.27	6.10	3.47	2.467	

3.2 *Genetic Variation*

The proportions of phenotypic and genetic variances, and individual and family heritability were given in Table 5 for different characters. The components of genetic variance were similar over the years for height, *DBH* and volume while they were 12.90%, 11.05% and 6.91% for *CBH*, *SF* and *BR*, respectively. Family heritability ranged 62.13-85.75% for different characters studied. As family heritability was evidently higher than individual heritability, the backward selection of the best families for further breeding would be an effective approach to achieving genetic gain. It seems that the genetic variation was large enough to make conventional breeding and selection successful.

Table 5. Phenotypic and genetic variance components (%), and family (h_f^2) and individual (h_s^2) heritabilities (%) for different characters of *E. tereticornis* families.

Variance	3-year					4-year			
	H	*DBH*	*V*	*SF*	*BR*	*H*	*DBH*	*V*	*CBH*
Phenotypic	23.18	27.12	59.37	30.81	17.63	20.52	27.52	61.50	26.29
Genetic	8.75	10.04	22.54	11.05	6.91	8.33	10.24	23.25	12.90
h_f^2	75.00	79.01	77.93	80.02	62.13	81.84	79.72	80.02	85.75
h_s^2	35.74	44.09	43.92	44.17	18.01	51.51	45.99	46.41	66.17
Variance	**5-year**			**6-year**			**7-year**		
	H	*DBH*	*V*	*H*	*DBH*	*SV*	*H*	*DBH*	*V*
Phenotypic	18.34	25.17	58.25	16.35	24.90	57.28	18.58	24.64	60.71
Genetic	7.33	9.80	22.34	6.87	9.73	22.63	7.81	9.40	24.02
h_f^2	81.51	81.16	80.54	82.76	81.07	81.28	82.79	80.27	81.24
h_s^2	50.29	49.85	48.09	53.92	49.65	49.98	54.39	47.49	49.85

3.3 *Genetic Correlation between Characters*

Height, *DBH* and volume were highly correlated with *CBH* and *SF* (Table 6). Branch character (*BR*) was significantly correlated with height but not with *DBH* and volume. As expected, height, *DBH* and volume were highly correlated with each other. In addition, significant phenotypic or genetic correlations were found among *CBH*, *SF* and *BR*.

3.4 *Multi-trait Index and Selection*

Because multi-trait selection had a positive impact due to the genetic differential and correlation, un-restricted index selection was adopted to screen superior families. Economic weights for different characters were determined by taking the equal weights and actual experiential estimates of economic weights into account (Cotterill and Jackson 1985). Two numerical solutions of economic weights obtained for further analysis below were 1:0.910:94.946:1.419:2.673:5.361 and 1:0.924:5.585:3.076:2.080:3.076, respectively for *H*:*DBH*:*V*:*CBH*:*SF*:*BR*. Genetic gain for different traits following unrestricted index selection and the index heritability were given in Table 7 for the two economic weight solutions.

Table 6. Phenotypic (upper triangle) and genetic (lower triangle) correlation between characters of *E. tereticornis*.

	H	*DBH*	*V*	*CBH*	*SF*	*BR*
H		0.8579**	0.8982**	0.6019**	0.3680**	0.2091*
DBH	0.8610**		0.9742**	0.2596**	0.3880**	0.0303
V	0.9069**	0.9848**		0.3390**	0.3454**	0.0420
CBH	0.6803**	0.2892**	0.3854**		0.2314*	0.5159**
SF	0.4178**	0.4399**	0.3886**	0.2503*		0.5397**
BR	0.2371*	0.0346	0.0474	0.5590**	0.5397**	

Note: * and ** indicate significance at the 5% and 1% levels.

Table 7. Genetic gain of different characters following unrestricted index selection with two economic weight solutions.

Weight	Estimated genetic gain(ΔG)						Index
solution	*H*	*DBH*	*V*	*CBH*	*SF*	*BR*	Heritability
Solution 1	0.6941	0.6728	0.0068	0.3933	0.3121	0.2512	0.8797
Solution 2	0.6672	0.5403	0.0057	0.4813	0.5293	0.3124	0.8884

By considering genetic gain for individual character and the index heritability, unrestricted index selection with the second solution of economic weights was applied to the 100 *E. tereticornis* families. The standard deviation of the index was 2.8949, and the "aggregate" index gain was 2.8849. Therefore, index selection function was transformed into the following form:

$$I = 0.573H - 0.56DBH + 153.055V + 2.547CBH + 4.437SF + 4.763BR.$$

The above function was used to calculate index value for each family, and the mean index value (I_0) of all the families was 38.516. There were 26 poor families with bad stem form and index lower than I_0 minus one standard deviation of index. These poor families should not be included in any further breeding work. Two selection schemes were considered to select superior families: 1) selection of families with I larger than I_0 plus one standard deviation of index, 22 superior families were selected, and the selection proportion 22%, and 2) selection of families with I larger than I_0, 55 families were selected, and the selection proportion was 55%. Genetic gain (ΔG) for different traits under these two selection schemes was given in Table 8. Obviously, the first selection scheme resulted in higher genetic gain than the second selection scheme.

Table 8. Genetic gain of the selected superior families under two selection schemes.

Selection	Genetic gain (ΔG) (%)					
schemes	*H*	*DBH*	*V*	*CBH*	*SF*	*BR*
Scheme 1	7.3050	6.3839	18.0109	13.7743	8.2219	5.3556
Scheme 2	3.9109	3.4129	9.3608	7.1934	4.6495	2.6662

4 Discussion and Conclusions

Highly significant differences in growth and other characters were found among 100 families within 14 provenances of *E. tereticornis* at 7-year old, and the provenance component was larger than the family component in the trial. The provenances from QLD and PNG such as 13659, 13398 and 16547 etc. performed best with rapid growth and straight stems. Individual volume, height and DBH at 7-year old of the best provenance (13659) were 1.27, 1.32 and 2.14 times, respectively, of those of the poorest provenance (13308) from Wallaroo State For., NSW. There were 48 families exceeding the population mean in volume, and the individual volume of best family (No.17) was 0.06134 m^3, 3.6125 times of that of the poorest family (No.33), and 1.7446 times of the population mean. Both provenance and family selection could contribute to the improvement of genetic gain.

Individual heritability was evidently lower than family heritability for all charactered studied. While *BR* had the lowest family heritability (62.13%), *CBH* had the highest family heritability (85.75%). With high family heritability, backward or family selection is recommended to improve genetic gain. With multi-trait index as the selection criteria, 22 families were screened (22% selection proportion). Genetic gain was 7.31%, 6.38%, 18.01%, 13.77%, 8.22% and 5.36 % for *H*, *DBH*, *V*, *CBH*, *SF* and *BR*, respectively.

To enrich the genetic resources, we suggest further introduction or inclusion of families or provenances from Laura, QLD into the existing breeding population. With further selection of superior individuals from the best provenances and best families, clonal test, selection and propagation are needed to realize genetic gain in plantations. For the advanced breeding at present, we have planned a subline population including 22 superior families, each comprising 4 clones. Hybrid crossing (including hybrid with other species), progeny test, clonal test and selection are now also being carried out to cycle the breeding population. In addition, the provenance/family trial was thinned according to the test results and managed as a seedling seed orchard.

References

Boland, D.J., Brooker, M.I.H., Chippendale, G.M., and Turnbull, J.W. 1984. Forest trees of Australia. Nelson CSIRO, Melbourne, 687p.

Cotterill, P.P. and Jackson, N. 1985. On index selection I. methods of determining economic weight. Silave Genet. 34:56-63.

Jacobs, M.R . 1981. Eucalypts for planting, FAO, Rome, pp610-614.

Liang, K. and Zhou, W. 1995. Provenance/family selection trial of *Eucalyptus tereticornis*. For. Res. 8:252-257.

Mckenney, D.W., Davis, J.S., Turnbull, J.W. and Searle, S.D. 1991. The impact of Australian tree species research in China. Canberra:ACIAR Economic Assessment Series, 1991(12):6-7.

SAS Institute. 1988. SAS/STAT user's guide for personal computers. Release 6.03 edition. SAS Institute, Gary, NC, USA.
Xu, J., Wu, K., Wu, J. and Bai, J. 1993. Genetic variation in a provenance trial of *Eucalyptus tereticornis*. For. Res. 6:242-248.
Xu, J., Bai, J., Wu, K. and Fu, J. 1995. Early trials of *Eucalyptus tereticornis* families. For. Res. 8:500-505.
Ye, Z. 1991. Keystone of SPQG and using directory. Nanjing University Press, Nanjing, China.
Zhou, W. and Liang, K. 1994. Provenance trial of *Eucalyptus tereticornis*. For. Res. 7:481-487.

细叶桉种源-家系综合选择的研究

徐建民，陆钊华，李光友，白嘉雨

中国林业科学研究院热带林业研究所
510520 中国广州

对细叶桉 14 个种源的 100 个家系生长性状、形质性状进行种源间和家系间的差异分析和遗传分析，结果表明各性状的差异极显著，家系遗传力大于单株遗传力，除分枝性状外，生长性状与形质性状相关极显著。7 年生时，生长最好的 13659 种源的平均树高、胸径和单株材积分别是最差种源 13308 的 1.27、1.32 和 2.14 倍；最优家系为 No. 17 号家系，其单株材积达 0.0613 m^3，是最差家系 No. 33 号的 3.6 倍，是家系群体均值的 1.74 倍。运用指数选择法对家系进行多性状综合选择，初步选出 22 个优良家系，其树高、胸径、材积、枝下高、干形和分枝的遗传增益分别达到 7.31%、6.39%、18.01%、13.77%、8.22%和 5.36%。

ANALYSIS OF GENETIC EFFECTS FOR GROWTH TRAITS OF *EUCALYPTUS GLOBULUS* LABILL. IN A 6×6 DIALLEL DESIGN

DANQING LI [1], YONGPING LIU [2], DEXIAN ZENG [2],
KAIPENG ZHONG [3] AND LINYUAN FAN [2]

[1]*Yunnan Academy of Forestry, Kunming 650204, China*
[2]*Tree Seed and Seedling Station of Yunnan Province*
Kunming 650215, China
[3]*Yipingliang Forest Farm ,Lufeng 651214*
Yunnan, China

The genetic effects of general (GCA) and specific (SCA) combining ability, paternal (MAL) and maternal (FEM) parents, and specific reciprocal cross (SRE) were estimated for height, diameter at breast height (DBH), volume, crown-width, under branch height, crookedness and ratio in diameter of branch to stem in *Eucalyptus globulus*. Three wood growth characters, i.e. tree height, DBH, and volume, were focused, and their genetic effects were discussed in detail. GCA, SCA and SRE were most significant, but FEM and MAL were minor. Three parents with high GCA were good materials for establishing seed orchard. Eight combinations significantly contributed to the improvement of the wood growth characters. A breeding strategy and the corresponding methods were discussed.

1 Introduction

The natural range of distribution of *Eucalyptus globulus* Labill includes coastal areas of eastern Tasmania, coastal regions of southern Victoria and adjacent islands in Australia, approximately spanning from 38.5°S to 43.5°S. The climate is warm and wet with most rainfall in the winter season in the region (Boland et al. 1984). The species was early introduced to Western Europe including countries such as France, Britain and Italia. Currently, it is widely cultivated in Europe, America, Asia and Africa, especially in Spain, Portugal, Brazil, China, India and Ethiopia (Eldridge and Griffin 1990). *E. globulus* is a fast growing species with great adaptability. It has become an important fiber species, mainly for paper and pulp. By planting *E. globulus*, Spain switched from a pure pulp buyer to a pulp exporter within 10 years. Its history of introduction in China is more than 110 years, most successful in Yunnan where it is already a main species for afforestation. A seedling seed orchard was established 15 years ago with seeds collected from plus trees selected from the populations early planted overall in Yunnan (Li 1990; Li et al. 1993). To better understand its genetic effects and facilitate breeding, a 6×6 complete diallel mating was conducted in the seed orchard in 1993.

Complete diallel design enables estimation of genetic and environmental parameters of a population with satisfied statistic efficiency, precision and reliability. In China, it had been widely used in crop and tree breeding (Ma 1982; Gao 1986; Zhu 1997; Ye et al. 1991; Hu et al. 1998). For tree species, research has mainly been focused on Chinese fir and other conifers (Ye et al. 1991; Hu et al. 1998). The relevant work on eucalypt is relatively limited.

2 Materials and Methods

2.1 *Experimental Materials*

Six superior individuals were selected from the seedling seed orchard, which were growing vigorously, with clear and straight trunk and plenty of flower buds. With the selected individuals as parents, a complete 6 × 6 diallel mating was conducted to generate the offspring (Table 1). Seeds were separately collected, and seedlings were raised and out-planted in the field. Except G-89 × G-59 with little seed and without seedling, all full-sib families had enough seeds and seedlings for the field test.

Table1. Codes of parents and crosses of a 6 × 6 diallel mating for *Eucalyptus globulus.*

♂ / ♀	G-7	G-9	G-41	G-59	G-60'	G-89
G-7	1	7	13	19	25	31
G-9	2	8	14	20	26	32
G-41	3	9	15	21	27	33
G-59	4	10	16	22	28	34[a)]
G-60'	5	11	17	23	29	35
G-89	6	12	18	24	30	36

[a)] Only a little seed and no seedling for the field test.

2.2 *Field Experiment and Measurement*

In 1996, seedlings were out-planted in the trial at Yipinglan Forest Farm. The site with upland red soil developed from sandy rock is located at longitude of 101°55′E, latitude of 25°09′N and altitude of 1860m. Complete randomized block design was used, which had 8 replications with 5-tree row plot for each cross or a mixed-seed as the control. The planting spacing was 3 m × 2 m.

Measurement was done in January each year. The traits directly measured included height, DBH, crown width, under branch height (UBH), crookedness and diameter ratio of branch to stem (RBS). Volume was calculated using $V=\pi fHD^2/4$ (f=0.44, D=DBH and H=height). Crookedness of each tree was expressed as the ratio of arc height to the chord of the bentest crook (for calculation convenience, the results were multiplied by 100). Ratio of the diameter of the biggest branch to that of stem at the same height was also obtained for analysis.

2.3 *Statistical Analysis*

General liner model (GLM) of SAS System was used for variance analysis, which could reasonably handle the unbalanced data. The DIALL procedure of SPQG package developed by Nanjing Forestry University was used to estimate variance components and genetic effect values (Ye 1990). The linear model considered was:

$$y_{ijk} = \mu + A_i + A_j + D_{ij} + F_i + M_j + R_{ij} + B_k + e_{ijk}$$

where y_{ijk}=the observation of the cross generated from i^{th} maternal parent and the jth paternal parent in the kth block, μ= the mean of population, A_i (or A_j)= GCA, D_{ij}=SCA, F_i(or M_j)= general reciprocal cross effect (GRE) of the ith maternal (or jth paternal) parent (i.e. FEM and MAL), R_{ij}= SRE of the ith and j^{th} parents, B_k =block effect, and e_{ijk}= the residue.

3 Results and Analysis

E. globulus is mainly cultivated for fiber. The present study therefore focused on the relevant traits such as tree height, DBH and volume. Other traits were related to the wood quality and foliage production, and discussed when appropriate.

3.1 *Variance Analysis*

The means of the experimental stand at age of 3 year and 5 month were: height=10.8 m, DBH=10.2 cm, individual tree volume=0.0418 m^3, UBH=5.5 m, crown width=1.76 m, crookedness=3.18, and ratio of branch to stem=0.15. Results of the significant test for cross and block effects on different traits were presented in Table 2. Cross effect was significant on all traits at level of 0.01 or 0.05 except crookedness and ratio of branch to stem at level of 0.1. All traits differed among blocks at level of 0.05 or 0.01 level, which was mainly attributed to the varying test site.

Table 2. Results of variance analysis for different traits of *E. globules* (F-value).

Source	Height	DBH	Volume	UBH	Crown width	Crookedness	RBS
Cross	2.04**	1.54*	1.50*	4.29**	1.48*	1.39^+	1.38^+
Block	2.99**	2.71**	2.80**	8.33**	3.42**	2.36*	2.51*

Note: **, * and $^+$ indicate significance at level of 0.01, 0.05 and 0.10, respectively.

3.2 *Genetic Effect*

Griffing (1956) divided the mean value of a cross (combination) of a complete diallel mating into GCA, SCA, reciprocal effect (REC) and random error. Keuls and Garretsen (1977) and Keuls and Garretsen (1978) further divided REC into GRE (including MAL and FEM) and SRE, and developed the corresponding analytical model (Ye et al. 1991). Since least square method was applied, the model provided high precision and reliability for analyzing complete diallel crossing with missing data. By using least square method, the expected mean squares of different effects were expressed as follows:

GCA: $V_E+V_{SRE}+2.9447V_{MAL}+2.9447V_{FEM}+1.666V_{SCA}+11.726V_{GCA}$

SCA: $V_E+V_{SRE}+0.0\,2796V_{MAL}+0.0\,2796V_{FEM}+1.669V_{SCA}$

FEM: $V_E+0.9999997V_{SRE}+0.02857V_{MAL}+5.829V_{FEM}+0.874V_{SCA}+5.926V_{GCA}$

MAL: $V_E+0.9999997V_{SRE}+5.828V_{MAL}+0.02857V_{FEM}+0.874V_{SCA}+5.926V_{GCA}$

SRE: $V_E+V_{SRE}-0.04659V_{MAL}-0.04659V_{FEM}-0.02.107V_{SCA}$

Error: V_E

where V_E, V_{SRE}, V_{MAL}, V_{FEM}, V_{SCA} and V_{GCA} were the variance components of the error, SRE, MAL, FEL, SCA and GCA effects, respectively. Their corresponding freedoms of degree were 168, 9, 5, 5, 15 and 5.

As shown in Table 3, GCA affected all traits studied, accounting for most of variation in RBS, UBH and crookedness (63.1-75.3%%), medium in height, DBH and volume (20-32.3%), and smallest part in crown width (5.9%). SCA controlled most of variation in crown width (72%), medium in height, UBH, crookedness, and DBH (15.2-39.7%), but none in volume and RBS. While MAL had no effect on all traits, FEM only affected crookedness and UBH. Except in UBH and crookedness, SRE contributed to the variation in all traits, particularly in DBH (64.8%) and volume (67.8%). Narrow- (h_N^2) and broad-sense heritability (h_B^2) were calculated according to Wang (1989). Table 3 showed that h_N^2 was high (0.32-0.56) for height, UBH, crookedness and RBS, medium for DBH (0.12) and volume, and low for crow width. Broad-sense heritability (h_B^2) was much higher than (h_N^2) for all traits. The difference and other results indicated the potential of using different genetic variations.

Table 3. Variance components and their contributions (in parenthesis) of different genetic effects.

Effect	Height	DBH	Volume	UBH	Crown width	Crookedness	RBS
2GCA	0.1768	0.1642	0.000023	0.3157	0.000884	0.4490	0.000137
	(53.5%)	(20.0%)	(32.2%)	(69.0%)	(5.9%)	(63.1%)	(75.3%)
SCA	0.1302	0.1253	-	0.1377	0.010795	0.1172	-
	(39.7%)	(15.2%)	(0)	(30.1%)	(72.0%)	(16.5%)	-
FEM	-	-	-	0.0044	-	0.1458	-
	-	-	-	(1.0%)	-	(20.5%)	-
MAL	-	-	-	-	-	-	-
	-	-	-	-	-	-	-
SRE	0.0223	0.5323	0.000049	-	0.003318	-	0.000045
	(6.8%)	(64.8%)	(67.8%)	-	(22.1%)	0	(24.7%)
h_N^2	0.32	0.12	0.18	0.56	0.03	0.35	0.40
h_B^2	0.59	0.60	0.56	0.81	0.48	0.55	0.53

Note: negative variance (-) was set to zero in calculating other components' contributions.

3.3 *General Combining Ability*

Parents G-9 and G-89 had great contribution to the improvement on most of the traits studied (Table 4), which could improve both wood growth and trunk quality traits. G-7 exclusively significantly improved trunk quality and crown traits. Other three parents had either significant (0.01) or non-significant negative effect on wood growth traits and UBH. Obviously, they should not be included as parents in breeding to cumulate additive genes. They may still be used to explore non-additive effects, depending on genetic effects rather than GCA (see below).

Table 4. Estimate of GCA for 6 parents.

Parents	Height	DBH	Volume	UBH	Crown width	Crookedness	RBS
G-7	0.078	-0.127	-0.0011	0.375**	-0.103**	-0.407**	-0.0046**
G-9	0.391**	0.181*	0.0033**	0.166**	-0.014	0.904**	-0.0022
G-41	-0.138**	-0.272**	-0.0023**	-0.138**	-0.010	0.117	0.0154*
G-59	-0.196**	-0.039	-0.0016	-0.175**	0.044**	-0.126	0.0027
G-60'	-0.472**	-0.313**	-0.0039**	-0.676**	0.054**	0.101	-0.0091**
G-89	0.350**	0.619**	0.0059**	0.473**	0.035*	-0.655**	-0.0022

Note: ** and * and indicate significance at level of 0.01, and 0.05, respectively.

3.4 *Specific Combining Ability*

While depression with negative SCA was significant at level of 0.01 in all wood growth traits of self-crosses 2 x 2, 3 x 3, and 6 x 6, superiority with positive SCA was significant at level of 0.01 in all wood growth traits of 5 x 5, DBH and volume of 1 x 1 and height of 4 x 4 (Table 5). It should be pointed out that crosses with selfing superiority are not necessarily superior genotypes and phenotypes. For instance, cross 5 x 5 had positive SCA, significant at level of 0.01, in all wood growth traits, but their phenotype values were all below the population mean. The poor performance of 5 x 5 resulted from the parent's negative and significant GCA (Table 4), other genetic effects such as SCA and SRE as well their interaction with environment. Nevertheless, significant positive SCA with growth genotype and phenotype superior to the mean population was also found among some self-crosses (Table 5). For example, cross 1 x 1 not only had selfing superiority but also was a genotype with performance in height, DBH and volume better than the whole population studied. Therefore, it is likely to explore selfing in breeding.

SCA was significant at level of 0.01 at least in one, two or even three wood growth traits of crosses 2 x 6, 3 x 6, 4 x 6, 2 x 3, 2 x 4, 1 x 3 and 1 x 2 (Table 5). Among these superior crosses, at least a parent had significant favorable GCA (0.01) in all wood growth traits (e.g. parents 2 and 6) or in trunk trait (e.g. parent 1). Therefore, it is possible to breed superior crosses with parents of high GCA and having high SCA as well. However, GCA is not necessarily related to SCA. For instance, SCA was significantly negative (0.05 or 0.01) in all wood growth traits of

1 x 6 and 5 x 6, but very significantly positive (0.01 level) in all wood growth traits of 1 x 3.

Table 5. Estimates of SCA for different crosses.

Cross [a)]	Height	DBH	Volume	UBH	Crown width	Crooked -ness	RBS
1 x 1	-0.0012	0.3699**	0.0038**	-0.2994**	0.1246**	0.4854**	-0.0077**
2 x 2	-1.1975**	-1.6935**	-0.0132**	-0.2668**	-0.1079**	-0.0602	0.0007
3 x 3	-0.8371**	-2.4297**	-0.0189**	0.6022**	-0.4857**	-0.5669**	-0.0108**
4 x 4	0.4432**	0.1370	-0.0005	0.2168**	-0.0096	0.2475	0.0080**
5 x 5	0.8712**	0.4841**	0.0048**	0.9233**	-0.0684**	1.2687**	0.0017
6 x 6	-1.0442**	-1.6820	-0.0165**	-0.2412**	-0.1588**	0.2732**	-0.0188**
1 x 2	0.3314**	-0.0320	-0.0014	0.2575**	-0.0321	-0.1445	-0.0123**
1 x 3	0.3741**	0.6631**	0.0063**	0.1890**	0.0653**	-0.0742	0.0040*
1 x 4	-0.2714**	-0.3471**	-0.0051**	-0.3342**	-0.0353*	0.1872	0.0092*
1 x 5	-0.2500**	-0.2001*	0.0011	0.1337**	-0.0405*	-0.3758**	-0.0027
1 x 6	-0.1826**	-0.5006**	-0.0054**	0.0414	-0.0878**	-0.0323	0.0085**
2 x 3	0.0850	0.5860**	0.0028**	-0.4064**	0.0631**	1.4706**	0.0171**
2 x 4	0.0734	0.5673**	0.0053**	0.2769**	0.0648**	-0.9656**	-0.0008
2 x 5	-0.0136	-0.0613	-0.0007	-0.4069**	0.0700**	-0.3162**	-0.0047**
2 x 6	0.6947**	0.5870**	0.0065**	0.5337**	-0.0638**	0.0616	-0.0009
3 x 4	-0.0489	-0.1231	-0.0011	-0.2403**	-0.0160	-0.1538	-0.0075**
3 x 5	-0.1089	0.0852	-0.0013	-0.2363**	0.0548**	-0.5889**	-0.00002
3 x 6	0.5361**	1.1718**	0.0115**	0.0799	0.3126**	-0.0511	-0.0037*
4 x 5	-0.3990**	-0.5734**	-0.0042**	0.0295	-0.0334*	0.4528**	-0.0092**
4 x 6	0.4317**	0.7320**	0.0125**	0.1401	0.0390	0.2422	0.0046
5 x 6	-0.1264*	0.2188*	-0.0003	-0.4552**	0.0116	-0.4047**	0.0140**

Note: 1, 2, 3, 4, 5 and 6 represent G-7, G-9, G-41, G-59, G-60'and G-89, respectively; ** and * indicate significance at level of 0.01, and 0.05, respectively.

3.5 *Specific Reciprocal Cross Effect*

SCA does not reflect the paternal or maternal effect on offspring (i.e. SRE), which is usually necessarily considered in breeding. In the following, only crosses with significant SRE (* 0.05 and ** 0.01) on wood growth traits are listed.

Height: G-9 x G-41, -0.6484*
DBH: G-7 x G-9, -0.6475*; G-9 x G-41, 0.9720**; G-9 x G-59, -1.3720**
Volume: G-7 x G-9, -0.0064*; G-9 x G-41, 0.0091**; G-9 x G-59, -0.0128**

In general, SRE effect on DBH and volume was more significant than on height. The significant SRE of G-9 x G-41, G-9 x G-7 and G-59 x G-9 indicated high hybrid vigor among these crosses. In particular, cross G-59 x G-9 had most significant hybrid vigor, not only with the highest SRE, but also with DBH and volume 25.5% and 63.6%, respectively higher than the population means. For G-9 x G-41, volume was increased while crookedness was low. To simultaneously improve both wood growth and trunk form is usually the objective of breeding for commercial plantations.

4 Discussion and Conclusion

Significant GCA effects on all traits except crown width in the present study suggests that the improvement of *E. globulus* can be obtained by using selective breeding. Seed orchard is a necessary practice for *E. globules* because of the difficulties to asexually propagate the seedlings for plantations. To establish second-generation seed orchard in Yunnan, breeding materials should be selected from the breeding populations with high GCA to increase genetic gain. In addition, superior individuals selected from the existing basic populations can be infused to broaden the genetic base of the seed orchard. When selecting crosses in breeding, both SCA and GCA should be considered. At least one parent's GCA should be high and acceptable.

Eight crosses (2, 10, 12, 13, 14, 18, 32 and 33) are superior considering the phenotypic values of height, DBH and volume. They are 5.6-14.8% in height, 4.9-25.5% in DBH and 14.1-63.6% in volume higher than the whole population. Without regard to the risk associated with the narrow genetic base, what deserves to be tried includes single parent seed orchard with G-7, cross seed orchards with two parents G-9 x G-59 or G-41 x G-89, and seed orchard with G-89 and G-9, G-41, G-59 etc as the materials.

Volume is the final objective of managing plantations, which is a function of height and DBH but mainly decided by the later. The variance components of GCA in height and DBH account for only 20-30%, but the variance components of SRE account for 65-68%, about 2-3 times of the former. Asexual propagation of individuals selected from superior crosses such as those eight crosses mentioned above (e.g. G-59 x G-9) is an efficient method to utilize non-additive genetic variance. To maximize genetic gain in these two traits of *E. globulus* in Yunnan, therefore, clonal forestry that combines hybridization breeding and vegetative propagation is essential. For successfully developing *E. globulus* plantations in the region, asexual mass propagation is a direction of research at present as well as in the future.

Acknowledgements

This study was financially supported by Forestry Department of Yunnan Province, China. Thanks are also due to B. Zhang and Y. Gao of the Forestry Bureau of Chengjiang County, for their participation in crossing pollinations for the present study.

References

Boland, D.J., Brooker, M.I.H., Chippendale, G.M., Hall, N., Hyland, B.P.M., Johnston, R.D., Kleinig, D.A. and Turner, J.D. 1984. Forest trees of Australia, Fourth edition. Thomas Nelson Australia, pp452-453.

Eldridge, K.G. and Griffin, A.R. 1990. Genetic improvement of *Eucalyptus globulus* and *E. nitens* - a review of the world scene in blue gum breeding and its relevance to China. In: S. Qi (ed), Proc. of International *Eucalyptus* Symposium, Zhanjiang, China, 20-30 November 1990, pp43-50.

Gao, Z. 1986. Qantitative genetics. Sichuan University Press, Chengdu, pp310-367.

Garretsen, F. and Keuls, M. 1978, A general method for the analysis of genetic variation in complete and incomplete diallels and North Carolina II (NC II) designs. Part II. procedures and general formulas for the fixed model. Euphytica 27:49-68.

Griffing, B. 1956. Concept of general and specific combining ability in relation to diallel crossing systems. Aust. Biol. Sci. 9:463-493.

Hu, D., Ruan, Z., Wu, Q., Lin, Y., Ling, W., Wei, Z., Qin, J., Liang, R. and Zhuang, X. 1998. Analysis for combining ability of growth characters in Chinese Fir. Guangdong For. Sci. Tech. 14:7-13.

Keuls, M. and Garretsen, F. 1977, A general method for the analysis of genetic variation in complete and incomplete diallels and North Carolina II (NC II) designs. Part I. procedures and general formulas for the random model. Euphytica 26:537-551.

Li, D. 1990. A study on plus tree selection in *Eucalyptus globulus* Labill. and *E. maidenii* F. Muell. Sci. Silvae Sini. 26:167-174.

Li, D., Zhang, B., Gao, Y., Li, F. and Zhao, H. 1993. Genetic parameters and gains expected in *Eucalyptus globulus* Labill. and *E. maidenii* F. Muell. Yunnan For. Sci. Tech. 1:1-9.

Ma, Y. 1982. The fundamental quantitative genetics of plant breeding. Jiangshu Science and Technology Publishing House, Nanjing, pp391-410.

Wang, M. 1989. Forest genetics and breeding. China Forest Publishing House, Beijing, pp230-244.

Ye, Z., Shi, J., Weng, Y. and Yu, R. 1991. Genetic analysis for eleven-parent incomplete diallel mating of Chinese fir. For. Res. 4:380-385.

Ye, Z. 1990. Guide of SPQG package. Nanjing Forestry University, pp178-188 and pp303-308.

Zhu, J. 1997. Analysis methods for genetic models. China Agricultural Publishing House, Beijing, pp66-74.

蓝桉 6×6 全双列交配生长性状的遗传效应分析

李淡清[1]，刘永平[2]，曾德贤[2]，钟开鹏[3]，范林元[2]

[1]云南省林业科学院，中国云南昆明 650204
[2]云南省林木种苗站，中国云南昆明 650215
[3]云南省禄丰县一平浪林场，中国云南禄丰 651214

本研究以蓝桉 6×6 完全双列交配的子代测定林为试验材料，对林木树高、胸径、材积、枝下高、冠幅、弯曲度和枝径比 7 个性状的遗传效应进行分析。结果表明，这些性状主要受一般配合力（GCA）、特殊配合力（SCA）和特殊正反交效应（SRE）的控制，母本效应（FEM）对弯曲度有较小作用，而父本效应（MAL）对所有性状影响甚微。有 3 个亲本对改良数量性状或树干品质有显著的作用，有 8 个组合对提高数量性状遗传增益的贡献极显著或显著。蓝桉的数量性状既有自交衰退，也有自交优势。我们对蓝桉育种的策略和方法也提出了建议。

ANALYSIS OF GENETIC EFFECTS ON GROWTH TRAITS IN *EUCALYPTUS MAIDENII* F. MUELL.

DANQING LI [1], YONGPING LIU [2], HANGSHENG ZHENG [2], RENGANG ZHU [2] AND YONGXIANG HUANG [3]

[1]*Yunnan Academy of Forestry, Kunming 650204, China*
[2]*Tree Seed and Seedling Station of Yunnan Province Kunming 650215, China*
[3]*Yipingliang Forest Farm ,Lufeng 651214 Yunnan, China*

The genetic effects of general (GCA) and specific (SCA) combining ability, parental (MAL) and maternal (FEM) parents, and specific reciprocal cross (SRE) were analyzed for height, dbh, volume, crown-width, under-branch height (UBH), crookedness and ratio in diameter of branch to stem (RBS) in *Eucalyptus maidenii*. Three wood growth characters, i.e. height, dbh, volume were focused, and their genetic effects were discussed in detail. SRE accounted for most of the total variance in wood growth characters, followed by MAL. GCA accounted for 39.4% of the total variance in UBH, but contributed almost nothing in other traits although two parents tested had high value of GCA. While SCA affected crookedness and UBH, FEM explained much of the total variance in crown-width, UBH and RBS. Selfing incurred serious depression. Eight crosses performed better than the population mean in wood growth characters. The best strategy to improve *E. maidenii* is to breed superior crosses, select superior trees from the superior crosses and to asexually mass-propagate the superior individuals for production (plantations). With the materials available in the region, seed orchard is an inefficient method in terms of genetic gain.

1 Introduction

Eucalyptus maidenii F. Meull is native to the southeast of New South Wales and the east of Victoria in Australia, found in slopes with altitude ranging from 200 to 900 m. The mean annual rainfall is about 800-1200 mm in its natural habitats and evenly distributed over the year, and the annual frost days are 20-80. It can usually grow to about 30-40 m in height with recorded maximum height of 75 m and maximum dbh of 2.5 m (Boland et al. 1984). It is usually of good form with clear, straight trunk often to two-thirds of total height. *E. maidenii* F. Meull has been introduced to many places with best growth performance found in the tropics and subtropics such as the uplands in Tanzania, Rwanda and Burundi (Eldridge and Griffin 1990). *E. maidenii* F. Meull was first introduced to Yunnan about 50 years ago. Now it is widely planted in landscapes, reforestation, and industry plantations for pulp, paper and extracted oil. In comparison with *E. globulus, E. maidenii* showed wider adaptability to varying site conditions, and more tolerance to draught and barren soil. In terms of timber density, fiber length and extracted oils, *E. maidenii* also has higher economic value than *E. globulus* (Li 1990; Li et al. 1992; Li et al. 2001). This species has been introduced to the provinces of Sichuan, Guizhou and Guanxi in China. The total plantation area is about 10,000 ha, which makes China the biggest country of planting *E. maidenii* in the world.

A seedling seed orchard of *E. maidenii* was established 15 years ago with seeds from plus trees selected from the populations early introduced overall in Yunnan, and a complete diallel mating was conducted among 6 selected individuals in 1993. In China, complete diallel mating is a good design to estimate genetic and environmental parameters of population, which is common in crop and tree breeding in China (Ma 1982; Gao 1986; Zhu 1997; Ye et al. 1991; Hu et al. 1998). For tree species, research is mainly focused to Chinese fir (Ye et al. 1991; Cheng et al. 1994; Hu et al. 1998). In the other parts of world, most of the efforts have contributed to *Pinus*, *Picea*, *Populus*, *Psoudotsuga* etc (e.g. Yanchuk 1996).

2 Materials and Methods

2.1 *Experimental Materials*

Six individuals chosen as the mating parents from the seedling seed orchard had high growth yield and vigor, clear and straight pole, and plenty of flower buds. Controlled pollination was conducted among the parents according to complete diallel mating design. Seeds were collected, and seedlings were raised with the seeds and out-planted in the field trial. The mating design, and codes of parents and crosses were described in Table 1.

Table1. Codes of parents and crosses of a 6 × 6 diallel mating for *E. maidenii.*

♂	♀					
	M-26	M-32	M-33	M-60	M-JJ	M-78
M-26	1[b)]	7[b)]	13	18	24	28
M-32	2	8[b)]	14	19	25	[a)]
M-33	3[b)]	9	[a)]	20	23	[a)]
M-60	4[b)]	10	15	[a)]	26	29
M-JJ	5	11	16	21	[a)]	30
M-78	6	12	17	22	27	[a)]

Note: [a)] no seed; [b)] only a little seed but, no seedling for the field test.

2.2 *Field Experiment and Measurement*

In 1996, seedlings were out-planted in the trial at Yipinglan Forest Farm. The experimental site with upland red soil developed from sandy rock was located at longitude of 101°55′E, latitude of 25°09′N and altitude of 1860 m. The randomized block design for the trial had 7 blocks, each with five-tree row plot for each cross or mixed seed orchard seed-lot that was the control. The spacing was 3 m × 2 m. Due to insufficient seedlings for the field experiment, some crosses might be missing in one or few blocks.

Measurement was done in January each year. The traits directly measured included height, DBH, crown width, under branch height (UBH), crookedness and diameter ratio of branch to stem (RBS). Volume was calculated using $V=\pi fHD^2/4$ (f=0.44, D=DBH and H=height). Crookedness of each tree was expressed as the ratio of arc height to the chord of the bentest crook (for calculation convenience, the results were multiplied by 100). Ratio of the diameter of the biggest branch to that of the stem at the same height was also obtained for analysis.

2.3 *Statistical Analysis*

RBD procedure of SPQG Soft Package developed by Nanjing Forestry University (Ye 1990) was used for variance analysis of data with missing crosses. The DIALL procedure of SPQG Soft Package was used to estimate variance components and genetic effect values of the mating. The linear model considered was:

$$y_{ijk} = \mu + A_i + A_j + D_{ij} + F_i + M_j + R_{ij} + B_k + e_{ijk}$$

where y_{ijk}=the observation of the cross generated from the ith maternal parent and the jth paternal parent in the kth block, μ= the mean of population, A_i (or A_j)= GCA, D_{ij}=SCA, F_i(or M_j)= general reciprocal cross effect (GRE) of the ith maternal (or jth paternal) parent (i.e. FEM and MAL), R_{ij}= SRE of the ith and j^{th} parents, B_k =k^{th} block effect, and e_{ijk}= the residue.

3 Results and Analysis

3.1 *Variance Analysis*

The means of the experimental stand at age of 3 year and 5 month were: height=7.8 m, DBH=7.4 cm, individual tree volume=0.0138 m^3, UBH=2.2 m, crown width=1.76 m, crookedness=2.18, and ratio of branch to stem=0.177. Variance analysis showed that the differences of most traits studied among different crosses are significant at level of 0.01 or 0.05, with crown width and RBS as exceptions (Table 2). Difference among blocks was significant due to varying site conditions of the experiment site.

Table 2. Results of variance analysis for different traits of *E. maidenii* (F-value).

Source	Height	DBH	Volume	UBH	Crown width	Crookedness	RBS
Cross	1.569^*	1.573^*	1.771^*	1.923^{**}	0.988^{ns}	2.110^{**}	0.836^{ns}
Block	7.838^{**}	8.001^{**}	9.538^{**}	3.909^{**}	4.313^{**}	3.315^{**}	1.910^+

Note: **, * and $^+$ indicate significance level of 0.01, 0.05 and 0.1, respectively.

3.2 *Genetic Effect*

Considering that *E. maidenii* is mainly cultivated for fiber. We mainly focused on wood growth traits such as tree height, DBH and volume that are closely related to fiber production. In addition, UBH, crookedness and RBS affected stem quality, and crown width, UBH and RBS affected to oil production. They may be mentioned when necessary. Following the theories and models by Griffing (1956), Keuls and Garretsen (1977) and Garretson and Keuls (1978), DIALL program decomposed the cross mean into GCA, SCA, GRE (including FEM and MAL), and SRE. Least-squared method was applied in the program to obtain satisfactory precision and reliability of complete diallel mating analysis with missing data (Ye 1991).

Different genetic effects on traits studied were analyzed using the randomized model. Variance components and their contributions to the total genetic variance were shown in Table 3. It was showed in that three wood growth traits were mainly controlled by SRE, accounting for 73% of the total genetic variance. While significantly affecting UBH, accounting for 40% of the total genetic variance, GCA had trivial effect on tree height (0.6%) and other traits. SCA only contributed most of genetic variation to crookedness (64.8%) and part to UBH (15.3%). The contribution of MAL to all wood growth traits was all above 20% while the contribution of FEM to UBH, crown width and RBS ranged from 38.4% to 54%.

Table 3. Variance components and their percentage (in parenthesis) of different genetic effects.

Effect	Height	DBH	Volume	UBH	Crown width	Crookedness	RBS
2GCA	0.00865 (0.6)	-	-	0.0828 (39.4)	-	-	-
SCA	-	-	-	0.0322 (15.3)	-	0.12137 (64.8)	-
FEM	-	-	-	0.0949 (45.2)	0.01475 (54.0)	0.00568 (3.0)	0.000096 (38.4)
MAL	0.2977 (21.1)	0.5428 (26.6)	0.000018 (26.1)		0.00883 (32.3)	0.0008 (0.4)	0.000034 (13.6)
SRE	1.1029 (78.3)	1.4971 (73.4)	0.000051 (73.9)	-	0.00376 (13.8)	0.05937 (31.7)	0.0001197 (47.9)

Note: negative variance (-) was set to zero in calculating other components' contributions.

It can be concluded that little improvement for the traits studied except for UBH could be expected through selection breeding. On the other hand, hybridization breeding by properly mating among parents can bring about efficient improvement, especially for wood growth traits.

3.3 *General Combing Ability*

General combining ability (GCA) was estimated for the six parents (Table 4). The value of GCA was compared with 0, and undergone t-test for significance. As shown in the table, GCA values of height, DBH, volume and crown width for M-JJ were all positive and statistically significant. GCA value of the crookedness for this parent was negative but also statistically significant. In seedling seeds orchard, with M-JJ as one of the parents can improve both growth volume and stem quality. M-78 had significant (positive) effect on tree height but not on other traits. GCAs in wood growth traits of other four parents were either significantly negative or non-significantly positive. The values of GCA in other traits were also not very well. Therefore, these four parents are not recommended for use in establishing second-generation seed orchards.

Table 4. Estimate of GCA for 6 parents.

Parents	Height	DBH	Volume	UBH	Crown width	Crookednes	RBS
M-26	-0.331*	-0.304*	-0.0021*	-0.038	-0.402**	-0.072	-0.0076*
M-32	-0.062	0.042	0.00092	0.203	0.315**	-0.070	0.0016
M-33	-0.291*	0.096	0.00027	-0.069	-0.424**	0.232*	-0.0026
M-60	-0.285*	-0.399**	-0.0023*	-0.074	0.045	0.295**	0.0036
M-JJ	0.661**	0.624**	0.00423**	0.076	0.329**	-0.337**	0.0018
M-78	0.184*	-0.113	-0.0016	-0.113	-0.014	-0.024	0.0010

Note: ** and * and indicate significance at level of 0.01, and 0.05, respectively.

3.4 *Paternal and Maternal Effects*

As shown in Table 5, the best maternal effect (FEM) was found with M-60, which was significantly positive for three wood growth traits (0.01) and UBH (0.05). It was non-significantly negative for crookedness and RBS. Hybridization breeding using M-60 as parent can improve both wood growth and stem quality traits. On the other hand, this parent may lead to the wide crown of the offspring, which does not favor wood production. Obviously, the other five parents were not suitable as maternal parents to improve wood production in hybridization breeding (Table 5).

Table 5. Estimates of FEM and MAL effects for 6 parents.

Parents	Height	DBH	Volume	UBH	Crown width	Crookedness	RBS
				FEM			
M-26	-0.014	-0.052	-0.0005	-0.101	-0.360**	0.039	-0.0063
M-32	-0.187	-0.058	0.0011	0.215**	0.206**	-0.163*	0.0091**
M-33	0.150	0.145	0.0006	0.098	0.113*	0.017	-0.0017
M-60	0.641**	0.844**	0.0047**	0.126*	0.146**	-0.017	-0.0034
M-JJ	-0.387**	-0.736**	-0.0050**	-0.354**	0.011	-0.015	-0.0047
M-78	-0.539**	-0.537**	-0.0032**	-0.123	-0.363**	0.197**	0.0088*
				MAL			
M-26	0.010	0.039	0.0004	0.076	0.2670*	-0.030	0.0047
M-32	0.187	0.058	-0.0011	-0.242**	-0.206**	0.163*	-0.0091**
M-33	-0.374	-0.362	-0.0016	-0.245	-0.283*	-0.042	0.0042
M-60	-0.802**	-1.055**	-0.0059**	-0.157*	-0.183**	0.021	0.0042
M-JJ	0.309**	0.589**	0.0040**	0.283**	-0.009	0.012	0.0038
M-78	0.323**	0.322**	0.0019**	0.074	0.217**	-0.118*	-0.0053*

Note: ** and * and indicate significance at level of 0.01, and 0.05, respectively.

The paternal effect (MAL) of M-JJ was favorably significant in wood growth traits and UBH (0.01), but was not significant in crown width, crookedness and RBS. M-78 also had significant and favorable MAL effect in wood growth traits (0.01), crookedness (0.05) and RBS (0.05). However, M-78 had significant and unfavorable MAL in crown width (0.01). Superior crosses and further individual (clone) selection for timber production plantation can therefore be obtained through hybridization breeding using M-JJ and M-78 as male parents. Other four parents either had significant and unfavorable MAL or non-significant MAL in the main wood growth and quality traits, thus not suitable as male parents of the crossbreeding. If a specific trait is considered however the genes of certain parents may still be useful, e.g. use of M-32 as male parent to obtain narrow crown width varieties.

3.5 *Specific Combining Ability and Specific Reciprocal Cross Effect*

Specific combining ability (SCA) for the two reciprocal crosses of the two parents was the difference between their mean value and the mean of all crosses involved (Ma 1982). The influence of a male or female parent was not included in SCA but in specific reciprocal cross effect (SRE). In the following, analysis was only conducted for crosses with significant SCA (and SRE) (0.01 or 0.05) at least in one of wood growth traits (Table 6).

Table 6. Estimates of SCA and SRE for crosses with significant and positive SCA and SRE.

Cross	Height	DBH	Volume	UBH	Crown width	Crookedness	RBS
			SCA				
M26×M32	0.416*	0.296	0.0019	0.146	0.093	0.823**	0.0023
M32×M33	0.632**	0.836**	0.0049**	0.188**	0.235**	-0.026	-0.0023
M32×MJJ	0.298**	0.298**	0.0026**	0.340**	-0.110*	0.227**	0.0104**
M32×M78	0.018	-0.003	0.0034**	-0.467**	-0.134	-0.433**	-0.0178**
M33×M78	-0.250	0.251	0.0026**	0.178	-0.462**	-0.034	-0.0124**
M60×MJJ	0.780**	1.154**	0.0071**	0.193**	0.098*	-0.116*	-0.0012
M60×M78	0.465**	0.631**	0.0038**	0.186**	-0.045	0.635**	0.0044*
			SRE				
M32×M33	-0.629**	-0.656**	-0.0032*	0.030	-0.042	-0.302*	0.0006
M32×M60	1.136**	1.162**	0.0050**	0.275*	0.031	0.436**	-0.0085
M32×MJJ	-0.507**	-0.506*	-0.0018	-0.305*	0.011	-0.134	0.0079
M33×M60	-0.629**	-0.653**	-0.0032*	0.030	-0.042	-0.302*	0.0006
M60×M78	0.333	0.564*	0.0033*	0.150	-0.060	0.469**	-0.0220**
MJJ×M78	-0.542**	-0.833**	-0.0058**	-0.155	-0.030	-0.386**	-0.1050*

Note: ** and * and indicate significance at level of 0.01, and 0.05, respectively; the mating sequence of crosses for SRE was ♀×♂.

As shown in the table, SCA in volume was significant for all crosses listed except for M26×M32. SCAs in wood growth traits of M32×M33, M32×MJJ, M60×MJJ and M60×M78 were all significant at level of 0.01. Some crosses, e.g. M32×M33, simultaneously had valuable SCA in wood growth and quality traits (height, DBH, volume, UBH, crookedness, and RBS). The findings in this study suggested that SCA should be paid more attention than GCA in *E. maidenii* breeding.

There were 25 crosses in total, and 18 pairs of reciprocal crosses. SRE was the same but different direction (positive and negative) for a pair of reciprocal crosses. Crosses with significant SRE in at least one of the wood growth traits were M32×M60, M78×MJJ, M60×33, M60×M78, and M33×M32 and M32×MJJ (Table 6). It should be mentioned that the genetic effect of a cross is jointly decided by GCA, SCA, MAL, FEM, SRE etc. In breeding, one or two effects may be favorable while others are unfavorable. Therefore, the performance of a cross depends on the combination of all effects.

By comparing with the population mean, 8 crosses were selected (Table 7). Tree height, DBH and volume of the selected crosses surpassed the population mean by 3.1-26.2%, 7.7-35.6% and 45.4-160.0%, respectively. Among them, 3 crosses had MJJ and M78 as the male parent and two crosses had M60 as the female parent, which was consistent with results presented in Table 5 that FEM of M60 and MAL of MJJ and M78 were greatest.

Table 7. Phenotypic values of height, DBH and volume for 8 selected crosses.

Cross	Parents		Height	DBH	Volume
	Female	Male	(m)	(cm)	(m^3)
5	M26	MJJ	8.39	8.54	0.02268
9	M32	M33	8.27	8.53	0.02556
11	M32	MJJ	9.47	9.73	0.03475
17	M33	M78	8.07	8.43	0.02406
21	M60	MJJ	9.88	10.17	0.03586
22	M60	M78	8.92	8.04	0.02005
23	MJJ	M60	8.81	7.96	0.02328
27	MJJ	M78	9.10	8.21	0.02260
Mean			7.83	7.39	0.01379

4 Discussion and Conclusions

The history of *E. maidenii* introduction in Yunnan is more than fifty years. Early planting of the species was limited, mainly for the roadsides and settlement surroundings. The seed origins and collection were not clear. Seeds for the early planting in Yunnan might originate from a single or few trees. *E. maidenii* population for plus tree selection in 1985 was quite small and the offspring of a small number of the founders. Therefore, genetic basis of *E. maidenii* for selection and breeding is quite limited. About 1% of the population was selected as plus trees. In the seedling seed orchards with seeds collected from plus trees, 95% of the individuals were thinned according to the family and individual performance in height, DBH and volume (Li 1992). Narrow genetic basis and intensive selection should explain low GCA found with parents studied (Ma 1982). With the current resources in Yunnan, *E. maidenii* genetic improvement should mainly use the non-additive effect. Our suggestion therefore is that the superior individuals selected from the superior crosses are asexually mass-propagated for clonal plantations.

In recent years, remarkable progress has been made in the clonal breeding of *E. maidenii* in Yunnan. The volume of the best cross in the present study was 2.6 times as much as that of the whole population (25 crosses); the 8 selected crosses (selection proportion is 32%) produced the volume 45% higher than that of the whole population. This suggests that by involving more parents, useful crosses can be obtained in relatively short period of time through double-pair mating (including orthogonal and reciprocal crossing).

The way for *E. maidenii* improvement through reproduction propagation is to establish hybrid seed orchard using parents of crosses with high SCA values. However, the genetic basis of the seed orchard or resultant plantations may further be narrowed. For efficient conventional breeding or seed orchard, it is necessary to increase the genetic diversity and reconstruct the breeding population with new genes introduced from aboard or other regions, or materials generated using other methods, which was suggested 10 years ago (Li 1992).

In terms of height, DBH and volume, MJJ has the highest GCA, followed by M78. Among the 8 selected crosses (Table 7), 7 crosses' parents (female or male) were MJJ or M78. In other words, parents with both high GCA and high SCA can be found in *E. maidenii* population. However, some poor parents (both male and female) with low GCA may still generate well-performing crosses such as cross No. 9 (Table 7). On the other hand, some parents (both male and female) with high GCA may generate poor crosses such as cross No. 30 (M78×MJJ). This indicates that GCA and SCA of *E. maidenii* may or may not be related to each other.

Among all the crosses studied, M60×MJJ performed best in terms of height, DBH and volume. In comparison to the reciprocal cross (MJJ×M60), the height, DBH and volume of M60×MJJ were 12.1%, 27.8% and 64.0% higher, respectively, indicating that besides SCA, SRE (choice of male and female parents) should carefully be considered in mating. To obtain superior crosses of *E. maidenii*, analysis of MAL, FEM and SRE together with SCA is necessary.

E. maidenii has low GCA and displays severe selfing depression in comparison with *E. globulus*. As a result, four of the 6 selfing crosses had no seed. The other two crosses had little seeds, but generated in no seedlings. That was why there was no selfing cross of *E. maidenii* included in the trial. In contrast, seeds were collected from all 6 selfing crosses of *E. globulus*, and all resulted in seedlings; selfing superiority were even found with some of the *E. globulus* crosses (Li 2002). In planning the respective breeding strategies, the performance of selfing crosses may be a factor to consider.

Acknowledgements

This study was financially supported by the Forestry Department of Yunnan Province, China. Thanks are also due to B. Zhang and Y. Gao of the Forestry Bureau of Chengjiang County, for their participation in crossing pollination for the present study, and to L. Fan for data analysis.

References

Boland, D.J., Brooker, M.I.H., Chippendale, G.M., Hall, N., Hyland, B.P.M., Johnston, R.D., Kleinig, D.A. and Turner, J.D.1984. Forest trees of Australia. Fourth edition. Thomas Nelson Australia, pp452-453.

Cheng, Z., Xu, Q., He, G. and Liu, S. 1994. Growth analysis of young stand with complete diallel mating design and selection of superior propagation materials of Chinese fir. In: "Seed orchard techniques for high genetic quality and ample production of seeds", X. Shen (ed), China Forestry Publishing House, Beijing, pp180-185.

Eldridge, K.G. and Griffin, A.R. 1990. Genetic improvement of *Eucalyptus globulus* and *E. nitens* - a review of the world scene in blue gum breeding and its relevance to China. In: S. Qi (ed), Proc. of International Eucalyptus Symposium, Zhanjiang, China, 20-30 November 1990, pp43-50.

Gao, Z. 1986. Quantitative genetics. Sichuan University Press, Chengdu, pp310-367.

Garretsen, F. and Keuls, M. 1978. A general method for the analysis of genetic variation in complete and incomplete diallels and North Carolina II (NC II) designs. Part II. procedures and general formulas for the fixed model. Euphytica 27:49-68.

Griffing, B. 1956. Concept of general and specific combining ability in relation to diallel crossing systems. Aust. Biol. Sci. 9:463-493.

Hu, D., Ruan, Z., Wu, Q., Lin, Y., Ling, W., Wei, Z., Qin, J., Liang, R. and Zhuang, X. 1998. Analysis for combining ability of growth characters in Chinese Fir. Guangdong For. Sci. Tech. 14:7-13.

Keuls, M. and Garretsen, F. 1977. A general method for the analysis of genetic variation in complete and incomplete diallels and North Carolina II (NC II) designs. Part I. procedures and general formulas for the random model. Euphytica 26:537-551.

Li, D. 1990. A study on plus tree selection in *Eucalyptus globulus* Labill. and *E. maidenii* F. Muell. Sci. Silvae Sini. 26:167-174.

Li, D., Zhang, B., Gao, Y., Li, F. and Zhao, H. 1993. Genetic parameters and gains expected in *Eucalyptus globulus* Labill. and *E. maidenii* F. Muell. Yunnan For. Sci. Tech. 1:1-9.

Li, D., Liu,J., Zhang, B. and Gao, Y. 2001. Genetic parameters and gains expected of important wood properties in *Eucalyptus globulus* Labill. and *E. maidenii* F. Muell. Sichuan For. Sci. and Tech. 22:4-9.

Li, D., Liu, Y., Zeng, D., Zhong, K. and Fan, L. 2002. Analysis of genetic effects for growth traits of *Eucalyptus globulus* Labill. in a 6×6 diallel design. Acta Genet. Sini. 29:385-840.

Ma, Y. 1982. The fundamental quantitative genetics of plant breeding. Jiangshu Science and Technology Publishing House, Nanjing, pp391-410.

Ye, Z. 1990. Guide of SPQG package. Nanjing Forestry University.1990, pp178-188 and pp303-308.

Ye, Z., Shi, J., Weng, Y. and Yu, R. 1991. Genetic analysis for eleven-parent incomplete diallel mating of Chinese fir. For. Res. 4:380-385.

Yanchuk, A.D. 1996. General and specific combining ability from disconnected partial diallels of coastal Douglas-fir. Silvae Genet. 45:37-45.

Zhu, J. 1997. Analysis methods for genetic models. China Agricultural Publishing House, Beijing, pp66-74.

直干桉生长性状的遗传效应分析

李淡清[1]，刘永平[2]，郑行生[2]，朱仁刚[2]，黄永祥[3]

[1]云南省林业科学院，中国云南昆明 650204

[2]云南省林木种苗站，中国云南昆明 650215

[3]云南省禄丰县一平浪林场，中国云南禄丰 651214

对直干桉的树高、胸径、材积、冠幅、枝下高、弯曲度和枝径比等 7 个性状的一般配合力（GCA）、特殊配合力（SCA）、母本效应（FEM）、父本效应（MAL）和特殊正反交效应（GRE）进行了分析，重点讨论了树高、胸径和材积 3 个性状的遗传效应。发现这些性状主要受 SRE 和 MAL 的控制，GCA 仅对树高和枝下高有较小的影响。有两个亲本数量性状的 GCA 值高，可用于建立种子园。直干桉自交衰退非常严重，有 8 个组合的数量性状显著高于平均值。对直干桉进行遗传改良的最好途径是：通过杂交育种，从中再选出优异个体进行无性繁殖，培育出优良无性系后应用于生产。若要以建立种子园的传统方式来改良直干桉，必需引进或诱变新的育种材料才有效果，如仅从研究的抽样群体中来进行选择育种，则效果不佳。

IV. Nutrition and Site Management

MICRONUTRIENT DISORDERS IN EUCALYPT PLANTATIONS: CAUSES, SYMPTOMS, IDENTIFICATION, IMPACT AND MANAGEMENT

BERNIE DELL [1], DAPING XU [2], CATHERINE ROGERS [1]
AND LONGBIN HUANG [1]

[1] *Plant Nutrition Laboratory, Murdoch University*
Perth, WA 6150, Australia
[2] *Research Institute of Tropical Forestry, Chinese Academy of Forestry*
Guangzhou 510520, China

Micronutrient disorders, especially B, Cu, Fe, Mn and Zn, have been recorded for eucalypts in nearly all the geographical regions where commercial plantations have been established. Whilst micronutrient disorders are often induced by the application of fertilisers containing only macronutrients, instances of primary B deficiency in China and Cu deficiency in Australia have been recently documented. Increasing records of micronutrient disorders in plantation eucalypts suggests that the capacity of micronutrients to limit productivity has not been adequately recognised in the past. Dramatic growth responses can occur with corrective fertilisation particularly where a micronutrient deficiency has been debilitating for growth. However, where deficiencies of elements have resulted in severe crown dieback and loss of form, it is very difficult to restart growth with good form. This paper describes the visual impact of micronutrient disorders on eucalypts in plantations, their effect on wood production and wood quality, and explores strategies for micronutrient management.

1 Introduction

Although micronutrient disorders have been reported for eucalypts in many of the countries where commercial plantations have been established, there is not a wide understanding of micronutrient deficiencies by plantation managers. Further, the reporting of micronutrient deficiencies in eucalypt plantations (Dell and Bywaters 1989; Dell 1994; Dell and Malajczuk 1994; Turnbull et al. 1994; Dell 1996) suggests that the capacity of micronutrients to limit productivity has not been adequately recognized in the past (Xu and Dell 1998). Dell et al. (2001) suggest that this has led in the past to operational fertilizer practices that exclude micronutrients and hence many fertilizer operations may not be optimal for productivity. The extra growth stimulated by macronutrient fertilization often induces deficiencies in micronutrients beyond the capacity of residual soil supply (Dell 1996), and this is typically the situation in Western Australia and parts of south China.

The effective management of eucalypt micronutrition requires an appreciation of how silvicultural and harvesting practices impact on the nutrient cycle in a plantation. Decrease in nutrient capital can result from common practices in south China such as poor site preparation leading to accelerated loss of soil, or from the excessive removal of organic matter including harvest residue. The selection of sites with soils that are naturally low in nutrients for plantations has consequences for fertilizer prescriptions. Knowledge of tree nutrient requirements, soil science and nutrient loss through harvesting are prerequisites to good prescriptions. The purpose of this paper is to address some of these topics with regard to micronutrients. Comparable information on macronutrients is readily available for most eucalypt-producing regions and are therefore not covered here.

2 Factors Affecting the Incidence of Micronutrient Disorders

2.1 *Soil Factors*

Nutrient retention and supply potentials of a soil are very much determined by their chemical and physical characteristics and these affect the frequency of nutrient disorders and fertiliser effectiveness. The detailed soil processes and mechanisms are beyond the scope of this paper. Micronutrient deficiencies in eucalypt plantations are more common in sandy soils (Table 1) which are prone to leaching (Dell and Robinson 1993; Dell and Malajczuk 1994; Dell et al. 2001). Boron deficiency in eucalypts in China and the Philippines is associated with soils which contain low available B. Copper deficiency is commonly observed in coarse-textured sandy soils formed on igneous or non-calcareous sedimentary materials (Gartrell 1981). Manganese deficiency is generally associated with impoverished soils where the parent material is either inherently low in Mn (arenosols) or from which the Mn has

Table 1. The main soil types and countries where micronutrient deficiencies have been reported in eucalypt plantations.

Element	Soil type	Where reported	Selected references
B	Sandy soils, especially those derived from granite and sandstones; quaternary deposits; peaty soils; some soils of high iron content; some serpentine soils	Australia, Brazil, Chile, China, Indonesia, New Zealand, Philippines, Portugal, Zambia	Dell and Malajczuk 1994; Dell and Xu 1995
Cu	Sands; sandy gravels; calcareous soils; peaty soils; soils rich in soluble nitrogen	Australia, Chile, Indonesia, South Africa	Dell and Bywaters 1989; Turnbull et al. 1994
Fe	Sands; peaty soils; soils over limestone; alkaline soils; some serpentine soils	Australia, China, New Zealand, Philippines	Stewart et al. 1981; Will 1985
Mn	Sands; sandy gravels; calcareous soils	Australia	Dell et al. 2001
Zn	Acid and calcareous sands; lateritic soils; black earths; some volcanic soils	Australia, Brazil, China, Indonesia, Philippines, South Africa	Dell and Xu 1995; Dell and Xu 1998; Dell et al. 2001

been depleted through weathering and leaching (solonetz), or with soils of high pH (Reuter et al. 1988).

Soil pH not only affects the growth of roots, but also the chemical forms and therefore bioavailability of micronutrients to plants. As a result, soil pH is one of the most significant causes of changes in the availability of micronutrients to plant roots. For example, extensive Cu deficiency often occurs in soils formed from calcareous dune sands or limestone from littoral zones of varying ages which are distributed in the south east of south Australia and the coastal plain of Western Australia (Gartrell 1981). Copper and Zn deficiencies are common in peaty soils (Gartrell 1981; Takkar and Walker 1993). Copper forms stable complexes with carboxyl and phenolic groups, but Zn-organic matter bonds are not so strong. Iron also forms complexes with soil organic matter, but its availability to plants is more affected by soil pH and redox conditions. Both native soil B and available B are positively correlated with organic matter content in soils (Goldberg 1997). In soils with long fertiliser history, increasing organic matter contents in general, is beneficial to the improvement of soil nutrient supplying capacity. However, in sandy soils, the decomposition of soil organic matter may eventually lead to non-wetting conditions, leading to water deficit in plants and surface erosion due to surface water movement.

2.2 *Fertilizer History*

Micronutrient disorders can be induced by heavy fertilization with macronutrients which promote rapid growth of eucalypts that exceeds the supply characteristics of the plantation soil. This has been observed by us in the case of B and Zn in China, and Cu, Mn, Zn and B in Australia. In southern Australia, land that previously was used for agricultural pursuits of cropping and animal production on annual pasture is being turned over to short-rotation (10y) eucalypt plantations. Application of superphosphate over many decades, and the use of N-fixing forage plants, has led to the build up of a bank of P and N in these soils that are of low natural fertility. Even though some micronutrients have been applied in the past, the amounts, though adequate for annual crops, appear to be inadequate for fast growing eucalypts.

2.3 *Herbicide-Induced Nutrient Deficiencies*

Inappropriate use of some herbicides can lead to micronutrient deficiencies in eucalypts. The sulfonylurea family need further study as recent field observations in Australia indicate that operational prescriptions may have to be modified to prevent post-transplant toxicity on some soils (Dell and Huang, unpublished data). Severely stunted growth of newly planted *E. globulus* seedlings was reported by growers in Victoria (Australia) in soils sprayed with a herbicide from this family. In southern Australia, the incidence of little leaf in young *E. globulus* (Figure 1) has increased recently and this may be a consequence of slow break down of herbicide in some soils or inappropriate herbicide rates. It is likely that some residual herbicides can restrict root performance of eucalypts as earlier research on chlorsulfuron showed that the herbicide decreased root length and total root volume, resulting in a decrease in the net uptake of Zn and Cu in wheat plants (Dong et al. 1995).

Figure 1. Micronutrient disorders in *Eucalyptus globulus* plantations in Western Australia. A. Weeping habit on low Cu soil; B. Crown dieback caused by soil low in either Cu or B; C. Zn deficiency induced by the herbicide Oust.

2.4 *Genotype*

There has been little attempt to assess differences in genotypes of plantation eucalypts for their capacity to do well on soils where micronutrient disorders are problematic. Indeed, most eucalypt screening trials in the past have been undertaken without regard for either the fertilizer inputs that are possible in well-managed plantations or have failed to recognise the importance of micronutrients by the application of so-called complete fertilizers that are often inadequate in their contents of micronutrients. Yet, there are a few field trials (B, Cu) where differences in growth appear to be related either to improved micronutrient acquisition or more efficient micronutrient utilisation (Figure 2). This suggests possibilities for selection and breeding in the future.

Figure 2. A. Example of the genetic differences in tolerance of eucalypts to low B soil. Left is ABL 12W5 and right is Leizhou No. 133; B. The response of B-deficient eucalypts to topdressing with B can be dramatic as shown by the two rows of trees on the right in the background.

In general, eucalypts have evolved on acid soils and perform poorly on calcareous soils due to Fe, Mn and other nutrient disorders. A small number of taxa, including one timber species that is slow-growing (*E. gomphocephala*) naturally occurs on shallow soils over limestone. Clearly, these species have overcome problems of nutrient acquisition inherent in this environment. Whilst slow growth is no doubt a factor that enables greater nutrient efficiency, the presence of beneficial fungi and bacteria in the root environment is important for the acquisition of some micronutrients at high pH.

3 Symptoms of Micronutrient Disorders

Symptoms of nutrient disorders in plantation eucalypts have been described in detail for the main plantation species (Dell et al. 2001) and are summarised in Table 2. The pattern of symptom expression and the nature of the symptom reflect the capacity of the nutrient to be redistributed in the tree and its function.

4 Identification

Where symptoms occur in trees in the field, these can be used to make a preliminary diagnosis. This can then be confirmed by foliar analysis and corrective fertilizer treatment. Soil analysis should also be undertaken as sometimes the underlying cause may not be apparent from foliar data. An example is the incidence of Fe deficiency symptoms in trees growing on serpentine soils high in Ni and other heavy metals in the Philippines (Malajczuk and Dell 1995).

Table 2. Typical symptoms of micronutrient deficiencies in eucalypts in plantations.

Element	Symptom
B	Leaves: Young leaves accumulate anthocyanins and turn purple or lose green pigmentation from the tips and margins, margins of leaves roll longitudinally, leaves fail to expand properly resulting in windows and uneven margins, veins become raised and corky, leaves become brittle. Tree form: Branches weep or become prostrate due to reduced lignification of stem wood, shoots die back, multi-leaders form, excess kino is produced and stems bleed. Trees may die.
Cu	Leaves: Young leaves become twisted, cupped or recurved and the margins may be irregular in out line. Leaves may be dark green or become chlorotic unevenly from the margins. Tree form: Symptoms vary depending on severity and environment. Branches may become pendulous due to impaired lignification of the stem wood especially where soil N is high. Alternatively, branches with few leaves develop and the canopy is visibly thin near the top of the crown. Shoots die back, axillary buds proliferate then shed and dwarf multi-stemmed trees form without a main leader. Branches become prone to infection by canker and other fungi.
Fe	Leaves: Interveinal areas of expanding leaves become pale green then yellow. Chlorosis spreads from young to old leaves. Leaves may become fully bleached in severe cases. Sometimes, symptoms are only expressed on young leaves low in the canopy. Tree form: Apart from reducing growth when severe, tree form is not affected.
Mn	Leaves: Margins of juvenile and adult leaves become pale green, chlorosis extends between the lateral veins towards the midrib. Tree form: When severe, crown thinning occurs, branches die back and some branches may weep due to reduced lignification of the wood.
Zn	Leaves: Young leaves fail to develop to their full size, a condition known as little leaf. Leaves become crowded and pockets of dead tissue form. Tree form: Because the internodes fail to extend, the branches are crowded and trees have a dwarf habit.

4.1 *Field Identification*

The pattern and type of symptom development in the field are useful in recognising micronutrient problems (Dell 1996; Dell et al. 2001). The following tree parts should be examined:

- Canopy - Look for multiple leaders (B, Cu, sometimes Ca), branches dying back (B, Cu, Mn), canopy thinning at the apex (Cu, Mn), and branches without leaves (Cu, Mn).
- Trunk and branches - Look for poor form such as wobbles (Cu), pendulous branches (B, Cu), enlarged nodes (Cu), stem bleeding (B).
- Leaves - Look for change in young leaf colour from green to yellow (Fe, Mn, S), misshapen leaves (B, Cu), small leaves (Zn), leaves with dead areas (K, Zn).

A key is provided (Table 3). Note that symptoms similar to some of the above can be caused by non-nutrition factors such as frost damage, pollutants, fungal diseases, herbicide damage, etc. (Dell 1996). This is one reason why foliar analysis is required to confirm a suspected disorder.

Table 3. Key to micronutrient deficiency symptoms in plantation eucalypts.

A1. Leaf symptoms appearing first on mature leaves	(N, P, K, Mg)
A2. Leaf symptoms appearing first on young leaves	
B1. Trees with tip dieback, weeping branches or trees prostrate	
C1. Leaves malformed or undulate, trunk often twisted, sub-terminal branches sometimes nearly leafless	**Cu**
C2. Leaves malformed, veins often raised and corky, sometimes windows in leaves, leaves often crisp	**B**
C3. Leaves drying before fully expanded	(Ca)
B2. Trees without tip dieback, erect though sometimes stunted, sometimes sub-terminal branches with few leaves	(Mn)
C1. Leaves normal in size	
D1. Leaves uniformally pale green to yellow	(S)
D2. Leaves yellow with green veins	**Fe**
D3. Leaves yellowing from the margins, main veins broadly flanked green	**Mn**
C2. Leaves small and crowded	**Zn**

4.2 *Laboratory Analysis*

- Soil analysis - Apart for B, soil analysis, that relates tree growth to some measure of the assumed available micronutrient pool in soil, has so far had little application in eucalypt plantation forestry. This is because of limitations including the following: trees may access micronutrients in a soil horizon that is not sampled, there is often a poor correlation between micronutrient concentrations in the extractant solution and tree growth, and there are no reliable data that plantation managers can use in site selection. Furthermore, soil tests are highly specific and plantation compartments sometimes extend across extremes in soil types from sand to clays. Of the micronutrients, perhaps the only one where soil analysis preplanting may predict tree performance is for B. Hot-water extractable B is well correlated with plant growth and foliar B concentrations.

- Foliar analysis - Plant analysis is the preferred method of diagnosing micronutrient deficiencies, as concentrations of nutrients in plant tissues are determined by the combined affects of all the processes affecting the supply of nutrients in soil and uptake by roots. The establishment of foliar concentration ranges in healthy and deficient trees for three commercial species (*E. globulus, E. grandis* and *E. urophylla*) for a specified sampling position (youngest fully expanded leaf) by Dell et al. (2001) greatly facilitates the diagnosis and prognosis of micronutrient deficiencies in young plantations (Table 4). Where tree growth or form is affected, healthy and unaffected trees can be sampled and the nutrient concentration ranges in the two groups compared. More generally, monitoring programs should be established and the micronutrient (and macronutrient) concentrations in sampling plots measured over time to follow the progress of a plantation. In this way, any decline in micronutrient levels can be addressed before it impacts on tree yield or wood quality. Ideally, growth plots can be used for this purpose. In southwestern Australia, foliar analysis at age 1, 2 and again towards mid-rotation is proving useful in the management of Cu deficiency. There is, however, a need to better define the concentration ranges in older trees to facilitate fertiliser management.
- Other tests - As mentioned earlier, the synthesis of lignin is impaired by low tissue concentrations of B, Cu and sometimes Mn. In B-deficient *E. globulus* in Yunnan, the debarked stem changes in colour from yellowish green to dark brown (Dell and Huang 1997), probably due to the oxidation of excess phenols whose precursors would normally be utilised for lignin synthesis in healthy trees. Dell (1994) used a lignin stain to distinguish between Cu-deficient and Cu-adequate *E. maculata*. Whilst this approach can not distinguish between woods formed in trees that are Cu or B or Mn deficient, it is a useful technique nevertheless and can be applied in the field. Whilst the activity of certain metallo-enzymes is reduced in affected trees (Dell 1996), it is presently too difficult to assess in the field. However, in our laboratory, the activity of phenolase was used to define functional Cu requirement in three eucalypts (Gherardi et al. 1999).

Table 4. Microutrient concentration ($\mu g{\cdot}g^{-1}$dwt) ranges observed in the youngest fully expanded leaves of one- to two-year old *Eucalyptus* trees in plantations (adapted from Dell et al. 2001).

	E. globulus			*E. grandis*			*E. urophylla*		
	Def	Marg	Adeq	Def	Marg	Adeq	Def	Marg	Adeq
Fe	8-10 *	10-25	25-700	10-14	14-25	25-130	?<20	20-25	25-100
Zn	8-11	11-15	15-50	5-9	9-15	15-50	9-12	12-16	16-47
Mn	2-19	19-30	30-2000	?<15	15-60	60-2300	?<15	15-130	130-4000
Cu	0.5-2.0	2.0-3.0	3.0-24	0.4-1.5	1.5-2.0	2.0-11	0.8-1.5	1.5-3.0	3.0-19
B	4-10	10-14	14-38	5-8	8-15	15-27	4-12	12-16	16-69

* < 20 on calcareous soils.

5 Impact on Plantation Yield

In parts of Yunnan (Dell and Malajczuk 1994) and Mindanao (Figure 2), B is the nutrient that is most limiting productivity. Severely B-deficient trees fail to form a trunk and remain as dwarf or prostrate shrubs. On sandy soils with a peaty surface layer in southwestern Australia, eucalypts are stunted and are non-commercial. The full impact of marginal micronutrient deficiencies has, so far, not been measured.

6 Management of Micronutrient Deficiencies

6.1 *Applying Fertilizer*

The key to managing the micronutrient status of eucalypts in plantations is to have reliable information on the micronutrient status of the trees and the soil. In our experience, correction of micronutrient deficiencies is not difficult once the problem has been identified and its cause is understood. Generally, as micronutrient deficiencies result from tree growth outstripping the supply capacities of plantation soil, application of fertilizer is recommended. Dramatic growth recovery has been achieved by topdressing early with B (Figure 2) in China (Yunnan) and in the Philippines where soils are acutely low in B. Likewise, addition of Cu has improved tree form and growth in southwestern Australia. However, where micronutrient disorders have resulted in severe crown dieback or poor stem form, it is very difficult and often impossible to restart growth of good quality. In this situation, replanting is preferred.

A range of micronutrient salts are available and are effective. Whilst rates will vary with soil type, climate, tree age, application method, and other factors, general guidelines arise from field trials conducted in Australia and China (Table 5). The oxide forms of Mn and Fe, which are not water soluble, may be effective in

Table 5. Fertilizer compounds and rates for the correction of micronutrient deficiencies in eucalypt plantations.

Element	Rate (kg/ha)	Suggested compound to apply to the soil
B	0.5-1.5	Colemanite, Borate, Borax, Boric acid, Ulexite
Cu	1-4	Copper chloride, Copper sulfate, Cupric oxide
Fe	?	Ferrous chloride, Ferrous sulfate
Mn	1-10	Manganous chloride, Manganous nitrate, Manganous sulfate
Zn	0.5-2	Zinc oxide, Zinc sulfate

replenishing soil reserves but are only slowly available to roots following reduction under acid conditions. Some micronutrients can be useful contaminants of macronutrient fertilizers produced from some natural deposits such as rock phosphate. The B level in fertilizer produced from rock phosphate varies with source and can be as high as 50 mg B/kg in single super to over 100 mg B/kg in triple super. In addition, there is a plethora of commercial formulations on the market, including chelated forms which are effective but costly. Once the need for a micronutrient is understood, the appropriate element can be incorporated into compound fertiliser at the required rate (e.g. super-Cu-Zn, N-P-K-Mn). Of the micronutrients, B is the most difficult to blend into a compound fertilizer.

Deficiencies of B and Cu often result in impaired stem form within a few months of outplanting or they can appear several years later. We recommend that micronutrients should be applied at plantation establishment. They can be incorporated into rip lines, broadcast, banded or spot applied. Depending on leaf analysis, a follow-up application may be required during the rotation. Aerial application of Cu has been effective in preventing loss of stem form and in improving growth in Australia. However, the foliage can often only absorb enough of the element to support growth for one season. Therefore, a follow-up application to the soil may be required if aerial application is insufficient to increase the available pools in soil. On peaty soils in southwestern Australia, it has not been possible to correct acute Cu deficiency even where high rates of Cu are applied (>10 kg Cu/ha) due to the binding of Cu to organic matter. The problem is further exacerbated by the drying of surface soil, which contains the majority of the Cu, over summer when tree growth is maximal.

6.2 *Nutrient Retention and Recycling*

Because soils available for eucalypt plantations in China are usually low in organic matter, are subject to erosion, and there is little slash retention following harvest (Xu and Dell 2002), the soil micronutrient capital is not being sustained. As for macronutrients (Xu and Dell 2003), practices that can increase nutrient retention on-site, and can increase nutrient cycling within eucalypt plantations, need to be encouraged. Silvicultural practices such as site preparation, slash and weed management, increasing soil microbial and fungal diversity (including functional ectomycorrhizal associations) need to be considered.

7 Conclusions

To date, most of the research effort on micronutrients in eucalypt plantations has been concerned with the diagnosis of disorders in the field, often where tree growth has been severely impaired. There is now reliable data on the micronutrient requirements of trees in the early years following establishment from seed or clonal plantlets. Further, corrective fertilizer schedules are now available for many soil and site types. However, there remain significant gaps in knowledge that need to be filled if micronutrients are to be managed in a sustainable manner in plantations across rotations. Some of the key issues concern the micronutrient requirements during mid to late rotation, and for coppice in subsequent rotations. The loss of micronutrients off-site with harvesting remains a concern in those areas where soil nutrient capital is limited.

Acknowledgements

Plant Nutrition Services Pty and the Australian Centre for International Agricultural Research (ACIAR) provided financial support for conference attendance. The research reported here was funded by ACIAR and the private sector in Australia, Indonesia, Philippines and the USA.

References

Dell, B. 1994. Copper nutrition of *Eucalyptus maculata* Hook. seedlings: requirements for growth, distribution of copper and the diagnosis of copper deficiency. Plant and Soil 167: 181-187.

Dell, B. 1996. Diagnosis of nutrient disorders in eucalypts. In: "Nutrition of eucalypts", P.M. Attiwill, and M.A. Adams (eds), CSIRO Publications, Melbourne, pp417-440.

Dell, B. and Bywaters, T. 1989. Copper deficiency in young *Eucalyptus maculata* plantations. Can. J. For. Res. 19:427-431.

Dell, B. and Huang, L. 1997. Physiological response of plants to low boron. Plant and Soil 193:103-120.

Dell, B. and Malajczuk, N. 1994. Boron deficiency in eucalypt plantations in China. Can. J. For. Res. 24:2409-2416.

Dell, B. and Robinson, J.M. 1993. Symptoms of mineral nutrient deficiencies and the concentration ranges in seedlings of *Eucalyptus maculata* Hook. Plant and Soil 155/156: 255-261.

Dell, B. and Xu, D. 1995. Diagnosis of zinc deficiency in seedlings of a tropical eucalypt (*Eucalyptus urophylla* S.T. Blake). Plant and Soil 176:329-332.

Dell, B. and Xu, D. 1998. Managing nutrients for plantation eucalypts in southern China. ACIAR Research Notes 20, 6p.

Dell, B., Malajczuk, N., Xu, D. and Grove, T.S. 2001. Nutrient disorders in plantation eucalypts, 2nd edition. Canberra, ACIAR Monograph No.74, 188p.

Dong, B., Rengel, Z. and Graham, R.D. 1995. Effects of herbicide chlosulfuron on growth and nutrient uptake parameters of wheat genotypes differing in Zn-efficiency. Plant and Soil 173:275-282.

Gartrell, J.W. 1981. Distribution and correction of copper deficiency in crops and pastures. In: "Copper in soils and plants", J.F. Loneragan, A.D. Robson and R.D. Graham (eds), Academic Press, Sydney, pp313.

Gherardi, M.J., Dell, B. and Huang, L. 1999. Functional copper requirement for catechol oxidase activity in plantation *Eucalyptus* species. Plant and Soil 210:75-81.

Malajczuk, N. and Dell, B. 1995. Heavy metals in Philippines soils affect growth and establishment of trees. ACIAR For. Newsl. 18:2-3.

Reuter, D.J., Alston, A.M. and McFarlane, J.D. 1988. Occurrence and correction of manganese deficiency in plants. In: "Manganese in soils and plants", A.D. Graham, R.J. Hannam and N.C. Uren (eds), Kluwer Academic, pp205.

Stewart, H.T.L., Flinn, D.W., Baldwin, P.J. and James, J.M. 1981. Diagnosis and correction of iron deficiency in planted eucalypts in north-west Victoria. Aust. J. For. Res. 11:185-190.
Takkar, P.N. and Walker, C.D. 1993. The distribution and correction of zinc deficiency. In: "Zinc in soils and plants", A.D. Robson (ed), Kluwer Academic Press, Dordrecht, pp151-165.
Turnbull, C.R.A., Beadle, C.L., West, P.W. and Cromer, N. 1994. Copper deficiency - a probable cause of stem deformity in fertilised *Eucalyptus nitens*. Can. J. For. Res. 24:1434-1439.
Xu, D. and Dell, B. 1988. Importance of micronutrients for productivity of plantation eucalypts in east Asia. In: "Overcoming impediments to reforestation", J. Kikkawa et al. (eds), Proc. of the 6th International Workshop of BIO-REFOR, Brisbane, Australia, 2-5 December 1997, University of Tokyo, pp133-138.
Xu, D. and Dell, B. 2002. Nutrient management for eucalypt plantations in South China. Chinese For. Sci. Tech. 1:29-41.
Xu, D. and Dell, B. 2003. Nutrient management of eucalypt plantations in South China. In: "*Eucalyptus* plantations – research, management and development", R.-P. Wei and D. Xu (eds), Proc. of International Symposium on "*Eucalyptus* plantations", Guangzhou/Zhaoqing, China, 1-6 September 2002, World Scientific, Singapore, pp269-289.
Will, G.M. 1985. Nutrient deficiencies and fertilizer use in New Zealand exotic forests. New Zealand For. Res. Inst. Bull.97, 53p.

桉树人工林微量营养元素失调：起因、症状、诊断、影响和管理

BERNIE DELL[1]，徐大平[2]，CATHERINE ROGERS[1]，
LONGBIN HUANG[1]

[1] *Plant Nutrition Laboratory, Murdoch University*
Perth, WA 6150, Australia
[2] *Research Institute of Tropical Forestry, Chinese Academy of Forestry*
Guangzhou 510520, China

桉树微量营养元素失调，特别是 B、Cu、Fe、Mn 和 Zn 等元素失调，在世界各地营造商业人工林的地区几乎都有报道。微量元素失调多数由于施用了仅含有大量元素的化学肥料引致，例如最近中国报道的主要缺乏 B 素，澳大利亚报道的主要缺乏 Cu。桉树人工林微量营养元素失调的报道越来越多，表明过去人们在微量元素对产量的影响方面没有适当的认识。在微量元素缺乏已经导致生长衰弱的林地，改善施肥状况，可以产生出人意料的生长结果。而在缺素已经导致严重顶端坏死和畸形的林地，则很难恢复树木原有的较好的树形。本文描述了微量营养元素失调对桉树人工林影响的外部症状和对木材生产及木材质量的影响，探索一些微量营养元素管理的策略。

AN OVERVIEW OF THE RESPONSE OF EUCALYPTS TO FERTILISER AT DONGMEN, SOUTHERN CHINA

JOHN A. SIMPSON [1], RICHARD E. PEGG [2] AND ZHENG BAI [3]

[1] *Queensland Forestry Research Institute, Locked Bag 16*
Gympie, Queensland 4570 Australia
[2] *3 Thulimbah Street, Coopers Plains, Brisbane 4108 Australia*
[3] *Dongmen Forest Farm, Fusui County, Guangxi, China*

Since the 1980's, eucalypts have been extensively planted mainly in the tropics and subtropics to help meet the worlds need for timber and fiber. The Dongmen project (1981-89) was established under the Australia-China Program of Technical Cooperation and represented the first major attempt in China to systematically test a wide range of fast growing species under a range of silvicultural regimes on degraded soils. During the life of the project more than 60 silvicultural experiments covering in excess of 1000 ha were established and maintained. The results of experiments to assess the eucalypt species/provenance by fertiliser interactions are reported. The experiments were established on Dongmen Forest Farm on degraded Acrisols representative of large areas of land potentially available for reforestation.

The most productive species/provenances were: *Eucalyptus urophylla*, *E. cloeziana* (Gympie provenance) and the hybrid between *E. grandis* and *E. urophylla*. Applications of mixed fertiliser containing: nitrogen (N) at 100 kg/ha, phosphorus (P) at 50 kg/ha, potassium (K) at 50 kg/ha and a trace element mix (TE) resulted in an improvement in total volume mean annual increment (for experiments aged between 5.6 and 10.9 years) of 6.1 m^3/ha/yr for the best three performing taxa. All species/provenances responded in an essenially similar fashion to fertilising. The variability in species performance and fertiliser responses between trials highlights the need for replication in space and time and care to be exercised in the extrapolation of results to different climatic/edaphic regions. It was estimated that the internal rate of return by investing in fertilising was of the order of 30-35%.

Extensive use was made of foliar nutrient analysis to help interpret fertiliser responses and estimated 'critical ranges' for the major nutrients and 'adequacy levels' for the elements not limiting tree growth at Dongmen were summarized.

1 Introduction

Prior to the early 1980's large areas of degraded wastelands or poorly managed low productivity forests were common in southern China. As a result of undesirable land use practices, large areas had been severely degraded by uncontrolled erosion with resultant excessive loss of the A horizon. The Dongmen project was established on Dongmen Forest Farm (80 km south west of the provincial capital – Nanning) in Guangxi Zhuzng Autonomous Region in 1981 under the Australia-China Program of Technical Cooperation Agreement with the goal of contributing to the national goal of increasing wood production through supporting the development of fast growing species on the degraded lands in south China (Carter et al. 1981). Mean annual increments for existing industrial plantations were commonly in the range 6-10 m^3/ha/yr or lower (Turnbull 1981). The project was implemented by the Guangxi Forestry Department and the Queensland Department of Forestry and represented the first major attempt in China to systematically test a wide range of fast growing species under a range of silvicultural regimes on degraded soils. The project was formally terminated in 1989 but much experimentation continued at Dongmen after the close of the project. During the life of the project more than 60 silvicultural experiments covering in excess of 1000 ha were established and maintained.

Since the 1980's, eucalypts have been extensively planted mainly in the tropics and subtropics to help meet the worlds need for timber and fiber. Extensive plantations have been established in China particularly at lower elevations south of latitude 27°N in the provinces of Hainan, Guangdong, Guangxi and Yunnan (Davidson 1995). Today the estimated plantation estate of eucalypts in China exceeds one million hectares (Xu et al. 2002) and constitutes the major timber resource of the south.

Although more than 300 eucalypt species have been introduced to China (Wang et al. 1994) only about 200 have survived and of these only about 15 species are of importance for commercial plantations. The most widely planted species include *Eucalyptus camaldulensis*, *E. citriodora*, *E. exserta*, *E. globulus* ssp. *globulus*, *E. globulus* ssp. *maidenii*, *E. grandis*, *E. urophylla*, *E.* '12 ABL' and the hybrid between *E. urophylla* and *E. grandis*. The taxon *E.* '12 ABL' is a land race of *E. tereticornis* introduced to China in the 1970s from the Congo. Today clones of the hybrid between *E grandis* and *E. urophylla* are extensively planted and in future will comprise a large proportion of eucalypt resource in China.

A major focus of the experimental work undertaken by the Dongmen project involved the testing of a wide range of species (in excess of 20 eucalypt species involving >34 provenances under a range of fertiliser regimes. This paper provides an overview of the results of these trials including a brief summary of foliar nutrient data and comments on the economics of fertiliser use at Dongmen.

2 Methods and Materials

2.1 *Site Details*

Dongmen Forest Farm is in the south western part of Guangxi Zhuang Autonomous Region at latitude 22°22'N and longitude 107°30'E about 80 km south-west of the capital city of Nanning. It is situated in the subtropical lowlands with altitude ranging from 90 to 340 m ASL. Rainfall is ca 1200 mm/yr and the monsoon climate has a well defined dry period of 5-6 months. Evaporation ranges from 1500 to 1800 mm/yr. Mean January temperature is 12.6°C, mean for July 28.5°C and the annual mean is 22.7°C. Absolute maximum temperature is 38.9°C and absolute minimum is –1.4°C.

The Dongmen region comprises low hilly tablelands. Extensive sheet erosion has occurred on upper slopes, and lower slopes and narrower bottoms are frequently gullied. Localised areas of karst occur due to limestone residuals. The massifs have developed colluvial fans about their bases (Carter et al. 1981). The parent materials of the soils at Dongmen were probably of Tertiary geological age, with some Quaternary interbedded sandstones, shales and conglomerates. Lesser areas of older limestone residuals and colluvial fans occur, together with mixed Recent age materials along the junction between the limestones and the sediments. Some transport and mixing of soil materials has taken place in the Recent. Broad bottoms have been filled by Recent alluvium due to local transport from surrounding ridges.

The major soil forming process is lateritisation which has given rise to a range of tropical latosolic soils on the sediments. Layers of iron – manganese nodules are common, particularly in undulating topography. These can reach the surface on the ridge tops because of peneplanation, giving rise to very shallow soils. Nodules occur commonly in the upper B soil horizon and increase in density with soil depth, in some instances making up a high proportion of the soil mass. Textures of surface soils vary from sandy loams to clay loams while subsoils are heavy clay loams to clays. The profiles are typical of laterites with gradually increasing clay content with depth, nodular layers in the upper B/lower A horizons and mottled B/C horizons. Depth and organic matter accumulation in the A_1 horizon depends on topographic position, degree of surface erosion and hydromorphic influence. The A_1 horizon varies from zero (old erosional surfaces) to 33 cm: 15-20 cm would appear to be average (*Ibid*). The plantation soils of the region are Ferric and Orthic Acrisols (Anon. 1978) (Red/Yellow Tenosols, Isbell 1996) derived from sedimentary material (siltstone/mudstone). Soil erosion, both sheet and gully, is a major feature of the area. The range of physicochemical characteristics of the soils used for plantation purposes is presented in Table 1. Composite soil samples were collected from the surface 0-10 cm layer in 82 plantation areas (aged between 8 and 18 years) in 1984.

Generally the soils available for plantations are well drained, highly structured light medium clays with good moisture holding capacity. They are acid with low available phosphorus and potassium and with the cation exchange dominated by aluminum.

2.2 *Experimental Background*

From the design stages of the Dongmen project it was realised that soils and nutrition studies would play an important role in developing modern effective forestry management practices on the extensive areas of degraded soils in southern China. Because of the importance of understanding the basic resources of the area, work first commenced on developing soil survey techniques applicable to the situation. This work formed the platform on which the experimental sites were selected and results of experiments are extrapolated.

The basic premise for testing species suitability, a major project aim, included doing so under a range of nutrient regimes. Species most suited to nutrient limiting situations would not necessarily be the most productive or provide the best return on investment if grown under non nutrient limiting conditions. At the commencement of the project there was very limited fertiliser response data available for tree species and edaphic environments of interest. Because of time constraints, it was therefore necessary to start with a simple robust fertiliser treatment array for testing species under a range of nutrient regimes. This involved testing N_{100} and P_{50}

individually and in combination then including a nutrient mixture containing K_{50}, Mg_{30}, Mn_5, $Mo_{0.2}$, Cu_5, Zn_5 and B_1 (where subscript numerals denote the quantity of element in kg/ha). This array was used in the major series of species by fertiliser trials (9 in total). Supplementary fertiliser trials involving additive, omission or factorial arrangement of treatments were carried out to provide more detailed nutrient response information. A series of later age fertiliser trials was carried out to determine if repeat fertilising was required.

Table 1. Physicochemical properties for Dongmen plantation soils.

Variable*	Mean	Range
Soil Physical Properties		
Sand (%)	18.4	4.1 – 39.5
Silt (%)	18.4	3.4 – 37.7
Clay (%)	63.0	36.1 – 92.5
Depth to 65% Clay	16.2	0 – 100
Soil Chemical Properties		
PH	4.6	4.1 – 5.0
OM (%)	2.94	0.66 – 3.01
CEC (meq/100 g)	13.15	7.18 – 16.60
N-NO_3 (mg/kg)	10.3	3 – 33
N total (%)	0.142	0.051 – 0.410
P total (mg/kg)	502	227 – 965
P avail (mg/kg)	1.3	0.4 – 2.7
Ca exch (meq/100g)	0.56	0.23 – 1.75
Mg exch (meq/100g)	0.22	0.07 – 0.72
K exch (meq/100g)	0.16	0.07 – 0.56
Na exch (meq/100g)	0.26	0.10 – 0.73
Al exch (meq/100g)	3.92	1.37 – 5.30

* Particle size analysis by the hydrometer method, pH on 1-2 soil water extract, OM by Wakley Black method, CEC by sum of exchangeable cations, NO3-N colorimitry on a 1:2 soil water extract, N total by Kjeldahl, P total Na_2CO_3 fusion and then colorimetry, P avail by Mehlich extract then colorimetry, Ca exch, Mg exch, K exch, Na exch, Al exch all by overnight 1mNH_4OAc extract and flame atomic absorption spectrophotometry.

The vegetation on the experimental sites prior to preparation for trial establishment was either sparse native shrubs, bamboo or low yielding plantings of *E. citriodora*, *E. rudis* or *E. exserta*. Prior to planting the vegetation, including stumps, was removed from the site and the area ploughed to a depth of 20-25 cm. The experimental areas remained relatively free of weeds.

Trials had either four or five replications. Common size for unit plots was 48 trees (6 rows at 3.0 m by 8 trees at 2.5 m) surrounded by an isolation of 4 rows or 5 trees (12 m minimum).

Foliar analysis was used extensively in the program and provided an important tool in:

- interpreting visual symptoms of nutrient deficiencies,
- determining the effectiveness of various fertiliser additions,
- determining the general nutrient status of the plot/crop,
- developing critical/acceptable concentrations that could be used extensively, and
- recommending fertiliser prescriptions.

Apart from the main species by fertiliser trials, a range of other soils and nutrient studies were carried out during the life of the project. These studies were designed to investigate specific issues. A comprehensive trial investigated the seasonal and positional changes in foliar nutrient concentrations in the crowns of fertilised and unfertilised *E. grandis*. Other studies established nutrient balance sheets for *E. grandis* plantations and provided base data for interpreting the nutritional consequences of various silvicultural and land use options. A complete listing of soils and nutrition experiments is shown in Appendix 1.

2.3 *Measures and Foliar Sampling*

Regular measures of individual tree heights and diameters at breast height (DBH) were recorded and foliar samples of selected treatments were regularly collected during winter according to a standard procedure. Foliar samples consisted of a sample of the most recently fully expanded leaves from the upper one third of the exposed crown (on the southern side) of the tree. Equal sub-samples (numbers of leaves) from four representative trees per plot were bulked to give a composite sample. Total N was determined by the micro-Kjeldahl method and P determination was determined by spectrophotometrically on the same digest. Samples for cations were dry ashed and dissolved in a mixture of hydrochloric and nitric acids, and cation concentrations were determined by atomic absorption spectrophotometry.

3 Results

There has been an enormous amount of data collected during the life of the Dongmen project and in the continuation of work initiated during the project. Only a brief summary of the results of the most relevant species/fertilizer trials is presented.

3.1 *Species Selection*

A total of ca 20 species of eucalypts involving 34+ provenances were included in the range of species by fertiliser trials. Based on the growth performance of the species at ages between 5.6 and 10.9 years, the following ranking of species suitability for Dongmen conditions was drawn up in Table 2:

Table 2. The ranking of species suitability for Dongmen site conditions.

Species	+ Fertiliser [a] Total volume MAI [b] (m^3/ha)
E. urophylla	24.3
E. cloeziana	21.4
Eg x Eu hybrid [c]	18.9
E. grandis	15.9
E. propinqua	15.6
E. tereticornis	13.2
E. pellita	12.6
E. camaldulensis	11.5
E. exserta	10.4
E. citriodora	10.2

[a] Fertiliser comprised N_{100}, P_{50}, K_{50} and a trace element mix.
[b] Total volume MAI=Total volume mean annual increment. Volume calculated as: TV = 0.42 basal area by predominant height.
[c] The Eg x Eu hybrid was imported from Aracruz in Brazil.

Some care must be exercised in the interpretation of the above data as there are different numbers and locations of trials used to generate species means, age differences between trials and a varying degree of replication within trials as a result of the different number of species/provenances tested in the individual trials. There were significant differences in growth between the provenances. While these difficulties in the interpretation of the data exist, there is still a high degree of confidence that the species with a large advantage are the more suited to the Dongmen environment.

The highest mean annual increments (MAI) recorded within individual trials were achieved by *E. urophylla* (30.4 and 29.2 m^3/ha in the fertilised treatments in Expts 54A and 59), Eg x Eu hybrid (25.3 m^3/ha in the fertilised treatment in Expt 59) and *E. cloeziana* (23.0 m^3/ha in the fertilised treatment in Expt 1).

Subsequent testing of Eg x Eu hybrids made from local selections of parental material at Dongmen elevated this taxon to head the productivity ranking. A clonal demonstration area of ca 50 ha at Dongmen involving a number of Eg x Eu hybrids and some pure *E. urophylla* when felled at six years yielded an average MAI of 35.31 m^3/ha (over-bark measure) and 28.25 m^3/ha (under-bark measure). This area was fertilised at planting and again at half rotation age. Profit calculated by Dongmen staff was ¥ 10,928 per ha. Coppice is being used to produce the second rotation crop.

Species which may have some potential include: *E. grandis*, *E. propinqua*, *E. tereticornis*, *E. camaldulensis*, *E. pellita*, *E. citriodora* and *E. exserta*.

Other species tested but because of poor overall performance not considered to be of sufficient interest to warrant further attention are: *E. siderophloia, E. drepanophylla*, *E. brassiana*, *E. acmenioides*, *E. pilularis*, *E. saligna*, *E. leilin* No.1[1], *E. dunnii*, and *E. globulus* ssp *maidenii.*

[1] *E. leilin* No.1 is an advanced generation of a hybrid between *E. exserta* and a broadleaved species (Turnbull 1981).

3.2 *Fertiliser Responses*

There was a major response to fertilising in all species by fertiliser trials. The largest fertiliser response has been to the 'All On' treatment (N_{100} P_{50} K_{50}, Mg_{30}, Mn_5, $Mo_{0.2}$, Cu_5, Zn_5 and B_1). Data from four large species by fertiliser trials illustrate the response to N and P applied individually and in combination and the additional response to a fertiliser mix containing K and trace elements (Table 3).

In all cases the species by fertiliser interaction was either non significant or small relative to the main effects.

Mean height after the first growing season was not improved in any of the experiments where N only was applied (Table 3). Phosphorus alone promoted a response (38 - 62%) in all trials. In two experiments (Expts 9 and 30) a significant NP interaction occurred resulting in a positive response to N where P had been applied. The addition of K and trace elements improved early growth (by 13.3 to 30.3%).

Table 3. Responses of eucalypts at Dongmen to fertiliser applied at planting.

Expt No.	1		9		19		30	
No. of taxa	15	15	9	9	9	9	4	8
Age (yrs)	1.0	7.8	1.2	7.2	1.1	6.9	1.2	5.9
Fertiliser Treatment	Ht[a] (m)	TV[b] (m^3/ha)	Ht (m)	TV (m^3/ha)	Ht (m)	TV (m^3/ha)	Ht (m)	TV (m^3/ha)
Nil	2.1	75.2	2.4	55.8	2.4	64.2	2.2	38.1
N	2.2	71.2	2.3	56.1	2.5	68.2	1.9	41.1
P	3.4	79.6	3.3	63.8	3.3	76.6	3.0	42.0
NP	3.0	85.5	3.6	72.7	3.3	74.4	3.4	55.2
NPKT	3.4	105.4	4.5	85.5	4.3	93.0	3.9	63.5
Mean	2.6	83.4	3.2	66.8	3.19	75.2	2.9	48.0
LSDP=0.05	0.34	9.3	0.38	13.0	0.45	13.0	0.58	13.4

[a] Ht = mean height of all stems.
[b] TV = Total volume calculated as TV = 0.42 basal area by predominant height.

By age 6 - 8 years total volume production was not significantly increased by additions of either N or P alone. The early response to P alone had lost significance in all cases. Where both N and P were applied, growth was significantly improved over the nil fertiliser treatment by between 10 and 17 m^3/ha (or 14 and 45%). The early response to K plus trace elements persisted and at age 6 - 8 years productivity had been improved by 8 to 20 m^3/ha (or 15.0 to 25%) over the N + P treatment. These data suggest that a mixed fertiliser is required to obtain optimum response. Results from other trials and foliar nutrient data suggest that the response is due essentially to the K component in the fertiliser mix.

A summary of the later age responses to the application of the 'All On' fertilizer treatment, compared to the Nil fertilizer treatment, for a wider range of trials is presented for species of interest in Table 4.

Table 4. Summary of fertiliser responses of selected eucalypt species in Dongmen trials.

Expt No.	Planted	Age (yrs)	Total Volume[a] (m^3/ha)		Volume Response (m^3/ha)	Response MAI (m^3/ha/yr)	MAI +Fert (m^3/ha)
			Nil[b]	NPKT[c]			
E. cloeziana							
1	9.83	10.9	229.6	250.6	21.0	1.9	23.0
59	5.89	6.8	102.8	134.5	31.7	4.7	19.8
Eg x Eu hybrid							
9	4.84	7.2	80.6	122.5	41.9	5.8	17.0
30	5.86	8.3	88.2	119.0	30.8	3.7	14.3
59	5.89	6.8	100.2	172.2	72.0	10.6	25.3
E. propinqua							
9	4.84	7.2	84.0	107.4	23.4	3.2	14.9
59	5.89	6.8	54.2	111.3	57.1	8.4	16.4
E. urophyll							
19	5.85	8.9	152.0	178.0	26.0	2.9	20.0
51	5.88	7.6	108.0	133.0	25.0	3.3	17.5
54A	5.89	6.4	148.6	194.7	46.1	7.2	30.4
59	7.89	6.6	92.8	192.5	99.7	15.1	29.2
E. grandis							
1	9.83	10.9	111.1	172.8	61.7	5.6	15.8
19	5.85	8.9	104.0	155.0	51.0	0.7	17.4
30	5.86	8.3	80.6	124.1	43.5	5.2	14.9
59	5.89	6.8	33.8	106.5	72.7	10.7	15.7
E. tereticornis							
1	9.83	10.9	113.9	152.4	38.5	3.6	14.0
19	5.85	8.9	89.0	117.0	28.0	3.1	13.1
30	5.86	8.3	93.9	100.5	6.6	0.8	12.1
59	5.89	6.8	37.9	92.3	54.4	7.9	13.5
E. camaldulensis							
1	9.83	10.9	88.4	113.6	25.2	2.3	10.4
9	4.84	7.2	57.2	76.9	19.7	2.8	10.7
19	5.85	8.9	66.0	103.0	37.0	4 .2	11.6
30	5.86	8.3	80.2		9.6	1.1	10.8
36	4.87	5.6	57.3	77.6	20.3	3.6	13.8
59	5.89	6.8	16.4	80.4	64.0	9.4	11.8

[a] Total volume calculated as 0.42 basal area by predominant height (or 0.50 basal area by mean height).

[b] Nil = no fertiliser treatment.

[c] All On fertiliser treatment (N_{100}, P_{50}, K_{50} + trace elements).

Table 4 (*continued*). Summary of fertiliser responses of selected eucalypt species in Dongmen trials.

Expt No.	Planted	Age (yrs)	Total Volume[a] (m^3/ha)		Volume Response	Response MAI	MAI +Fert
			Nil[b]	NPKT[c]	(m^3/ha)	(m^3/ha/yr)	(m^3/ha)
E. citriodora							
1	9.83	10.9	98.1	119.5	21.4	2.0	11.0
59	5.89	6.8	25.1	65.0	39.9	5.8	9.5
E. exserta							
1	9.83	10.9	93.5	113.2	19.7	1.8	10.4
9	4.84	7.2	52.0	84.3	32.3	4.5	11.7
19	5.85	8.9	74.0	119.0	45.0	5.1	13.4
30	5.86	8.3	60.6	82.3	21.7	2.6	9.9
59	5.89	6.8	27.6	44.4	16.8	2.5	6.5
E. pellita							
59	5.89	6.8	50.1	79.6	29.5	4.3	11.7
53A	4.89	6.6	55.9	89.1	33.2	5.0	13.5
Exptl Mean						4.8	15.4

[a] Total volume calculated as 0.42 basal area by predominant height (or 0.50 basal area by mean height).

[b] Nil = no fertiliser treatment.

[c] All On fertiliser treatment (N_{100}, P_{50}, K_{50} + trace elements).

The fertiliser response varies considerably (Table 4). The total volume MAI response (at ages 5.6 to 10.9 years) to fertilising varies from 0.7 to 15.1 m^3/ha/yr for the experiments/species considered. Overall mean volume MAI response is 4.8 m^3/ha/yr but for the species considered to have most promise (*E. cloeziana*, *E. urophylla* and the Eg x Eu hybrid) this response is 6.1 m^3/ha/yr (3.3, 6.7 and 7.1 m^3/ha/yr respectively). The precise magnitude of the response is confounded to some extent by age differences and the degree of replication but the data provide a good general guide as to what can be expected. The range in responses within a species illustrates the importance of replicating trials over a range of sites and seasons if reliable data which are capable of extrapolation beyond a particular experimental site are to be collected.

There is a trend in the data for the fertiliser response to decline with age. There has been a wide range in the magnitude of the fertiliser responses between ages 6 and 8 years with much smaller responses recorded in the few points at age 10-11 years. The age effect is confounded with experiment differences so it is not possible to say other than it appears to be a trend. A more detailed investigation of the data for individual trials is required to further elucidate this question. This trend suggests that the initial fertiliser is inadequate to satisfy the needs of the plantation for the length of the rotation.

Results from later age fertiliser trials are variable. Neither *E. exserta* nor *E. citriodora* (Expts 3 and 7, Dongmen) stands aged 3.8 or 4.8 years responded by four to five years after treatments were applied to applications of N (four rates between nil and 200 kg/ha) or P (two rates nil or 50 kg/ha) either alone or in combination. The addition of 100 kg N/ha to *E. grandis* at age two years (Expt 47, Dongmen) improved volume increment by 23% five years after application but no response to either P or K was apparent. On the other hand *E. camaldulensis* responded to K (Expt 48, Dongmen), total volume increment being increased by 18% over a 5.5 years period but no response to N or P was recorded. Obviously further work is required to better define the need for refertilising/later age fertilising of established stands on the more productive species.

3.3 *Foliar Nutrient Data*

For the species by fertiliser trials reported, foliar N concentrations ranged from 1.34 to 2.03%. While significant differences between species and fertiliser treatments were demonstrated in some instances, the interaction has not attained statistical significance. The apparent differences were not consistent so detailed interpretation of age, species or fertiliser effects on foliar N concentrations is difficult.

Foliar P concentrations ranged from 727 mg/kg to 1560 mg/kg. While there is a trend for higher concentrations to occur in the material from older trees, it is not possible to determine if this difference is an age, site or seasonal effect. Species differences exist but again no consistent pattern is apparent in the data. An increase in foliar P concentration resulting from P fertilising is apparent in the youngest sampling of all experiments but this response has diminished or is not present in the latter age samplings.

Foliar K concentrations vary enormously - from 3180 to 15100 mg/kg. In *E. camaldulensis,* the material from older trees had higher foliar K concentrations than material from the younger trees, and concentrations far above those found in the other species (*E. grandis* and the hybrid). Foliar K concentrations have been consistently increased by the addition of K fertiliser. In experiments 1, 9 and 30 the increase in foliar K concentrations as a result of fertilising is more apparent in the older trees.

The foliar nutrient data available from the Dongmen trials and estimated 'critical ranges' for the major nutrients and 'adequacy levels' for the elements not limiting tree growth at Dongmen were summarized by Simpson et al. (1997) (Table 5).

4 Discussion

It is difficult to summarise the large quantity of data available from the trials carried out at Dongmen but sufficient data have been presented to identify preferred species and fertiliser regimes. However, care must be exercised in the extrapolation of these findings to areas with dissimilar climatic or edaphic conditions. Climatic regime and seasonal extremes are critical to species selection and edaphic conditions influence not only species performance but have a large impact on fertiliser responses. The general degraded condition of the soils at Dongmen would be expected to increase the responses to nutrient additions more than if the sites had not been degraded. Whilst the soils of the plantation areas at Dongmen have been derived from common parent materials and exhibit some degree of similarity, the relative species performance and fertiliser responses do vary to a surprising extent. The fact that there has been a degree of replication in time and space lends support to the overall findings reported in this paper. It does however highlight the need to be very cautious in the interpretation of data and extrapolation of results from a single trial located at a single site.

Table 5. Foliar Nutrient Concentrations of Eucalypts at Dongmen. Based on >10,000 trees samples and >2,500 composite samples analysed at Dongmen 1983-89. Trees aged 1 - 7 years old. Species examined include: *E camaldulensis*, *E. citriodora*, *E. cloeziana*, *E. drepanophylla*, *E. dunnii*, *E. exserta*, *E. grandis*, *E. grandis* × *E. urophylla*, *E. pilularis*, *E. propinqua*, *E. tereticornis*, and *E. urophylla*.

	Estimated critical range [a]		Approx. range of all data	
N (%)	1.2	1.8	0.92	2.4
P (mg/kg)	800	1100	442	1560
K (mg/kg)	4000	6000	1025	15100
	'Adequacy Levels' [b]			
Al (mg/kg)	40		40	219
Ca (mg/kg)	5000		1132	24120
Mg (mg/kg)	1000		709	3313
Na (mg/kg)	-		3	3175
Mn (mg/kg)	<3000 [c]		634	5392
Zn (mg/kg)	14		9	48
Fe (mg/kg)	50		28	108
B (mg/kg)	12		6	52
Cu (mg/kg)	6		5	16
S (mg/kg)	1000		946	1901
SO_4-S (mg/kg)	200		201	808
Mo (mg/kg)	0.06		0.06	0.14

[a] Based on levels recorded in fertilised plots.

[b] From lower end of range where no relationship with growth could be established.

[c] Based on samples of *E. cloeziana* where Mn toxicity symptoms were apparent coupled with Fe deficiency.

Notes: Species, age and sampling methods must be taken into account when interpreting foliar nutrient data. For example, striking differences occur between species in foliar Na levels. In Experiment 1, Dongmen the winter foliar Na levels (1985-89) for *E. grandis* varied between 920 and 1953 mg/kg whereas in *E. camaldulensis* the corresponding levels were 93-146 mg/kg.

Foliar nutrient data has been summarised and used for interpreting the results from fertiliser trials. Simpson (1986) after reviewing the use of foliar nutrient analysis concluded that for eucalypts it is useful as an aid in interpreting fertiliser responses but it does not provide all the answers expected of the technique. Foliar nutrient analysis is far less suited for use with eucalypts than for *Pinus* species. The inability to establish widely accepted critical foliar nutrient concentrations for eucalypts limits the usefulness for diagnosing marginal nutrient deficiencies or as a crop logging technique for eucalypts. The 'estimated critical ranges' for foliar N, P and K concentration established from this work at Dongmen are low when compared to the critical levels published for eucalypts (Lamb 1977; Schonau 1983) but are in accord with the data reported by Judd et al. (1996).

The foliar N and P concentrations in the trials reported lie generally within the 'estimated critical range' so heavier/repeated dressings of N and P fertiliser could be expected to elicit a growth response. For *E. grandis* and the Eg x Eu hybrid, the foliar K concentrations for the nil fertiliser treatment is generally at the lower end of the estimated 'critical range' suggesting a more severe K deficiency in these species. Additions of K fertiliser have improved foliar K concentrations but not to the extent that K concentrations have been increased to 'acceptable' levels. Foliar cation concentrations in *E. camaldulensis*, and to a lesser extent *E. tereticornis*, are somewhat atypical, in that relative to other species of interest in China, high foliar K concentrations are recorded whereas foliar Na concentrations tend to be very low. It follows that the 'estimated critical range' for these species will therefore be at the higher end or above the range shown.

There has been a major response in all trials to fertilising. In all trials the best fertiliser response has been to the 'All On' treatment. There is evidence to suggest that at planting application of a fertiliser treatment containing N_{100}, P_{50} and K_{50} will be adequate initially for at least most sites. Overall the eucalypt species by fertiliser interaction has failed to attain statistical significance indicating that all species have responded in a similar fashion to fertilising. Fertiliser response varies considerably depending on the experiment, age and species. Overall mean volume MAI response to fertilising is 5.0 m^3/ha/yr but for the species considered to have most promise (*E. cloeziana*, *E. urophylla* and the Eg × Eu hybrid) this response averages 6.1 m^3/ha/yr. There is a trend in the data for the fertiliser response to decline with age.

In order to put the benefits attained from fertilising into an economic perspective, simple economic comparisons were made. The basic cost and value data was taken from Xu and Bai (1990). The relevant figures used in calculations by these authors were:

Initial fertiliser costs per ha	-	¥580
Value of wood per m^3	-	¥195
Plantation establishment costs per ha	-	¥1500

A range of scenarios were examined with the above data used as a datum. Sensitivity analysis were performed in order to illustrate the effects of changing conditions (fertiliser response 30, 60 and 90 m^3/ha; plantation establishment costs ¥1500, ¥3000 and ¥4500 per ha; fertiliser costs ¥500, ¥1000, ¥2000 and ¥3000 per ha; wood value ¥100, ¥200 and ¥ 300 per m^3; interest 5%, 10% and 15%) of the profitability of fertiliser application.

Internal rates of return on costs of fertilising reach zero only when most pessimistic scenarios (low wood prices, low fertiliser responses and extremely high fertiliser costs) are considered. The most optimistic scenario yield internal rates of return of 50%. A reasonable estimate of returns from investing in fertilising might be in the order of 30-35%, which is quite a worthwhile investment. Returns of this order have a net present worth in the order of ¥ 6000 /ha (assuming an interest rate of 5% and fertiliser costs of <¥ 1000).

An alternative scenario is to reduce rotation age to meet a predetermined yield (say 160 m^3/ha). Reduced rotation ages results in lower interest charges and attendant improved financial viability. In addition, where land resource is limited, higher long-term productivity is achieved. Assuming fertiliser costs of either nil or ¥ 1000/ha, plantation establishment costs of ¥ 3000/ha, wood value of ¥ 200/m^3 and interest rate of 5%, then the fertilised plantation option has a future value of ¥ 1832 above the nil fertilised plantation. This type of economic analysis can be a valuable aid in making investment decisions.

5 Conclusions

The species which have demonstrated the best growth potential at Dongmen are: *E. urophylla*, *E. grandis* by *E. urophylla* hybrid and *E. cloeziana*. Other species which have some potential are: *E. grandis*, *E. propinqua*, *E. tereticornis*, *E. camaldulensis*, *E. exserta and E. citriodora*. Species which would appear to have little potential are: *E. siderophloia*, *E. drepanophylla*, *E. acmenioides*, *E. brassiana*, *E. pilularis*, *E. saligna*, *E. leilin* No.1 and *E. dunnii*. The application of mixed fertiliser at planting improved volume production by 6.1 m^3/ha/yr.

The soils and nutrition program carried out during the life of the Dongmen project and continued after the projects completion has provided a sound scientific basis for optimising the species selection and fertiliser management for forest plantations at Dongmen and with due care results have application to similar area in sub-tropical southern China.

Acknowledgements

The work reported in this paper has been carried out as part of the China - Australia Afforestation Project at Dongmen Forest Farm. The project was funded by AIDAB and the Chinese Government and managed by the Australian Project Director, Mr. R.E. Pegg. Field and office work associated with the trials has been supervised by the permanent advisors under the direction of Mr. E.G. Mannion and Mr. K.E. Gould and staff of the Dongmen Forest Farm. Chemical analyses were undertaken in the project laboratory at the Guangxi Forest Research Institute, Nanning, under the direction of Mr. Wang Yong Liang and Mr. Paul Milham. Assistance with statistical analyses from Mr. He Sanzhong and Mr. Zhou Wei in China, and Mr. M. Nester of the Queensland Forest Research Institute is acknowledged.

References

Anon. 1978. FAO Uneso soil map of the World 1: 5 000 000. Vol.VIII. North and Central Asia. Prepared by the Food and Agriculture Organisation of the United Nations, Unesco - Paris 1978, 165p+maps.

Carter, W.G., Bevege, D.I., Curtin, R.A., Ferguson, I.S. and Pegg, R.E. 1981. Eucalypt afforestation project at Dongmen State Forest Farm, People's Republic of China. A report on feasibility study of a proposed Australian/Chinese technical co-operation project at Dongmen State Forest Farm, Guangxi Zhuang Autonomous Region, People's Republic of China, 115p.

Davidson, J. 1995. Ecological aspects of *Eucalyptus* plantations. In: Proc. of the Regional Expert Consultations on Eucalypts, FAO Regional Office for Asia and the Pacific, Bangkok, 4-8 October 1993.

Isbell, R. 1996. The Australian soil classification. CSIRO, Collingwood, 143p.

Judd, T.S., Attiwill, P.M. and Adams, M.A. 1996. Nutrient concentrations in *Eucalyptus*: A synthesis in relation to differences between taxa, sites and components. In: "Nutrition of eucalypts", P.M. Attiwill and M.A. Adams (eds), CSIRO, Australia, pp123-153.

Lamb, D. 1977. Relationships between growth and foliar nutrient concentrations in *Eucalyptus deglupta*. Plant and Soil 47:495-508.

Simpson, J.A. 1986. Foliar analysis and its application to the Dongmen fertiliser trials. Paper presented at the Third Technical Exchange Seminar, China-Australia Afforestation Project, Dongmen, People's Republic of China, December 1986, Queensland Department of Forestry, Brisbane, 15p.

Simpson, J.A., Pegg, R.E. and Xiang, D. 1997. Fertiliser responses of eucalypts in Dongmen Forest, PR China. In: Proc. of the IUFRO Conference on "Silviculture and improvement of eucalypts", Salvador, Brazil, August 1997, pp219-227.

Schonau, A.P.G. 1983. Fertilisation in South African forestry. South African J. For. 125:1-19.

Turnbull, J.W. 1981. Eucalypts in China. Aust. For. 44:222-234.

Wang, H., Jiang, Z. and Yan, H. 1994. Australian trees grown in China. In: "Australian tree species research in China", A.G. Brown (ed.), Proc. of an international workshop, Zhangzhou, Fujian Province, PRC, 2-5 November 1992, ACIAR Proc. No.48, pp19-25.

Xu, D., Dell, B., Malajezuk, N. and Gong, M. 2002. Effects of P fertilisation on productivity and nutrient accumulation in a *Eucalyptus grandis* × *E. urophylla* plantation in southern China. For. Ecol. and Mgt. 161:89-100.

Xu, J. and Bai, J. 1990. The successful experiences, economic analysis and extension value of the Dongmen project. In: "Project completion report, Dongmen State Forest Farm eucalypt afforestation project, People's Republic of China", R.E. Pegg, R.E. (ed), Queensland Forest Service, 31p.

Appendix 1. Listing of Dongmen Project soils and nutrition experiments which include eucalypt species.

No.	Experiment Name	Year	Location	Spp	Prov	Fert Tmts	Reps
1E	Species x Fertiliser trial – *Eucalyptus* spp	1983	Dongmen	8	17	5	5
2	NP Factorial Fertiliser trial – *E. exserta*, *E. grandis*, PCH	1983	Dongmen	3	3	12	3
3	Mid Rotation Fertiliser trial – *E. exserta*	1983	Dongmen	1	1	8	3
5	Site Productivity Study – *E. exserta, E. citriodora*						
7	Mid Rotation Fertiliser trial – *E. citriodora*	1984	Dongmen	1	1	8	3
9	Species x Fertiliser trial – *Eucalyptus* spp	1984	Dongmen	8	9	5	3
16	NP Factorial Fertiliser trial – *E. camaldulensis*	1984	Dongmen	1	1	32	3
18	Provenance x Fertiliser trial – *E. cloeziana*	1983	Dongmen	1	2	5	3
19	Species x Fertiliser trial – *Eucalyptus* & *Acacia* spp	1985	Dongmen	7	5	9	4
28	N, K and Trace Element Fertiliser trial – *E. grandis*	1985	Dongmen	1	1	12	3
29	K and Trace Element Fertiliser trial – *E. camaldulensis*	1985	Dongmen	1	1	6	4
30	Species x Fertiliser trial – *Eucalyptus* spp	1986	Dongmen	9	11	5	5
32	Long term Fertiliser trial – *E. grandis*	1986	Dongmen	1	1	13	3
35	Long term Fertiliser trial – *E. camaldulensis*	1987	Dongmen	1	50	1	6
38S	Species x Fertiliser trial	1987	Satang	6	6	2	3
39W	Species x Fertiliser trial	1987	Weidu	7	7	2	3
40	Foliar Nutrient Sampling Study – *E. grandis*	1987	Dongmn	1	1	2	2
47	Mid Rotation Fertiliser trial – *E. grandis*	1987	Dongen	1	2	8	4
48	Mid Rotation Fertiliser trial – *E. camaldulensis*	1987	Donmen	1	2	9	5
49	Nutrient Immobilization Study – *E. grandis*	1988	Dongmen				
1	Fertiliser trial – *E. urophylla*	1988	Dongmen	1	2	26	5
52	Thinning x Fertiliser trial – *E. urophylla*	1988	Dongmen	1	1	5	3
59	Species x Fertiliser trial	1989	Dongmen	12	12	3	4
87	Fertiliser by Spacing trial	1993	Dongmen	1*	-	6	4

* Single clone.

华南东门林场桉树施肥反应综述

JOHN A. SIMPSON [1]，RICHARD E. PEGG [2]，郑白 [3]

[1] *Queensland Forestry Research Institute, Locked Bag 16*
Gympie, Queensland 4570 Australia
[2] *3 Thulimbah Street, Coopers Plains, Brisbane 4108 Australia*
[3]中国广西扶掇县东门林场

二十世纪八十年代以来，随着世界各地对木材和纤维的需要量日益增加，桉树人工林的发展非常迅速，尤其在热带和亚热带地区，桉树的种植更为广泛。东门项目（1981-1989）是中澳技术合作子项目，也是中国在退化林地上多种造林条件下系统测试多个速生树种的第一个主要的尝试。在整个项目执行过程中建立并维持 60 片试验林，总面积超过 1000 公顷。本文报道了对桉树树种/种源进行施肥试验的结果，试验林地点设在东门林场具有代表意义、可供大面积造林的典型土壤 - 退化酸性土壤区域。

生产力最高的树种/种源有尾叶桉(*Eucalyptus urophylla*)、昆斯兰桉(*E. cloeziana*)、巨桉(*E. grandis*)和尾叶桉的杂交种。对三种生长最好的树种施混合肥，混合肥含氮（N）100 kg/ha、磷（P）50 kg/ha、钾（K）50 kg/ha 和多种微量元素，结果表明施肥可使年平均蓄积量增加 6.1 m^3/ha/yr（试验林为 5.6 - 10.9 年生）。所有树种/种源对施肥的反应在本质上都是相似的。试验林之间树种生长和对施肥反应的差异表明在空间和时间上需要进行重复试验，在不同气候/土壤条件下应用该结果时需要谨慎。据估计，投资施肥的内部收益率在 30-35%之间。

本研究广泛地应用了叶片营养诊断来解释施肥反应，估算了大量元素使用浓度的“临界点”，以及营养元素不影响林木生长的“适当浓度”。

NUTRIENT MANAGEMENT OF EUCALYPT PLANTATIONS IN SOUTH CHINA

DAPING XU [1] AND BERNIE DELL [2]

[1] *Research Institute of Tropical Forestry, Chinese Academy of Forestry Guangzhou 510520, China*
[2] *Plant Nutrition Laboratory, Murdoch University Perth, WA 6150, Australia*

Eucalypts are very popular for revegetation of many parts of South China because of their capacity to tolerate degraded sites and infertile soils, their fast growth potential and their ability to coppice. More than 1,500,000 ha of plantations have been established, increasing annually by about 150,000 ha. The productivity is very low, only about 25-50% of the average productivity of eucalypt plantations in the world. Most of the land available for eucalypts has been degraded over recent centuries with extensive loss of the A horizon. The soils are generally oxisols and utilsols, which are acidic, highly leached and deeply weathered. Phosphorus deficiency is very severe and its availability is reduced because of limited compatible ectomycorrhizal fungi. Nitrogen supply is limited because of low organic matter content in the topsoil. This paper reviews a decade of field trials in South China; undertaken as part of several bilateral research programs in plantation forestry, concerning the use of fertilisers, harvest residue management and inoculation with eucalypt-compatible ectomycorrhizal fungi. One of the key questions addressed is whether the productivity of plantation eucalypts in South China can be increased to the world average by nutrient management across a rotation.

Some of the major findings include: application of superphosphate at establishment substantially increased growth, tree survival and nutrient uptake. The P recovery by tree uptake was between 6-35% and decreased as the quantity of P applied increased. Phosphorus applied in the first rotation resulted in elevated levels of available, total and organic P in the topsoil in the second rotation. Coppice was less responsive to superphosphate than seedlings/cuttings to P in the first rotation, indicating the potential importance of P storage in below-ground biomass. Only a small number of the introduced ectomycorrhizal fungi increased stand volume and the effect was sometimes temporary, possibly due to nursery practices impairing mycorrhizal development and poor fungal persistence after outplanting. A diverse range of species and isolates of ectomycorrhizal fungi should be screened for specified site conditions in order to identify functionally superior commercial isolates.

Reduction in the amount of residue removed during harvest increased tree growth by increasing available nutrient supply and reducing the loss of soil organic C after tree planting. Intercropping with N-fixing trees enhanced tree growth and increased litterfall. More than 80% of nutrient accumulation by trees is lost via current harvesting practices.
We conclude that productivity of plantation eucalypts in south China can be improved to 10-20 m^3/ha/yr, the average productivity in the world, by nutrient management. Recommended site management (land preparation, weed control, fertilizer rates) options are discussed for the first rotation. Harvest residue retention and coppice regeneration are recommended as operational practices in the second rotation.

This research indicates that the productivity of well-managed plantations should be sustainable whereas poor management practices result in serious yield declines across rotations and continued soil degradation. Operational manuals are being developed that incorporate these findings.

1 Introduction

With a steady increase in world population and the reduction in wood production from natural forests, forest plantations represent the world's best hope for meeting global wood requirements in the twenty-first century (Powers 1999). Industrial use of eucalypts, the main plantation genus in the tropics and warmer subtropics, is the most rapidly expanding sector in world forestry (Turnbull 1999). There are more than 12 million hectares of eucalypt plantations in the world, and about half were established during the last decade (Turnbull 1999). Eucalypts are favoured for high-yielding plantations because: (i) they adapt well to different ecosystems, (ii) large natural genetic populations occur, (iii) they mature precociously, (iv) species may be crossed to produce vigorous hybrids and wood with new characteristics, (v) many commercial species and hybrids can be cloned, and (vi) they produce wood with suitable properties for the purposes of fibre, sawntimber, chipboard, charcoal, posts and civil construction (FAO 1988).

Eucalypts are very popular for revegetation of many parts of South China because of their capacity to tolerate degraded sites and infertile soils, their fast growth potential and their multi-purpose use (Turnbull 1994). One of the advantages of eucalypts is their ability to coppice, thus generating new forests from the same plantation without the investment of large amount of money. More than one and half million ha of eucalypt plantations have been established in South China, mainly including Guangdong, Hainan, Guangxi, Fujian and Yunnan since the 1980s (Xu et al. 2000) and the rate of establishment of new eucalypt plantations is more than 150,000 ha annually.

A convenient way to compare the productivity of plantations is by the net production of biomass/unit area/unit time (Powers 1999). Tropical plantations are among the most productive ecosystems in the world, with growth rates often exceeding 40 t/ha/yr of aboveground net primary productivity, and 15 t of wood production (Lugo et al. 1988). Binkley and Ryan (1998) reported 323 t/ha/yr aboveground biomass production for a 16 years-old *Eucalyptus saligna* plantation on young organic soil in Hawaii. With different fertiliser treatments, aboveground biomass production of eucalypt plantations is 8-15 t/ha/yr in Brazil (Fernandez et al. 2000), 7-18 t/ha/yr in Victoria (Bennett et al. 1997), and 15-18 t/ha/yr under irrigation in India (Hunter 2001). In Brazil, MAI of eucalypt plantations can reach 90 m^3/ha/yr (Eldridge et al. 1994). Since site productivity is a function of canopy light interception (Cannel 1989) and the canopy should be managed to maximise interception of light throughout the rotation (Beadle 1997). Fertilisation of poor sites can markedly increase leaf area or mass. For example, annual net primary production in *E. grandis* (1-3 years old) was 15-32 t/ha at leaf area index (LAI) 2-5 compared to only 3-9 t/ha at LAI <1 for unfertilised trees (Leuning et al. 1991).

Productivity of eucalypt plantations in South China is not only low (25-50% of the average in the world), but also uneven, from MAI 5 to 50 m^3/ha/yr (Xu et al. 2000) and is declining with successive rotations in some areas (Yu et al. 1999). A common phenomenon is that the LAI is lower than in some foreign countries (Table 1) and both LAI and growth slows down 3-4 years after planting (Xu et al. 1996). Most foresters in China blame the low productivity on either uneven precipitation or poor soil fertility. However, there is no evidence that precipitation (1100 to 2600 mm per annum; Xu et al. 1998) is restricting productivity. Furthermore, as most high-yielding eucalypt plantations in Leizhou, Dongmen and Hainan occur in locations where precipitation is lower than in areas where productivity is low, such as Yangxi, Yangjiang and Guangzhou. Since both biomass production and leaf area decrease when soil mineral nutrients are in limited supply, particularly P and N (Marschner 1995), soil infertility may restrain productivity in South China.

Table 1. Leaf mass and aboveground N content of eucalypt plantations in South China and selected plantations in the world.

Species	Leaf mass (t/ha)	N applied (kg/ha)	N content (kg/ha)	Age (year)
E. urophylla [1*]	2.7	444	162	4.6
E. grandis x *E. urophylla* [2]	1.9	194	96.1	6.3
E. grandis x *E. urophylla* [3]	2.8	393	129	3.5
E. grandis [4]	2.5	100	74	4.8
E. grandis [5]	3.1	0	246	9.5
E. grandis [6]	5.0	2180	331	9.5
E. urophylla [7]	10.6	11	371	3.5
E. grandis [8]	5.9	0	185	3.0
E. grandis [9]	7.2	320	203	3.0

*[1] Xu et al., unpublished data; [2] Xu et al. (2002); [3] Xu et al. (1996); [4] Simpson et al. (1997); [5] Birk and Turner (1992); [6] Harrison et al. (2000); [7] Harrison et al. (2000); [8] Hunter (2001); [9] Hunter (2001).

2 Effect of Soil Phosphorus Supply on Productivity of Plantation Eucalypts

Phosphorus is the most important element (after N) limiting agricultural production in most regions of the world (Holford 1997). It has been suggested that about 65% of tropical soils are strongly P-deficient and 27% are moderately deficient for cropping (Gonçalves et al. 1997). These soils are often low in total P and some have a high capacity to rapidly fix applied P (Willett et al. 1998). The soils available for eucalypt plantations in South China are generally lateritic and lateritic red earths (oxisols and ultisols) which are acidic in reaction (pH_{water} 4-6), highly leached, generally heavy in texture and deeply weathered (Li 1983; Wang and Zhou 1996). Oxisols and ultisols may contain toxic concentrations of Al and Mn, and are often low in available P and other nutrients (Stewart et al. 1991). Phosphorus may be the primary macronutrient limiting the growth of eucalypts in South China, since, without P application, *E. urophylla* did not respond to N fertiliser at Dongmen and Enping (Chen et al. 1996; Wang and Zhou 1996). Other data are available for Yangxi (Li et al. 1995), Yangdong (Wu et al. 1996) and Leizhou (Lin et al. 1999; Liang and Zhou 1999) in Guangdong, and Baoshan (He et al. 1999) in Yunnan. From these studies, it can be concluded that fertilisation with P is essential for fast-growing eucalypt plantations in South China. In general, low soil pH and high concentrations of Al^{+3} and Fe^{3+} in South China greatly reduce the availability of P to trees through adsorption processes (Li 1983). In eucalypt plantations outside China, P is also most often the nutrient that controls biomass increment (McLaughlin 1996).

In soils of high P-fixing capacity, there is low efficiency of P fertiliser use for the first crop and also a low availability, or residual effect, for subsequent crops. Thus, a high input of P might be necessary before a normal response to fertilisation is obtained (Ruaysoongnern and Keerati-kasikorn 1998). A key issue for improving P efficiency is whether a single application of a large amount of P is an efficient strategy for optimizing P uptake and crop yield or, alternatively, whether applying small amounts of P to alleviate low productivity to an intermediate level is a better strategy. The validity of applying large amounts of P requires that the applied P remains available over time. Dias et al. (2001) found that after a rotation of eucalypts, soil available P in an oxisol was elevated due to phosphorus application 9.5 years earlier. Complementary action on management of P, such as increasing P cycling in litterfall and organic residues, can minimise soil P-fixation and increase P effectiveness (Gonçalves et al. 1997). Phosphate management recommendations for high P-fixing soils often consider P fertilisation as a long-term investment that can be amortised over several cropping cycles (Fölster and Khanna 1997).

Our studies showed that the addition of 4, 13, 40, 120 and 300 kg P/ha increased stand volume by 15, 29, 55, 42 and 66%, respectively, at age 3 years on a site with a mildly P-deficient red ultisol at Chuxiong, Yunnan (Xu et al. 2001). Application of 40 kg P/ha was sufficient to maintain marginal growth in the short-term. The addition of 20, 50 and 200 kg P/ha increased stand volume by 5, 5 and 7 times the MAI at P0, respectively, at age 4.5 years on a site with a severely P-deficient oxisol at Gaoyao, Guangdong. Application of 200 kg P/ha was adequate at this site. The addition of 13, 52, 104, 208 and 312 kg P/ha increased stand volume by 2, 3, 3, 4 and 4 times the MAI at P0, respectively, at age 6.3 years on an oxisol at Kaiping, Guangdong. Application of 208 kg P/ha was adequate at this site.

From the above results, and other published studies from the region (Figure 1, Table 2), it is clear that P responses in the southeast are higher than those in the southwestern part of China. This is because soils, which are mainly oxisols (lateritic red soils), in the southeast have lower pH and available P than the ultisols (red soils) in the southwest. In the southeastern China, the responses to P application in our P-rate-trials are higher than in reports by Liang and Zhou (1999) and Lin et al. (1999). However, the highest P treatments in the latter two studies are lower than the adequate P level determined in our P-rate-trials. A further complicating factor is that these studies used factorial designs with P, N and K treatments. Therefore, N and K could have confounded the response of trees to P application as fertilisers were only applied once at plantation establishment and rates were generally low.

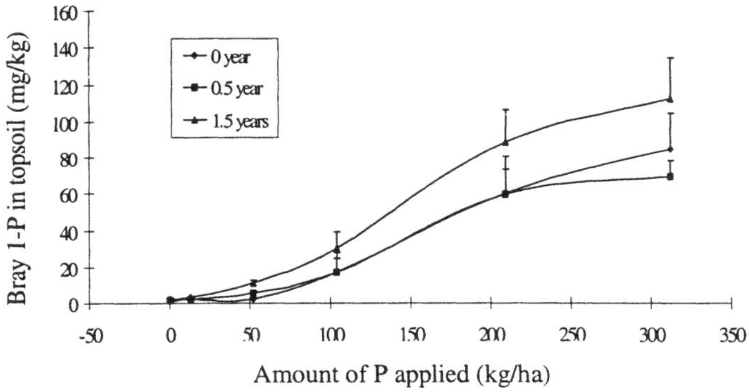

Figure 1. Effect of P application in first rotation on available P in the topsoil of second rotation coppice of *E. grandis* x *E. urophulla* in Zhenghai, Kaiping, Gaungdong.

Table 2. Summary of the results of P application trials in South China.

Location	Species and age (year)	P applied (kg/ha)	Increment over control	Soil pH	Trial design
Baoshan [1*]	*E. globulus*, 5	31-62	39-64%	5.8 (H_2O)	Factorial design with N and K
Chuxiong [2]	*E. globulus*, 3	13-300	15-66%	4.8 (KC1)	P rates trial
Dongmen [3]	*E. grandis x E. urophylla*, 6-8	50	40-67%	5.7 (H_2O)	Factorial design with N and K
Leizhou [4]	*E. urophylla*, 4.6	15-58	160-320%		Factorial design with N and K
Gaoyao [5]	*E. urophylla*, 4.5	20-200	470-740%	3.2 (KC1)	P rates trial
Yangxi [6]	*E. urophylla*, 3.7	32.5-65	130-460%	4.4 (H_2O)	With N and K
Kaiping [7]	*E. grandis x E. urophylla*, 6.3	13-312	160-370%	3.3 (KCl)	P rates trial
Kaiping [8]	*E. urophylla*, 4.5	43-130	130-210%	4.6 (H_2O)	Factorial design with N and K

* [1] He et al. (1999); [2] Xu et al. (2001); [3] Simpson et al. (1997); [4] Xu et al. (2001); [5] Lin et al. (1999); [6] Xu et al. (2000b); [7] Xu et al. (2000a); [8] Liang and Zhou (1999).

Phosphorus can not only increase tree growth and biomass production, but also increase accumulation of nutrients other than P (Table 3). The addition of P (20-200 kg P/ha) increased N uptake by 2-4 times, K uptake by 3-5 times, Ca uptake by 3-4 times and Mg uptake by 3-4 times the uptake by trees at P0, respectively, at age 4.5 years at Gaoyao. Similarly, it increased P uptake by 3-5 times of the uptake by trees at P0. The addition of P (13-312 kg P/ha) increased N uptake by 1-2 times and K uptake by 2-3 times of trees in P0, respectively, at age 6.3 years at Zhenhai, Kaiping. Further, it increased P uptake by 1-9 times the uptake by P0 trees. The percentage of P recovery was highest at low P application and decreased as the quantity of P applied increased.

Table 3. Effect of P fertilization on nutrient accumulation of a 4.5-year-old plantation of *E. urophylla* in Gaoyao, Guangdong (kg/ha).

Treatment	N	P	K	Ca	Mg	Biomass
P0	43.27 a	2.29 a	30.40 a	38.22 a	2.07 a	13.36 a
P20	165.71 c	7.89 b	142.73 b	155.02 b	8.42 b	73.36 b
P50	131.44 b	9.16 b	126.73 b	147.02 b	9.67 bc	70.28 b
P200	190.53 c	13.53 c	178.78 c	192.51 c	10.56 c	100.82 c

Note: Means in columns with the same letter are not significantly different at p=0.05 using the Newman-Keuls Critical Range Test.

In a P fertilisation trial in Zhenghai, Guangdong without good weed control, though P uptake by the trees, understorey and litter was increased by P application (Table 4), the proportion of P uptake by the understorey and litter decreased as the quantity of P applied increased. A similar pattern occurred in the proportion of N and K uptake by the understorey and litter, again indicating that the understorey was more tolerant to the low P supply in the soil than eucalypt trees. However, the increment in nutrient uptake by the understorey was much lower than by the trees. The percentage of N uptake by the trees to total N uptake was much lower than the percentage of P and K uptake by the trees to total uptake. Even in P312, N uptake by the trees was less than 50% of total N uptake.

Table 4. Nutrient accumulation (kg/ha) of a 75 months-plantation of *E. grandis* x *E. urophylla* in Kaiping, South China.

Treatment	Eucalypt trees			Understorey			Litterfall		
	P	N	K	P	N	K	P	N	K
P0	2.5 a	38.7 a	44.9 a	4.2 a	93.6 bc	84.4 bc	2.1 ab	55.2	14.7bc
P13	5.1 b	79.1 b	95.3 b	5.0 ab	104.4 c	91.5 c	1.9 a	55.2	12.2 ab
P52	10.4 c	109.9 d	119.9 c	6.3 bc	102.7 c	81.0 bc	2.2 ab	57.2	10.4 a
P104	12.7 d	102.0 c	114.4 c	5.3 ab	76.8 ab	68.5 a	3.2 c	63.2	18.6 c
P208	21.9 e	119.4 e	152.9 e	7.3 c	100.7 c	77.8 ab	2.5 b	49.6	16.7 c
P312	23.2 f	119.4 e	140.7 d	6.8 c	75.3 a	68.7 a	2.4 b	47.6	11.4 ab

Note: Means in columns with the same letter are not significantly different at p=0.05 using the Newman-Keuls Critical Range Test.

In a coppice trial in Kaiping, Guangdong, coppice in low P treatments (P0 and P13) had a faster growth in the second rotation than the trees in the first rotation, and coppice at high P treatments (P52, P104, P208 and P312) had slower growth than trees in the first rotation (Table 5). The adequate P fertiliser level was 208 kg P/ha in the first rotation and 52 kg P/ha in the early stage of the second rotation. It remains to be determined whether the optimal level changes with coppice age. The better performance of the coppice trees at low P supply than trees in the first rotation may be due to P in roots becoming available for shoot growth. In addition, the large root system would enable the coppice to access large soil volumes for nutrient absorption early in the rotation cycle. More research is needed to understand why the coppice at high P did not perform as well as in the first rotation. As expected, application of P at establishment of the first rotation significantly increased Bray-1 P following harvest, before slash burning and 0.5 and 1.5 years after slash burning (Figure 1).

Table 5. Comparison between tree growth in the first rotation and coppice growth in the second rotation of *E. grandis x E. urophylla* with re-fertilisation of N, K and B.

	Tree height (m)				Tree diameter (cm)			
Rotation	1st	1st	2nd	2nd	1st	1st	2nd	2nd
Age (month) / Treatment	19	75	17	30	19	27	17	30
P0	3.6	10.6	6.9	10.2	2.25	3.68	5.0	7.4
P13	6.9	14.2	7.8	11.1	5.06	7.13	5.6	8.6
P52	8.9	16.0	7.9	11.9	7.09	8.67	6.0	9.3
P104	8.8	15.7	7.9	11.4	7.14	8.67	5.7	8.7
P208	9.6	16.6	8.2	11.8	8.09	9.29	6.0	9.1
P312	9.4	16.8	8.1	11.2	8.45	9.71	6.0	8.9

3 Inoculation with Ectomycorrhizal Fungi

It has been suggested that the low productivity of plantation eucalypts in China is related to the absence of symbiotic fungi which are normally associated with eucalypts in their native habitat (Malajczuk et al. 1994). Ectomycorrhizal fungi are an essential component of eucalypt forests, where they are reputed to have key roles in the acquisition of nutrients by trees and in nutrient cycling (Vogt et al. 1997). Of their many functions, the most understood is the capacity of mycorrhizal fungi to gain access to limited supplies of P and to deliver P to the root (Vogt et al. 1997). Ectomycorrhizal fungi inoculation may be essential for the success of exotic trees introduced to places where indigenous fungi are unable to effectively form associations (Malajczuk et al. 1994). Although many studies have established the potential of ectomycorrhizal fungi inoculation to increase growth of eucalypts in glasshouse experiments under conditions of limiting P availability (Monson et al. 2000; Xu et al. 1994; Burgess et al. 1990), very few works have followed up these studies in the field either with young or mature trees. Further research is needed to determine more definitively how mycorrhiza benefit plantation eucalypts under field conditions. Studies on effects of fertiliser on plantation performance rarely consider the implications for the symbiont component. A notable exception is the work by Pampolina et al. (2001) on the effect of superphosphate on ectomycorrhizal fungi in *E. globulus* in Western Australia and in *E. urophylla* in the Philippines. Further research is therefore needed to determine what level of P fertilisation, for example, is optimal for enhancing tree growth while minimising the suppression of the mycorrhizal association.

A small number of field experiments have quantified effects of ectomycorrhizal fungi inoculation on tree growth (Table 6). With the exception of the field trial in Western Australia reported by Malajczuk (1987), where *E. diversicolor* (karri) was reestablished in native karri forest. The inoculation trials have been undertaken in plantations of exotic eucalypts. In the latter group, the most significant responses have been obtained in the Congo (Garbaye et al. 1988), and in East Asia (Aggangan et al. 1999). There was, for example, a 31% and 21% increase in stand volume of 5 years old *E. urophylla* on a marginal grassland in the Philippines inoculated with Australian isolates of *Pisolithus* (H445) and *Laccaria* (H4100), respectively (Aggangan et al. 1999). A few trials for ectomycorrhizal fungal inoculation in South China have demonstrated 15-33% increase in early tree growth compared to the control treatment. However, responses are isolate dependent. Moreover, only small number of ectomycorrhizal fungi has been screened in the field. More study is need for fungal screening study with a diverse range of species or isolates in order to identify functionally superior commercial isolates.

4 Nitrogen Management

Turning to other factors, most of the N uptake by eucalypt trees is from the mineralisation of organic N (Binkley and Giardina 1997). Because organic material in the topsoil and at the soil surface is usually low, N uptake by eucalypts in South China is driven closely by N fertilisation. Several experiments demonstrated that N fertilisation combined with P increased tree growth in south China (Chen et al. 1996; Wu et al. 1996; He et al. 1999). Therefore, N could be another constraint after P contributing to the small canopy of plantation eucalypts in the region. Nitrogen application increases productivity initially by increasing the photosynthetic efficiency of foliage (Reich et al. 1994), and later through increased foliage production (Xu et al. 1996).

Table 6. Published accounts of field trials where eucalypts have been inoculated in the nursery with ectomycorrhizal fungi and then out-planted.

Country	Eucalypt species	Results	Reference
Australia	*E. diversicolor*	Inoculation with ectomycorrhizal fungi increased tree growth by 50%	Malajczuk (1987)
Congo	*E. urophylla x E. tereticornis*	Inoculation with ectomycorrhizal fungi increased tree growth by 26-30%	Garbaye et al. (1988)
China	*E. urophylla*	Inoculation with ectomycorrhizal fungi increased tree growth by 33%	Zhong et al. (2000b)
China	*E. grandis x E. urophylla*	Inoculation with ectomycorrhizal fungi increased tree growth by 18%	Huang and Liang (2000)
China	*E. urophylla*	Inoculation with ectomycorrhizal fungi increased tree growth by 27%	Xu et al. (2001)
China	*E. globulus*	Inoculation with ectomycorrhizal fungi increased tree growth by 15%	Xu et al. (2001)
Philippines	*E. urophylla*	Inoculation with ectomycorrhizal fungi increased tree growth by 21-31%	Aggangan et al. (1999)
Philippines	*E. deglupta*	Inoculation with ectomycorrhizal fungi increased tree growth by 40-52%	De La Cruz et al. (1988)

Eucalypt plantations rarely respond to N fertilisation under tropical and subtropical conditions (Herbert and Schönau 1990; Gonçalves et al. 1997) because natural sources of N, mainly organic N mineralisation during the rotation, are enough to meet tree demand in the first rotation (Gonçalves and Barros 1999). Though plantation eucalypts in parts of South China did not response to N (Simpson 1989; Chen et al. 1996), N increased tree growth after P application (Zhou 1995). Therefore, N could be limiting productivity after P fertilisation in the first rotation, especially on sites with low organic matter content or where there is rapid growth of weeds competing for N. For example, the total N content (191 kg/ha) in tree biomass of a 4.5 years-old plantation of *E. urophylla* supplied with P (P200), was much higher than the N content (119 kg/ha) of a 6.3 years-old *E. grandis x E. urophylla* supplied with 208 kg P/ha (Xu et al. 2002). Leaf biomass in the former trial (2.7 t/ha) was also higher than the leaf biomass in the second (1.9 t/ha). However, more N fertiliser (444 kg/ha) was applied to the plantation of *E. urophylla* than to the hybrid plantation of *E. grandis x E. urophylla* (194 kg/ha). These results suggest that N could become limiting after P even on sites with relatively high fertility with 2-3% SOM. In general, nitrogen fertilisation in the first rotation should probably be more than 200 kg/ha. This should go earlier in this section. Compared with results from other countries, both leaf mass and N content in above-ground biomass of plantation eucalypts in South China are lower (Table 1). This means that N may still be limiting for tree growth even where high amounts of N (at least in Chinese silviculture) have been applied, perhaps because of some fertiliser N loss by erosion, leaching and volatilisation. It is important that the organic matter content and N:C ratio of soils are increased so as to guarantee adequate N supplies in the future.

Further research on other nutrient supply rather than P in second rotation coppice of *E. grandis x E. urophylla* shows that refertilization with N, K and B significantly increased coppice growth (Figure 2). More study is required to define the timing and rates of N fertiliser required in first rotation eucalypt plantations in South China.

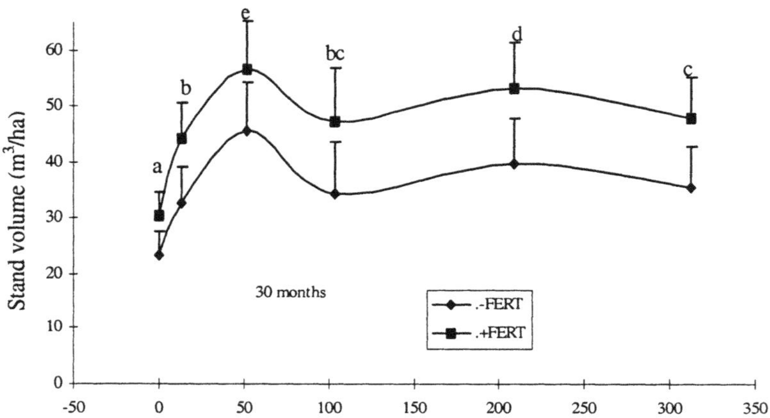

Figure 2. Effect of refertilization of N.K. and B in second rotation coppice of eucalypts on tree growth after P fertilization in first rotation. Means of the treatment with same letter are not significantly different at p=0.05 using Newman-Keuls Critical Range Test.

The long-term sustainability of high productivity in tropical plantations depends, in part, on a continued supply of nutrients. Rapid growth of tropical plantations leads to high rates of nutrient accumulation in biomass, and harvesting at short rotation intervals removes large quantities of nutrients and may deplete soil fertility (Binkley and Giardina 1997). Burning of slash could also deplete soil N pools. Therefore, N fertilisation could be very important in the second rotation following P fertilisation in the first rotation. However, more study is needed to determine the demand for N in the second, and later, rotations. The N use efficiency of eucalypt plantations in China is quite similar with eucalypt plantations in other countries (Table 7).

5 Slash Management

Another possible major constraint to productivity of eucalypt plantations in south China is low SOM. Most of the land available for eucalypts in south China has been degraded over recent centuries with extensive loss of the A horizon. This soil degradation was caused by population pressure, inadequate management, over-harvesting (litter, understorey and all tree components) and severe soil erosion (Xu 1997). As mentioned above, a common practice by rural people in South China is to harvest litter and understorey from forests to meet their fuel needs. The average amount of annually harvested litter and understorey is 3.3 t ha^{-1}, which can be as much as 55% of litter and understorey production (Brown et al. 1995). It has been calculated that the harvest of litter and understorey removes a substantial quantity (about 44-73%) of the nutrients that could be potentially recycled (Mo et al. 1995).

Table 7. Nutrient use efficiency of aboveground biomass for the 75-month-old plantation of *E. grandis* x *E. urophylla* in South China with data from tropical eucalypt plantations from the literature.

Species	Country	Age	NUE (kg/kg)	PUE (kg/kg)	KUE (kg/kg)
E. hybrid	Congo (Laclau et al. 2000)	7	592	2000	1475
E. hybrid	Congo (Laclau et al. 2000)	7	611	1638	1026
E. grandis	Brazil (Morais et al. 1990)	8	282-443	4269-8783	550-621
E. hybrid	India (Negi and Sharma 1996)	7-10	318-352	1370-4333	436-914
E. grandis	India (Negi and Sharma 1996)	9	204	2000	278
E. grandis	South Africa (Herbert 1996)	7	538	7604	471
E. hybrid	Kaiping, China	6.3	566-760	3422-8785	498-652
E. urophylla	Gaoyao, China	4.5	325-553	5589-9255	416-545

Slash management has a significant effect on the amount of inorganic P made available to the following rotation. Bekunda et al. (1990) showed, for example, that retention of slash of *P. radiata* increased Olsen inorganic P in the soil by 7.5 kg P/ha, which was adequate to sustain high early growth rates in the following crop. Above all, high P applications may increase productivity by improving P availability over a long time, through litterfall and slash retention.

A survey by Yu et al. (1999) in Leizhou, China, revealed that with the same clone and fertiliser rates, the yield of fourth rotation eucalypt plantations was only 50% of the yield of first rotation plantations established on land that previously supported secondary forest. Soil fertility of the first rotation site was much higher than that of the fourth rotation site. In general, it seems that productivity in the second rotation relates both to the management practices used in the previous rotation and practices in the current rotation, as well as other factors such as site quality and tree species characteristics.

As mentioned previously, the retention of unburnt harvest residue on site should increase nutrient supply to trees in the following rotation, especially N (Table 8). In addition, the study in Yangxi has established that slash retention improved soil available P. So far, there are only a few studies on slash management for second rotation eucalypts. Gonçalves et al. (2000) observed a significant improvement in tree growth in second rotation *E. grandis* at age 4 years following slash retention in the humid tropical climate in Brazil. However,

Jones et al. (1999) reported only a small, but insignificant, positive response to slash retention for a second rotation plantation of *E. globulus*. For the latter species, O'Connell et al. (2000) measured a significant response after second rotation plantation establishment (first 2 years), but it was not significant at year 4.

Table 8. Comparison of nutrient loss (kg/ha) from 4.5-year old *E. urophylla* trees with or without phosphorus fertiliser with four harvesting strategies.

Harvest treatment	Rate of P applied (200 kg/ha)					Rate of P applied (0 kg/ha)				
	N	P	K	Ca	Mg	N	P	K	Ca	Mg
Scenario 1[a]	77.6	4.6	51.6	38.6	4.2	12.8	0.6	5.4	4.2	0.5
	41[*]	34	29	20	40	30	26	18	11	25
Scenario 2	108.7	7.4	116.2	128.1	4.7	19.7	1.1	15.8	21.3	0.6
	57	54	65	67	44	45	48	52	56	28
Scenario 3	162.2	11.7	160.9	169.8	9.3	34.8	2.0	27.1	31.9	1.7
	85	86	90	88	88	80	88	89	83	83
Scenario 4	181.8	12.9	173.2	184.8	10.1	39.9	2.2	28.9	35.8	1.9
	95	95	97	96	96	92	95	95	94	92

* Percentage of total tree accumulation; [a] Scenario 1 = only stem-wood harvested, Scenario 2 = only bole harvested, Scenario 3 = all aboveground harvested (including leaves and branches), and Scenario 4 = all above ground plus tap-root harvested.

Again, the main reason for the insignificance of slash management in some studies may be related to site quality and the size of the nutrient demand of the second rotation trees. On some sites, plantation eucalypts in the second rotation, where soil reserves have become depleted, may respond to the pools of mineralisable nutrient resulting from slash decomposition. Of course, retaining slash will also be useful to reduce wind and water erosion, which contributes to nutrient loss, and to improve soil physical properties and water holding capacities. In some cases, slash retention suppresses weeds. On site where nutrients are not limiting for the growth, the beneficial effect of slash retention could be measured as long-term soil fertility improvement instead of short-term effects on the promotion of tree growth. The effect of slash retention was significant (Figure 3) in a study in Yangxi, China (Xu et al. 2000) because the soil fertility at the study site was very low. Also, there was an obvious yield decline in the second rotation eucalypts even with the same genetic material and similar fertilisation. Thus, on these sites, there remains the need to develop better silvicultural practices across rotations to prevent further yield decline. The dramatic response to fertilisation at the site in Yangxi indicates that nutrient deficiency was the main limitation to tree growth. One consequence of slash retention was an increase in litter fall, which can benefit long-term soil fertility maintenance. Although removal of all organic material improved tree growth in the early stage of the second rotation, compared to whole tree harvest, growth could not be sustained as soil fertility declined. So far, only early growth responses has been measured and further research should monitor long-term changes in soil fertility and site productivity in relation to organic matter management.

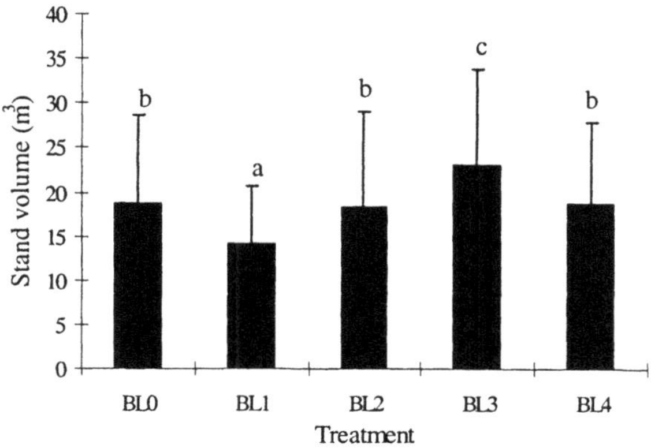

Figure 3. Stand volume of *E. urophylla* plantation with different slash retention treatment at Yangxi at 46 months. Means of the treatment with same letter are not significantly dnifferent at p=0.05 using Newman-Keuls Critical Range Test. Bars = one standard deviation.

As mentioned before, coppice grew better than replanted trees in a poor degraded site. Coppice was also better than replanted trees when the P application rate was low (Xu et al. 2000). However, coppice had poorer growth than trees in the first rotation in plots with relatively high P application. Gonçalves et al. (2000) reported that coppice was better in the beginning and poorer later in a second rotation of *E. grandis* in Brazil. In Congo, where P is not limiting for tree growth, coppice is much worse than replanted trees (Bouillet et al. 1999). In many cases, eucalypt stands in Brazil that are generated from coppice have a lower productivity than trees in the first rotation and the reduction can be up to 50% in stands growing on savanna soils. Barros et al. (1984) reported a 330% increase in productivity by NPK application to a second rotation coppice on a poor Brazilian savanna soil. In China, coppice was much better than replanting at low fertilisation and the gap became smaller as the amount of fertiliser applied increased (unpublished). Therefore, the effect of regeneration method on productivity of second rotation eucalypt plantations is related to soil fertility and fertiliser practice. For most eucalypt plantations in South China, coppice should be a good choice until improved genetic material is available and adequate fertilisation is guaranteed.

6 Management of Other Nutrients (K and B)

So far, only a few trials have demonstrated a positive response to K fertilization for first rotation plantation of eucalypts because P or N is the main limitation for tree growth. However, up to 133 kg K/ha may be lost by erosion in the first year after site preparation (Xu et al. 1999). Hence, further study is needed to monitor soil K status change in successive rotations of plantation eucalypts to ensure that soils do not deplete their K capital.

Boron deficiency has been found after adequate fertilization with N, P and K. This often occurs in sandy or granite-derived soils without organic topsoil after serious erosion. On sandstones in Yunnan, B may be the primary limiting nutrient for eucalypts. Boron deficiency is often expressed towards the end of the dry season.

7 Productivity Increment by Nutrient Management

Results from our trials in Guangdong and Yunnan show that MAI can readily be lifted to 20 m^3/ha/yr by better nutrient management. On a typical plain site where the soil was cultivated and P was not applied, the MAI of a 4.5-year old *E. urophylla* plantation was only 4 m^3/ha (Xu et al. 2001). The MAI was increased to 22 m^3/ha/yr by the application of 20 kg P/ha, and to 30 m^3/ha by the application of 200 kg P/ha. Similarly, on a relatively fertile hilly land (less disturbed topsoil) without P fertilisation, the MAI of a 6.3-year old *E. grandis x E. urophylla* plantation (Xu et al. 2002) was only 4 m^3/ha. At this site, the MAI was increased to 16 m^3/ha by the application of 52 kg P/ha and to 19 m^3/ha by the application of 209 kg P/ha. Moreover, the MAI of a 2.5-year old second rotation coppice was only 9 m^3/ha in the treatment without P fertilisation in the previous rotation (unpublished). The application of 52 kg P/ha in the previous rotation lifted the MAI to 18 m^3/ha. Re-fertilisation with N, K and B in the second rotation further increased the MAI to 23 m^3/ha. Even on a degraded site, the MAI of a 3.7-year old second rotation *E. urophylla* plantation was increased from 1 to 7 m^3/ha with replanted trees, and from 11 to 14 m^3/ha with coppice. Therefore, it is possible to increase the average MAI of eucalypt plantations in South China from 5-10 (current situation) to 10-20 m^3/ha in the near future and to 20 m^3/ha in the future. The challenge will be to determine whether further productivity increase will be driven by improvements in tree breeding or by refinement to silvicultural regimes.

Following the previous discussion on productivity increase by fertilisation, organic material management and ectomycorrhizal fungi inoculation, there is no doubt that the productivity of plantation eucalypts can be improved. However, the cost of plantation establishment is the largest cost for plantation eucalypts in South China. If the rotation period can be extended from the current 4-6 to 6-8 years (the average rotation period for plantation eucalypts in the world) without a decline in MAI, this will benefit both the economic position of the forest farmers and the maintenance of site fertility because there will be less disturbance to the soil if the rotation is longer. Moreover, it will be very helpful for external nutrient cycling and reduce the amount of nutrient loss due to harvesting in the current management model of eucalypt plantations in South China. As mentioned before, the key to maintaining productivity of plantation eucalypts is to manage tree canopy. Therefore, more research effort should be placed on managing nutrients so as to extend the length of the rotation period without a decline in MAI or bole-wood production.

Results from our study so far suggest that productivity of plantation eucalypts is very dependent on management. The low productivity of eucalypt plantations in South China is a consequence of poor management of forest soil fertility in the past, including the over-harvesting of organic matter from forest land and inadequate fertilisation. It is predictable that better management of eucalypt plantations today will improve soil fertility leading to higher productivity in the future. Lastly, more demonstration trials should be established to monitor changes in soil fertility with different long-term management schemes.

References

Aggangan, N.S., Dell, B., Malajczuk, N. and De La Cruz, R.E. 1999. Comparative effect of ectomycorrhizal inoculation and fertilization on *Eucalyptus urophylla* S.T. Blake five years after outplanting in a marginal soil in the Philippines. In: R.E. De La Cruz (ed), Proc. of seventh International BIO-REFOR Workshop, Manila, Philippines 3-5 November 1998, University of Tokyo, Tokyo, pp77-82.

Barros, N.F., Silva, O.M., Pereira, A.R., Braga, J.M. and Ludwig, A. 1984. Análise do crescimento de *Eucalyptus saligna* em solo de cerrado sob diferents niveís de N, P e K no vale do Jequitinhonha. MG. IPEF 26:13-17.

Beadle, C.L. 1997. Dynamics of leaf and canopy development. In: "Management of soil, nutrients and water in tropical plantation forests", E.K.S. Nambiar and A.G. Brown (eds), ACIAR, Canberra, Australian, pp169-212.

Bekunda, M.A., Smethurst, P.J., Khanna, P.K and Willett, I.R. 1990. Effects of post-harvest residue management on labile soil phosphorus in a Pinus radiate plantation. For. Ecol. Mgt. 38:13-25.

Binkley, D. and Giardina, C. 1997. Nitrogen fixation in tropical forest plantations. In: "Management of soil, nutrients and water in tropical plantation forests", E.K.S. Nambiar and A.G. Brown (eds), ACIAR, Canberra, Australian, pp297-328.

Binkley, D. and Ryan, M.G. 1998. Net primary production and nutrient cycling in replicated stands of *Eucalyptus saligna* and *Albizia facaltaria*. For. Ecol. Mgt. 112:79-85.

Bennett, L.T., Weston, C.J. and Attiwill, P.M. 1997. Biomass, nutrient content and growth response to fertilisers of six-year-old *Eucalyptus globulus* plantations at three contrasting sites in Gippsland, Victoria. Aust. J. Bot. 45:103-121.

Birk, E.M. and Turner, J. 1992. Response of flooded gum (*E. grandis*) to intensive cultural treatments: biomass and nutrient content of eucalypt plantations and native forests. For. Ecol. Mgt. 47:1-28.

Bouillet, J.P., Nzila, J.D., Ranger, J., Laclau, J.P. and Nizinski, G. 1999. *Eucalyptus* plantations in the equatorial zone, on the coastal plains of the Congo. In: "Site management and productivity in tropical plantation forests", E.K.S. Nambiar, C. Cossalter and A. Tiarks (eds), Workshop Proc., Pietermaritzburg, South Africa, CIFOR publishing, Bogor, pp13-22.

Brown, S., Lenart, M. and Mo, J. 1995. Structure and organic matter dynamics of a human-impacted pine forest in a MAB reserve of subtropical China. Biotropica 27:276-289.

Burgess, N.L., Grove, T.S. and Malajczuk, N. 1990. Growth and phosporus acquisition of karri (*Eucalyptus diversicolor* F. Muell) seedlings inoculated with ectomycorrhizal fungi in relation to phosphorus supply. New Phytol. 114:77-85.

Cannel, M.G.R. 1989. Physiological basis of wood production: a review. Scand. J. For. Res. 4: 459-490.

Chen, S., Wang, G. and Xiang, D. 1996. Fertilizer response of *Eucalypt urophylla*. For. Res. 9:573-578.

De La Cruz, R.E., Bartolome, H.T. and Aggangan, N.S. 1988. Pilot testing of mycorrhizal tablets for pines and *Eucalyptus* in the Philippines. In: Proc. of UNESCO Regional Workshop on "Development and production of mycorrhizal inoculation", Biotech, UPLB College, Lagum, Philippines.

Dias, L.E., Fernandez, J.Q.P, Barros, N.F. and Novais, R.F. 2001. Availability of phosphorus in a Brazilian oxisol cultivated with *Eucalyptus* after nine years as influenced by phosphorus fertilizer source, rate and placement. Commun. Soil Sci. Plant Anal. 32:837-847.

Eldridge, K.G., Davidson, J., Harwood, C. and van Wyk, G. 1994. Eucalypt domestication and breeding. Clarendon Press, Oxford.

FAO (Food and Agriculture Organization of The United Nations) 1988. The Eucalypt dilemma. FAO, Rome.

Fernandez, J.O.P., Dias, L.E., Barros, N.F., Novais, R.F. and Moraec, E.J. 2000. Productivity of *Eucalyptus camaldulensis* affected by rate and placement of two phosphorus fertilizers to a Brazilian Oxisol. For. Ecol. Mgt. 127:93-102.

Fölster, H. and Khanna, P.K. 1997. Dynamics of nutrient supply in plantation soils. In: "Management of soil, nutrients and water in tropical plantation forests", E.K.S. Nambiar and A.G. Brown. (eds), ACIAR, Canberra, Australian, pp339-378.

Garbaye, J., Delwaulle, J.C. and Diangana, D. 1988. Growth response of Eucalypts in the Congo to ectomycorrhizal inoculation. For Ecol. Mgt. 24:27-42.

Gonçalves, J.L.M. and Barros, N.F., 1999. Improvement of site productivity for short-rotation plantations in Brazil. Bosque 20: 89-106.

Gonçalves, N.F., Barros, N.F., Nambiar, E.K.S. and Novais, R.F. 1997. Soil and management for short rotation plantations. In: "Management of soil, nutrients and water in tropical plantation forests, E.K.S. Nambiar and A.G. Brown (eds), ACIAR, Canberra, Australia, pp379-417.

Gonçalves, J.L.M., Serrano, M.I.P., Mendes, K.C.F.S. and Gava, J.L. 2000. Effect of site management in a *Eucalyptus grandis* plantation in the humid tropics: Sao Paulo, Brazil. In: "Site management and productivity in tropical plantation forests", E.K.S. Nambiar, A. Tiarks, C. Cossalter and J. Ranger (eds), Workshop Proc., Kerala, India, CIFOR publishing, Bogor, pp3-10.

Harrison, R.B, Reis, G.G., Reis, M.D.G.F., Bernardo, A.L. and Firme, D.J. 2000. Effect of spacing and age on nitrogen and phosphorus distribution in biomass of *Eucalyptus camaldulensis*, *Eucalyptus pellita* and *Eucalyptus urophylla* plantations in southeastern Brazil. For. Ecol. Mgt. 133:167-177.

He, R., Jiang, Y., Zeng, P., Chen, D., Zhao, M. and Wang, P. 1999. Study on the effect of fertilization on *Eucalyptus globulus* plantation in Yunnan. For. Res. 12:474-478.

Herbert, M.A. 1996. Fertilizers and eucalypt plantations in South Africa. In: "Nutrition of eucalypts", P.M. Attiwill and M.A. Adams (eds), CSIRO Publishing, Melbourne, pp303-326.

Herbert, M.A. and Schönau, A.P.G. 1990. Fertilizing commercial forest species in Southern Africa: research progress and problems (part 2). South African For. J. 152:34-42.

Holford, I.C.R. 1997. Soil phosphorus: its measurement, and its uptake by plants. Aust. J. Soil Res. 35:227-239.

Huang, B.L. and Liang, X.T. 2000. The continued effects of *Pisolithus tinctoris* on the growth of *Eucalyptus grandis x E. urophylla*. In: "Mycorrhizal fungi biodiversity and applications of inoculation technology", China Forestry Publishing House, Beijing, pp161-167.

Hunter, I. 2001. Above ground biomass and nutrient uptake of three tree species (*Eucalyptus camaldulensis*, *Eucalyptus grandis* and *Dalbergia sissoo*) as affected by irrigation and fertiliser at 3 years of age in southern India. For. Ecol. Mgt. 144:189-199.

Jones, H.E., Madeira, M., Herraez, L., Dightom, J. and Fabiâo, A. 1999. The effect of organic-matter management on the productivity of *Eucalyptus globulus* stands in Spain and Portugal: tree growth and harvest residue decomposition in relation to site and treatment. For. Ecol. Mgt. 122:73-86.

Laclau, J.P., Bouillet, J.P. and Ranger, J. 2000. Dynamics of biomass and nutrient accumulation in a clonal plantation of *Eucalyptus* in Congo. For. Ecol. Mgt. 128:181-196.

Leuning, R., Wang, Y.P. and Cromer, R.N. 1991. Model simulation of spatial distribution and daily totals of photosynthesis in *Eucalyptus grandis* canopies. Oecologia 88:494-503.

Li, Q.K. 1983. Red soils in China. Science Publishing House, Beijing, pp299-302.

Li, Y., Hong, S., Zhou, W. and Chen, D. 1995. Two years' growth response to fertilization in young *Cunninghamia lanceolata, Pinus elliottii* and *Eucalyptus urophylla* plantations. For. Res. 8:241-246.

Liang, K. and Zhou, W. 1999. The effect of fertilization on young clone plantation for *Eucalyptus urophylla*. For. Res. 12:127-131.

Lin, S., Li, S., Liao, G., Lan, P. and Liao X. 1999. A study on fertilization of short rotation *Eucalyptus* plantations. For. Res. 12:275-282.

Lugo, A.E., Brown, S. and Chaoman, J. 1988. An analytical review of production rates and stemwood biomass of tropical plantations. For. Ecol. Mgt. 23:179-200.

Marschner, H. 1995. Mineral nutrition of higher plants, second edition. Academic Press, London.

Malajczuk, N. 1987. Ecology and management of ectomycorrhizal fungi in regenerating forest ecosystems in Australia. In: "Mycorrhizae in the next decade-practical application and research Priorities", D.M. Sylvia (ed), University of Florida, Gainesville, pp118-120.

Malajczuk, N., Grove, T.S., Bougher, N.L., Dell, B. and Gong, M.Q. 1994. Ectomycorrhizas and nutrients : their importance to eucalypts in China. In: "Australian tree species research in China", ACIAR, Canberra, ACIAR Proc. No.48, pp132-139.

McLaughlin, M.J. 1996. Phosphorus in Australian forest soil. In: "Nutrition of eucalypts", P.M. Attiwill and M.A. Adams (eds), CSIRO Publishing, Melbourne, pp1-30.

Mo, J., Brown, S. and Lenart, M. 1995. Nutrient dynamics of a human-impacted pine forest in a MAB reserve of subtropical China. Biotropica 27:290-304.

Morais, E.J., Barros, N.F., Novais, R.F. and Brandi, R.M. 1990. Biomassa e efficiência nutricional de espécies de eucalipto em duas regioes bioclimaticas de Minas Gerais. Revista Brasileira de Ciência do Solo 14:353-362.

Monson, P.A., Ingleby, K., Munro, R.C., Wilson, J. and Kamal, I. 2000. Interactions of nitrogen and phosphorus on mycorrhizal development and shoot growth of *Eucalyptus globulus* (Labill) seedlings inoculated with two different ectomycorrhizal fungi. For. Ecol. Mgt. 128:259-268.

Negi, J.D.S. and Sharma, S.C. 1996. Mineral nutrition and resource conservation in *Eucalyptus* plantations and other forest covers in India. In: "Nutrition of eucalypts", P.M. Attiwill and M.A. Adams (eds), CSIRO Publishing, Melbourne, Australia, pp399-416.

O'Connell, A.M., Grrove, T.S., Mendham, D. and Rance, S.J. 2000. Effect of site management in eucalypt plantations in southwestern Australia. In: "Site management and productivity in tropical plantation forests", E.K.S. Nambiar, A. Tiarks, C. Cossalter and J. Ranger (eds), Workshop Proc., Kerala, India, CIFOR publishing, Bogor, pp61-71.

Pampolina, N. 2001. Ecology of ectomycorrhizal fungi in eucalypt plantations in Australia and Philippines. PhD Thesis, Murdoch University.

Powers, R.F. 1999. On the sustainable productivity of planted forests. New For. 17:263-306.

Reich, P.B., Walters, M.B., Ellsworth, D.S. and Uhl, C. 1994. Photosynthesis-nitrogen relations in Amazonian tree species. I. Patterns among species and communities. Oecologia 97:62-72.

Ruaysoongnern, S. and Keerati-kasikorn, P. 1998. Role of phosphorus fertilization in improving the soil fertility of acid tropical and subtropical soil in Asia. In: "Nutrient management for sustainable crop production in Asia", A.E. Johnson and J.K. Syers (eds), CAB International, Wallingford, UK, pp61-72.

Ryan M.G. 1991. The effect of climate changes on plant respiration. Ecol. Appl. 1:157-167.

Simpson, J.A. 1989. Biomass production, nutrient pools and potential nutrient losses in young *Eucalyptus grandis* plantations at Dongmen. In: Proc. of China-Australia Afforestation Project at Dongmen, forth Technical Exchange Seminar, Nanning, China, pp40-52.

Simpson, J.A., Pegg, R.E. and Xiang, D.Y. 1997. Fertilizer response of eucalypts in Dongmen forest, P.R. China. In: Proc. of IUFRO Symposium on

"Silviculture and genetic improvement of eucalypts", Salvaador, Brazil, pp108-119.
Stewart, B.A., Lal, R. and El-Swaify, S.A. 1991. Sustaining the resource base of an expanding world agriculture. In: "Soil management for sustainability", R. Lal and F.J. Pierce (eds), Soil and Water Conservation Society, Ankeny, Iowa, pp125-144.
Turnbull, J.W., 1994. Eucalypts in China. Aust. For. 44: 222-234.
Turnbull, J.W. 1999. Eucalypt plantations. New For. 17:37-52.
Vogt, K., Asbjornsen, H., Ercelawn, A., Montagnini, F. and Valdes, M. 1997. Roots and mycorrhizas in plantation ecosystem. In: "Management of soil, nutrients and water in tropical plantation forests", E.K.S. Nambiar and A.G. Brown (eds), ACIAR, Canberra, Australia, pp1-19.
Wang, H. and Zhou, W. 1996. Fertilizer and eucalypt plantations in China. In: "Nutrition of eucalypts", P.M. Attiwill and M.A. Adams (eds), CSIRO, Melbourne, Australia, pp389-397.
Willett, I.R., Moody, P.W. and Blamey, F.P.C. 1998. The essential role of phosphorus in crop production. In: "Nutrient management for sustainable crop production in Asia", A.E. Johnson and J.K. Syers (eds), CAB International, Wallingford, UK, pp51-60.
Wu, Z., Ye, D., Li, T. and He, K. 1996. Effect of two-year's fertilization on the growth of *Eucalyptus urophylla* stand. For. Res. 10:167-176.
Xu, D. 1997. A approach to nutrient balance in tropical and south subtropical short rotation plantations, In: "Forest soils", C.D. Yang (ed), Science and Technology Publishing, Beijing, pp125-133.
Xu, D., Malajczuk, N. and Grove, T. 1994. Effects of ectomycorrhizal fungal inoculation on growth and phosphorus uptake of four provenance of *Eucalyptus urophylla*. In: "Mycorrhizas for plantation forestry in Asia", M. Brundett, B. Dell, N. Malajczuk and M. Gong (eds), ACIAR, Canberra, ACIAR Proc. No.62, pp91-94.
Xu, D., He, Q. and Yang, Z. 1996. Above-ground primary productivity and nutrient cycling of the plantation of *Eucalyptus grandis* x *E. urophylla*. In: Proc. of the Conference on "Tropical forestry in 21st century", Bangkok, Thailand, pp81-86.
Xu, D., Liu, Y. and Yang, Z. 1999. Soil erosion control by vegetation restoration in southern China. In: Proc. of International Symposium on "Soil erosion and dryland farming", Xian, China, September 1997, Soil Society of America, USA, pp256-268.
Xu, D., Bai, J. and Dell, B. 2000a. Overcoming the constraints to productivity of eucalypt plantations in southern China. In: Proc. of International Workshop on "Mycorrhizas in forestry", Guangzhou, China, September 1998, China Forestry Publishing House, Beijing, China, pp150-162.
Xu, D., Yang, Z., Chen, M. and Chen, Q. 1999. Effect of site management on productivity of second rotation plantation of *Eucalyptus urophylla* Stake in Guangdong province, China. In: "Site management and productivity of tropical plantations forests", E.K.S. Nambiar, C. Cossalter and A. Tiarks

(eds), Workshop Proc., Pietermaritzburg, South Africa, CIFOR publishing, Bogor, pp32-41.

Xu, D., Dell, B., Malajczuk, N. and Gong, M., 2001. Effect of phosphorus application and ectomycorrhizal fungal inoculation on productivity of a *Eucalyptus urophylla* plantation in South China. In: "Plant nutrient-food security and sustainability of agro-ecosystems through basic and applied research", W.J. Horst, M.K. Schenk, A. Bürkert, N. Claassen and H. Flessa (eds), Proc. of XIV International Plant Nutrition Colloquium, Hanover, Germany, Kluwer Academic Publishers, Dordrecht, pp646-647.

Xu, D., Dell, B., Malajczuk, N. and Gong, M. 2002. Effects of P fertilisation on productivity and nutrient accumulation in a *Eucalyptus grandis x E. urophylla* plantation in southern China. For. Ecol. Mgt. 161:89-100.

Xu, D., Dell, B., Malajczuk, N., Gong, M. and Grove, T. 2000b. Effects of P fertilization and ectomycorrhizal fungal inoculation on tree growth and productivity of *Eucalyptus grandis* x *E. urophylla* plantation in southern China. In: Proc. 8th International BIO-REFOR Workshop, Kathmandu, Nepal, 28 November - 2 December, University of Tokyo, Tokyo, pp164-171.

Yu, X., Xu, D. and Wang, S. 1999. Management of long term productivity of eucalypt plantations. China Forestry Publishing House, Beijing.

Zhong, C., Gong, M., Chen, Y., Wang, F., Malajczuk, N., Bougher, N., Grove, T. and Dell, B. 2000b. Inoculation of mycorrhizal fungi for *Eucalyptus* plantations in southern China. In: "Mycorrhizal fungi biodiversity and application of inoculation technology", M. Gong, D. Xu, C. Zhong, Y. Chen and B. Dell (eds), China Forestry Publishing House, Beijing, pp172-175.

Zhou, W. 1995. A study of fertilization effect on young trees of *Eucalyptus urophylla*. For. Res. 9:159-163.

华南地区桉树人工林的养分管理

徐大平[1], BERNIE DELL[2]

[1]中国林业科学研究院热带林业研究所，
510520 中国广州
[2]*Plant Nutrition Laboratory, Murdoch University*
Perth, WA 6150, Australia

桉类树种不仅速生和易于无性繁殖，而且能够适应各种退化和贫瘠的立地环境条件，在华南地区深受人们的喜爱，是植树造林的最佳树种之一。目前华南地区桉树人工林面积已经超过 150 万公顷，平均每年增加 15 万公顷，但这些桉树人工林的生产力非常低，仅达世界桉树人工林平均产量的 25-50%。近年来，大多数营造桉树人工林的土地都已经退化，大都呈酸性，土壤高度渗透和风化，造成了土壤营养的缺乏。磷元素严重缺乏，与桉树共生真菌数量有限，养分利用效率亦大大降低，而且由于表土层有机物质含量低，氮元素的供应亦非常有限。本文论述了几个国际林业合作项目在华南地区进行了十年的田间试验，包括林地施肥、采代剩余物的管理和桉树共生真菌的接种工作等，关键是想验证通过一个轮伐期的营养管理，能否使华南地区人工林的产量达到世界平均水平。

研究结果如下：在营林初期施用过磷酸钙可以大大提高林木成活率和生长量并增加营养吸收。施磷后的元素吸收率在 6-35%之间时，吸收率随施肥量的增加而降低。第一轮伐期施用磷肥，可以增加第二轮伐期表土层中有效磷、全磷和有机磷的含量。第一轮伐期后，萌芽林对过磷酸钙的增产作用低于重新造林的实生林和无性系林，这说明树木根系中磷的积累对萌芽林具有重要意义。在接种使用的共生真菌中，只有少数几个可以增加立木材积，其效果有时是暂时性的，这可能是苗期其他管理措施削弱了共生菌本身的发展，也可能是移植后，共生真菌的持效性较差。对此，要想获得具有特别效用的接种菌株进行商业性推广，就需要对大量可供菌株进行筛选，以确定那些适用于不同立地条件的菌株。

在采伐以后，保留大量剩余物可以增加新生林的营养供应，减少土壤有机碳的减少，从而提高林木生长量。林间间种固氮树种同样可以提高林木生长量，增加林间凋落物。目前的采伐措施使得林木积累的营养中 80%的营养都随收获物而损失掉了，必须对现有措施进行改进。可以肯定的是，通过营养管理，可以使华南地区桉树人工林的产量提高到世界平均水平，即每年每公顷 10-20 立方米。本研究推荐在第一轮伐期使用的立地管理措施包括整地、杂草控制和施肥等，在第二轮伐期采用保留采伐剩余物、进行萌芽更新等营林措施。

研究结果表明，经营管理条件好的林地的林分产量可以持续发展，而管理较差的林地在几个轮伐期以后其产量就会急骤下降而且土壤也不断退化。根据这些研究结果将制定出华南地区桉树人工林的经营管理模式。

DATABASE SYSTEM APPROACH FOR INTEGRATED PLANTATION NUTRITION MANAGEMENT

LONGBIN HUANG AND BERNIE DELL

Division of Science and Engineering, Murdoch University
Perth, WA 6150, Australia

Fertiliser management for optimal productivity in monoculture forest plantations involves key processes including site-specific prediction and on-going calibration (in terms of tree growth and nutrient requirements), and decision-making (in terms of fertiliser management plans). These processes are ultimately dependent on the integrity and representativeness of the databases on local conditions, tree growth potential and soil fertility indicators. Given the fact that data collection is a costly exercise, it is important to identify the scope of data entities that are economically achievable and scientifically sound, and logical processes that are required for mining the databases for decision making. Four key components of the system framework proposed are: (1) growth modelling of total and component biomass at the tree and stand level; (2) databases of internal nutrient requirements in trees and nutrient loading in biomass components; (3) databases of available nutrient pools in the soil; and (4) databases about fertilisers' effectiveness in terms of replenishing available nutrient pools in soil. The framework presented here is by no means complete, but serves as a starting point for detailed data and process modelling and field calibration.

1 Introduction

Since the 1980s, plantation forests have become important sources of timber and paper pulp (Evans 1999), due to the diminishing supply of natural forests and the protection of natural forests by governments and the general populace. By the early 1990s, 40-50 million ha of forest plantations had been planted in the tropics and subtropics (FAO 1993). Among the genera selected for plantations, *Eucalyptus* has been particularly favoured by commercial plantations, because of (1) the fast growth and high yielding nature, (2) adaptability to different ecosystems, (3) maturing precociously (short-rotation requirements), and (4) wood quality of excellent properties for fibre and saw timber (FAO 1988). During the last decade, more than 6 million ha of eucalypt plantations were established worldwide (Turnbull 1999). In Australia, there has been a rapid expansion of eucalypt plantations (in particular, *E. globulus*) and by the year 2020, it is visioned to reach about 3.3 million ha of plantations (MCFFA 1997).

In response to increasing demands for plantation timber and fibre and governments' encouragement to shift industrial dependence of wood/fibre away from natural forests onto plantations, there has been a steady flow of private/governmental investments into commercial plantation projects (with rotation terms of 7-10 years) on ex-farm (crop/pasture) and marginal land. As a result, investors expect profit stability on these commercial plantations, which are ultimately dependent on sustainable soil fertility management.

Plantation species like *E. globulus* and *E. urophylla* are fast growing, requiring adequate supply of nutrients and water to attain the profitable level of productivity. During the first couple of rotations, soil nutrient reserves built up from decades of farming activities (crop or pasture production) may be able to support a reasonable level of productivity. However, integrated fertiliser management is the only way to maintain the existing soil fertility for infinite rotations of short-term plantations. Previous lessons have been learned from eucalypt plantations in the tropics. For example, in Brazil plantation productivity can be severely damaged due to the depletion of soil K and the application of 80 kg K_2O almost doubled the volume of timber produced (Barros et al. 1990). In southern China, marginal lands of mountainous areas are usually allocated to eucalypt plantations and their low productivity is often associated with low soil nutrient supply, leading to the necessity of fertiliser (N, P, K) application in the field (Xu et al. 2002). Hence, without integrated nutrient management, the risks can be two fold: (1) the failure to reach the expected productivity, and (2) the degradation of site fertility that results in unsustainable plantation productivity within and between rotations.

The use of growth models for silviculture, production planning and wood quality grading have been well practiced in the forestry industry. Decision support software tools have been proposed for pine plantations, that link silviculture, wood quality and grading of structural timber and allow the evaluation of alternative silviculture scenarios in relation to internal rate of return and net present value (Catchpoole and Nester 2002). However, in most models, there is a lack of consideration of tree nutrient requirements and soil fertility (Roux et al. 2001). In some models, tree nutrition consideration is only given to the macronutrients – usually nitrogen and phosphorus (Roux et al. 2001). As a result, to achieve an integrated fertiliser management for sustainable plantation productivity, it is necessary to develop a modelling approach which links growth models with tree nutrition requirements for target productivity and nutrient behaviour in soils.

The objective of the present paper is to propose a framework for a database system that combines the existing knowledge of growth models, tree nutrition (including nutrient requirements, mobility, and partitioning among biomass components) and nutrient characteristics in soils. We will discuss the use of suitable growth models for the estimation of total and component biomass, the scope of data entities that are necessary for reliable growth projection, representation of soil fertility for plantation productivity, and fertiliser effectiveness for maintaining available nutrient pools in soil and which sustain tree growth in subsequent rotations. The decision-making processes for site-specific scenario analysis of fertiliser inputs and productivity output will be identified. Effective plantation management is dependent on a representative and sound database.

The framework of the database system is proposed to include modules for (1) at preplanting, predicting the maximum biomass increments during the term of rotation at the regional level, by considering climatic factors such as temperature, photosynthetic active radiation (PAR) and rainfall, and setting the benchmark of the upper limit of internal nutrient requirements and projected fertiliser input; (2) site-specific estimation of tree biomass (and component biomass) by taking into account site index that is the combined result of site conditions including soil properties, available nutrient pools and water availability; (3) estimating the effectiveness (the conversion factor of raw fertilisers into available nutrient pools) of fertilisers applied for maintaining or improving the available pools of nutrients in the soil and their supply to the roots; and (4) analysing scenarios of costs of various fertiliser management options and productivity outcome, in order to choose the optimal management solutions for achieving sustainable profitability.

2 The Role of Growth Models in Plantation Fertiliser Management

Detailed review of growth models is beyond the scope of the present paper. However, briefly speaking, tree growth models include two main types: process-based and empirical. Process models take into account the main physiological processes in carbon metabolism, including photosynthate production, respiration, reserve dynamics, allocation of assimilates and growth (Roux et al. 2001). Unlike empirical models which are only limited to specific plantation and site conditions and lack predictive power under different conditions, carbon-based process models provide a valuable means to predict the growth of fast-growing trees under different site, climatic and cultural conditions and can make predictions beyond the range of data used in model development. The carbon-based models of individual tree growth generally consist of sub-models, each describing one of the main carbon processes. The processes driving the carbon dynamics and growth remain fundamentally identical between different tree species, with the only difference in their species- and site-specific parameters (Roux et al. 2001). As a result, models developed for particular species can be applied to a range of tree species when suitably parameterised. However, process models are often complex and difficult to use due to the requirement of many input parameters. The combined used of process models and empirical models may provide solutions to plantation management issues (Adlard 1995).

3 Growth Model Selection

From the perspective of plantation management, it is important to estimate the maximum potential of productivity and associated fertiliser input requirements (after considering fertiliser effectiveness of conversion into available nutrients in soils concerned (to be discussed later) during the project planning phase. This estimation sets the benchmark for the upper limit of fertiliser costs and the most optimistic productivity of plantations to be achieved. To accomplish this task, it is necessary to utilise an appropriate growth modelling system, which produces accurate estimates of tree biomass at stand level after considering regional climatic factors (including PAR flux density, temperature and rainfall). It is assumed that species-related parameters have been built into the system, including morphological, physiological and phenological parameters characterising eucalypt species.

Tree species selected for commercial plantations (such as *E. globulus*) are usually fast growing and maturing within a short-rotation period (8-10 years).

Among various models published, the ECOPHYS model is the only individual tree level model for fast-growing trees, with detailed description of the architecture of the plant (Philippot 1996). The ECOPHYS model operates at the individual leaf level on an hourly time step, based on leaf photosynthesis process that is a function of hourly light and temperature. The photosynthate model parameters are genetically determined and vary with leaf age (Host et al. 1996). Outputs of the model include cumulative whole-tree photosynthate production, leaf area development, biomass by component (leaves, shoot, roots), diameter and height (Ceulemans 1996; Host et al. 1996). The model is used to develop morphological and physiological criteria and to assess the impacts of multiple environmental stresses on tree growth at the juvenile stage (Ceulemans 1996).

It is recognised that it is necessary to scale the model at the individual tree level up to the stand or plantation level, for producing growth simulation across the entire rotation years, to be useful for management planning and decision support. To overcome this weakness, it is proposed to link the ECOPHYS model with expert databases, and with a field scale model like EPIC (Host et al. 1996). EPIC is an empirical model at the field scale, operating on a daily time step using soil, weather, cultural, and crop-related parameters (Host et al. 1996). Biomass production in EPIC is based on a radiation use efficiency approach where the crop-specific coefficient describes the relationship of intercepted PAR to biomass. A method of interfacing ECOPHYS and EPIC has been proposed by Host et al. (1996), in which EPIC provides climatic variables (light, temperature) to the ECOPHYS model and ECOPHYS feedbacks data on leaf area index (LAI), total above ground biomass and leaf litter.

Coppice regeneration is an alternative to replanting in the second and subsequent rotations. While developing and calibrating biomass simulation models, considerations should also be given to the root system and morphology of coppiced trees. It is reasonable to expect that nutrient uptake and supply efficiency to the growth of the above-ground biomass would be elevated during the early growth phase, compared to newly planted juvenile trees. Time-course data about the development of leaf area and canopy in coppiced eucalypt trees are lacking and are worth particular attention if coppicing becomes a significant option for the next rotation.

4 The Importance of Permanent Plots for Site-Specific Calibration

Without calibrating growth models with site-specific data, regional scale growth modelling may not be able to generate the information required for on-going site-specific fertiliser management decisions. One common approach is to gather required data from permanent sampling plots (PSPs) that are specifically set up at pre-planting to represent soil conditions at each plantation site. Comprehensive databases generated from representative PSPs not only enable the development and calibration of growth models for biomass estimation, but also provide the basis for valid inferences on optimal silviculture regimes including planting density, thinning rates and timing, rotation length and regeneration method (Vanclay et al. 1995). As plantations are normally of mono-culture nature and relatively uniform (compared to natural forest ecosystem), the establishment of PSPs is usually achievable in terms of associated costs.

The data collected from the PSPs may include two major categories: (1) tree growth parameters such as BHD, basal area, height; and (2) soil variables such as soil physical/chemical properties, and available nutrient pool sizes. Site index is usually estimated from tree growth parameters (e.g. wood volume) of the PSPs and it is an indicator of site productivity (Nigh and Love 1999). In the case of short-term, fast-growing plantations, site index is the combined result of native site productivity and site management intervention.

It is proposed to establish two categories of PSPs (1) the least managed (LM-PSP) and (2) the best managed (in reality, it is determined by the financial capacity of a grower) (BM-PSP), in terms of fertiliser inputs and weed control. Site indices derived from the LM-PSPs and BM-PSPs will be used to set the benchmarks of the scenarios of the worst productivity (native productivity) and the best productivity achievable through the best (or the most affordable) site management practices. The LM-PSP site index would indicate the existing/inherited site productivity without any management input. The BM-PSP site index would indicate the management-achievable productivity under the site-specific climatic conditions. As a result, optimal management practices may be inferred from these two indices while reaching a balance of cost-profit.

However, the reliability of these site indices would depend on the quality of PSP databases, which should be structured according to geography, soil type, soil depth, and productivity levels (Vanclay et al. 1995). It is also important to sample trees with minimal or insignificant damage from the PSPs (Nigh and Love 1999).

5 Nutrient Requirements for Expected Productivity

After reaching an understanding of the site-specific growth potential of the plantations concerned, it is necessary to have a quantitative estimation of nutrient loading and partitioning in the biomass components and associated nutrient pool size in soil. For the convenience of data collection, the internal nutrient requirement may be simply categorized as the amount of all essential nutrients loaded in biomass components of healthy trees representative of plantations at a defined age. The external nutrient requirement may be defined as the available pool size of all nutrients required for supporting the expected productivity without depleting soil fertility for subsequent rotations. After obtaining these data, the application rates of raw fertilisers at a given site can then be calculated by considering the effectiveness of various chemical fertilisers in replenishing the available nutrients in soil under the site-specific conditions.

6 Internal Nutrient Requirements and Relative Risks of Acute Deficiency

The internal nutrient requirement for a given plantation may be estimated by field sampling of established plantations of various age and sampling trees in permanent sampling plots. Data on internal nutrient requirement or loading in biomass components of trees sampled form the basis of estimation of the minimal fertiliser input requirement for the site. Existing data on nutrient partitioning among biomass components in trees of representative age categories (e.g. year 1 to year 10 for eucalypt plantations) are hardly available in the literature, perhaps due to the commercial value of these data.

When research data are not available, field sampling is the quick way of establishing the initial database of internal nutrient requirements. The procedure of field survey for such a purpose includes: (1) identifying a series of plantations of the same species of various ages, which are established at sites with similar management history and soil and climatic conditions, (2) selecting best trees (without significant damage from insects, past nutrient disorders, and others) for destructive sub-sampling of biomass components for nutrient quantification, and (3) analysing the nutrient data critically by relating them to nutrient physiological functions, and then compiling them into a relational database. Due to variations in soil fertility and climatic conditions, the field sampling approach may be able to generate a relatively complete dataset that is representative of all sites under management. Permanent sampling plots set up for site index assessment are ideal for the collection of nutrient loading in biomass components and optimal nutrient requirements in relation to the best management practice.

Existing data from commercial plantations on nutrient concentrations in biomass components in a healthy tree are scarce and incomplete in terms of age profile and site conditions. Xu et al. (2002) published a set of data on biomass of tree components and nutrient concentrations, in response to optimal NPK fertiliser application treatments in a 75-month old clonal plantation of *E. grandis x E. urophylla*. These values certainly provide an initial guidance for estimating the total internal nutrient requirement for the desired productivity at the plantation site. However, they are not optimal as the plantation studied showed a decline in LAI over time. The extrapolation of these data to different plantations of the same species in different age groups requires great care. First of all, different plantations may have different management practices such as planting density and site preparation, which can determine the resource competition intensity, particularly after canopy closure. However, research data related to physiological functions of each biomass component will provide a reliable basis for critically screening and validating data gathered through field sampling.

Risk factors or rating may be assigned to each category of nutrients when evaluating nutrient requirements for the expected productivity. The risk of acute deficiency of individual nutrients is often associated with their mobility in the plant system. Acute deficiencies are those that occur abruptly in the new growth, and which often result in stunted growth or deformed morphology, leading to losses in productivity and quality. Chronic deficiencies of nutrients are those that take relatively longer time to develop into severely stunted growth and shoot deformation. Micronutrient deficiencies are usually acute and may cause severe growth retardation if they are not promptly diagnosed and corrected (Dell et al. 2003). Micronutrients (Mn, Zn, Cu, B, and Fe) have low re-utilisation efficiency when their supply is limited and deficiencies exhibit as stunted or discoloured new growth. Most macronutrients (particularly N, P, K) are mobile and re-utilised by the new growth when the external supply is low/limited through retranslocation from old leaves. As a result, the requirements of micronutrients should be factored in as an insurance policy, rather than pursuing quantitative optimum.

7 Soil Fertility and Sustainability

An equally important component of integrated fertiliser management is the quantitative estimation of soil nutrient supplying capacity at the plantation site. One approach to quantify changes in site fertility within rotation and between rotations is to document the size and distribution of biomass and macro- and micro-nutrient pools in soil and (living/dead) biomass. By measuring biomass distribution and nutrient concentrations, total site nutrient inventory and nutrient cycling data can be compared between years and rotations to reveal if site nutrient supplying capacity is being maintained.

The quantification of soil fertility changes and sustainability is the most complex and difficult component of plantation fertiliser management. There are abundant detailed information about soil chemistry and nutrient behaviour in soils in the literature, but simple and reliable indicators have to be developed for effective management decision-making. The quantification of all soil processes is unconceivable and unrealistic, in terms of management requirements and financial costs. A "black-box" approach may be adopted, by which the net consequences of natural influences and human interventions on soil fertility are quantified on a regular basis, particularly in the PSPs representative plantation sites. However, the existing information on soil chemistry and nutrient behaviour in soils can provide an important basis for critically interpreting the data acquired and inferences for fertiliser application decisions.

Gresham (2002) proposed to use the concept of site nutrient conservation to reflect changes in soil fertility and the sustainability of site productivity. The total site nutrient conservation indicators include the internal nutrient load of the standing plantation and nutrient pools (total and available pool) in the soil. If a plantation accumulates nutrient faster than predicted by age, it may suggest that the associated forest management practices not only conserve site nutrients, but also allow accumulation (Gresham 2002). By comparing the trend of changes in these pool sizes between seasons and rotations, the effectiveness of fertiliser management practices can be measured and the sustainability of the site for the expected productivity in the following rotations can then be assessed. Fertiliser management measures should be adequate to allow the maintenance of soil fertility with the possible trend of improvement and as a result sustainable production in future rotations.

Permanent sampling plots set up for the purposes of fertiliser management options (as discussed before) can play a significant role in the quantification of soil nutrient pools. Data generated from the PSPs reflect the consequences of fertiliser management practices in relation to expected productivity. To establish a long-term database on changes in soil nutrient pools, it is important to sample the site representative at preplanting, in order to set a baseline for subsequent comparison and inferences.

In practice, there are some difficulties in achieving the quantification of soil nutrient pools, due to the variability of soil conditions including soil properties, depth and fertiliser history. Prasolova et al. (2000) clearly demonstrated the spatial and temporal variability of soil moisture, nitrogen availability indices (total nitrate and ammonium, potentially mineralizable N) and other chemical properties including organic carbon, available P, pH, CEC, and exchangeable cations. The degree of variability in soil properties and total and labile pools of nutrients within a sampling plot increases with the sampling area (Prasolova et al. 2000). To accurately estimate the soil nutrient pools within a 1.36 ha pine plantation, up to 66 separate samples may be required to control the sample mean relative error within 10% (Prasolova et al. 2000). Fortunately, PSPs (300-400 m^2) and sub-plots (50-100 m^2) are usually relatively small and soil samples can be collected from the same plot annually during the wet season. The trend of changes is more important than the spatial variability. One approach to reduce the number of samples is to sample adequate number of spots to make up one composite soil sample for the analysis of nutrient pool indices.

Another important factor to consider is the sampling depth: which depth is reasonably reflecting the majority of root uptake activity? The common practice is to sample surface soil from 0 to 20 cm, on the assumption that most of the feeding roots are distributed within this layer of soil. Various sampling depths have been used in published studies, for example, 10-30 cm in pine plantations (Prasolova et al. 2000; Gresham et al. 2002), and 10-20 cm in eucalypt plantations (O'Connell and Rance 1999; Xu et al. 2002).

There are two important aspects to be considered when choosing a sampling depth that closely matches the distribution pattern of fine roots: soil types and plantation age. Soil type not only influences the movement of nutrients along the soil profile, but also the penetration depth of roots. It is reasonable to expect that the amount of nutrients moving downward increases with the proportion of the sandy components. The pattern of fine root distribution of young trees would be very different from the mature ones. The typical depth for the majority of fine roots can be estimated by field sampling of existing plantations of various age classes at representative sites. For soil nutrient budget and pool size estimation, it is preferable to sample as deep as 30-50 cm. Most of the fine roots (2 mm or less in diameter) are distributed within the first 50 cm soil layer, though varying with species (O'Connell and Sankaran 1997) and they not only determine the nutrient uptake efficiency, but also play a critical role in nutrient recycling through seasonal growth and decay (Vogt et al. 1997).

In summary, for the practicality of management decision without compromising too much of the data reliability, the top 50 cm soil layer should be sampled for the estimation of soil nutrient budget and total and available nutrient pools. The representativeness of the data is determined by the sample size. One approach to reduce the number of soil samples is to combine soils sampled from a number of spots within a PSP into a composite sample. The bottom line of this exercise is strongly influenced by the understanding of plantation managers as well as financial considerations.

8 Fertiliser Properties and Effectiveness

After estimating the internal and external nutrient requirements for the expected productivity of a plantation, it is important to select the right type of mixture of different types, suitable application methods, and application timing, to timely replenish the available nutrient pools, to prevent nutrient disorders from occurring, and effectively correct existing nutrient deficiencies if any. It is very useful for managers to refer to an expert database that contains data about (1) nutrient physical and chemical properties (including solubility and nutrient composition), and (2) their reactions with soils (e.g. nutrient releasing rate in soils). It is easy to compile data of the former through fertiliser manufacturers' product information. However, the challenges lie with the latter. The simple and quick way may be to incubate various fertilisers under defined conditions and *in situ*, to estimate the conversion factors into available nutrients. The estimate of site nutrient pools in PSPs is also an indirect measurement of fertiliser effectiveness in maintaining the nutrient pools. The estimate of fertiliser conversion factors would then allow the calculation of fertiliser costs for a plantation concerned.

9 Nutrient Disorder Diagnosis

Foliar sampling is a useful method to monitor the on-season growth response to soil nutrient supply. Trees may be randomly sampled from the PSPs to establish a database that allows the examination of historic trends of changes in relation to seasonal conditions (such as rainfall/water availability) and fertiliser input. Foliar analysis results also serve an initial alert for plantation managers to find out existing problems with nutrient status and correcting solutions to them. It is often the case that foliar results are sent to managers without proper interpretation due to a lack of expertise in commercial analytical labs or the lack of calibrated standards for correct interpretation. An automated nutrient deficiency diagnosis tool with a built-in data base of foliar standards will provide a great convenience to plantations managers to initiate follow-up measures and solutions, in response to identified problems.

10 Future Development

The present paper only sets the basic framework of main aspects to be considered in the development of a database system for integrated fertiliser management for short-term fast-growing tree plantations. Detailed database schema and functional processes are to be defined before any physical implementation of these concepts becomes possible. On-going calibration of growth models and fertiliser requirements is necessary, in order to strengthen the relationships between soil fertility parameters chosen and tree growth performance and therefore, improve the predicting capacity of the system and minimise investment risks for plantation shareholders and landowners. A database system for integrated fertiliser management should provide tools for: (1) a holistic examination of plantation input requirements and growth performance across regions and seasons, (2) multiple scenario analysis of management options and possible consequences for risk (mainly nutrient disorders) management, (3) mining historic data and on-going calibration, (4) forecasting forthcoming events of possible environmental stresses including pathogen and insects outbreaks, drought and nutrient disorders, and (5) tools for profit margin evaluation of fertiliser management inputs.

Acknowledgement

Plant Nutrition Services Pty Ltd (plantns@iinet.net.au) provided full financial support for the present research and conference attendance.

References

Adlard, P.G. 1995. Myth and reality in growth estimation. For. Ecol. Mgt. 71:171-176.

Barros, N.F., Novais, R.F. and Neves, J.C. 1990. Fertilizer correction of eucalypts plantations. In: "Relacao solo-eucalipto", N.F. de Barros and R.F. de Novais (eds), University of Federal de Vicos, Brazil, pp128-181.

Catchpoole, K.J. and Nester, M.R. 2002. Decision support software tools for pine plantations. NZ J. For. Feb:15-18.

Ceulemans, R. 1996. An inventory of tree and stand growth models with potential application in short-rotation forestry. Biomass and Bioenergy 11:95-107.

Dell, B., Xu D., Rogers C. and Huang L. 2003. Micronutrient disorders in eucalypt plantations: causes, symptoms, identification, impact and management. In: "*Eucalyptus* plantations – research, management and development", R.-P. Wei and D. Xu (eds), Proc. of International Symposium on "*Eucalyptus* plantations", Guangzhou/Zhaoqing, China, 1-6 September 2002, World Scientific, Singapore, pp241-252.

Evans, J. 1999. Planted forests of the wet and dry tropics: their variety, nature, and significance. New For. 17:25-36.

FAO 1988. The Eucalypt Dilemma. FAO, Rome.

FAO 1993. Forest Resource Assessment 1990: tropical countries. FAO Forestry Technical Report Paper 112, pp59-99.

Gresham, C.A. 2002. Sustainability of intensive loblolly pine plantation management in the South Carolina Coastal Plain, USA. For. Ecol. Mgt. 155:69-80.

Host, G.E., Isebrands, J.G., Theseira, G.W., Kiniry, J.R. and Grahams, R.L. 1996. Temporal and spatial scaling from individual trees to plantations: a modelling strategy. Biomass and Bioenergy 11:233-243.

MCFFA 1997. Plantations for Australia: The 2020 Vision. Implementation Committee, Ministerial Council for Forestry, Fisheries and Aquaculture, Department of Primary Industries and Energy, Canberra, pp88.

Nigh, G.D. and Love, B.A. 1999. How well can we select undamaged sited trees for estimating site index? Can. J. For. Res. 29:1989-1992.

O'Connell, A.M. and Sankaran, K.V. 1997. Organic matter accretion, decomposition and mineralisation. In: "Management of soil, nutrients and water in tropical plantation forests", E.K.S. Nambiar and A.G. Brown (eds), ACIAR Monograph No.43, pp443.

O'Connell, A.M. and Rance, S.J. 1999. Predicting nitrogen supply in plantation eucalypt forests. Soil Biol. Biochem. 31:1943-1951.

Philippot, S. 1996. Simulation models of short-rotation forestry production and coppice biology. Biomass and Bioenergy 11:85-93.

Prasolova, N.V., Xu, Z.H., Saffigna, P.G. and Dieters, M.J. 2000. Spatial-temporal variability of soil moisture, nitrogen availability indices and other chemical properties in hoop pine (*Araucaria cunninghamii*) plantations of subtropical Australia. For. Ecol. Mgt. 136:1-10.

Roux, X.L., Lacointe, A., Escobar-Gutierrez, A. and Le Dizes, S. 2001. Carbon-based models of individual tree growth: a critical appraisal. Ann. For. Sci. 58:469-506.

Turnbull, J.W. 1999. Eucalypt plantations. New For. 17:37-52.

Vanclay, J.K., Skovsgaard, J.P. and Hansen, C.P. 1995. Assessing the quality of permanent sampling plot databases for growth modelling in forest plantations. For. Ecol. Mgt. 71:177-186.

Vogt, K., Asbjornsen, H., Ercelawn, A., Montagnini, F. and Valdes, M. 1997. Roots and mycorrhizas in plantation ecosystems. In: "Management of soil, nutrients and water in tropical plantation forests", E.K.S. Nambiar and A.G. Brown (eds), ACIAR Monograph No.43, pp247.

Xu, D., Dell, B., Malajczuk, N. and Gong, M. 2002. Effects of P fertilisation on productivity and nutrient accumulation in a *Eucalypts grandis* x *E. urophylla* plantation in southern China. For. Ecol. Mgt. 161:89-100.

人工林营养综合管理的数据库系统法

LONGBIN HUANG，BERNIE DELL

Division of Science and Engineering, Murdoch University
Perth, WA 6150, Australia

单一品系人工林最佳生产力的施肥管理牵涉到几个关键过程，包括立地特有条件的预测、树木生长和养分需求的不断校正、施肥管理计划的决策。这些过程最终依赖于有关当地环境条件、树木生长潜势和土壤肥力指标方面数据库的完整性和代表性。考虑到数据收集的昂贵，重要的是界定所需要的数据范围，这些数据应该是在经济上不难获得、在科学上是合理的，此外，还需要为决策而开发的数据库确定逻辑流程。我们建议的系统框架包括四个关键组分：（1）个体和林分水平上总体和组分生物量的生长模型；（2）树木本身养分需求和生物量组分养分负荷的数据库；（3）土壤中有效养分库的数据库；（4）肥料效能，即通过施肥补充土壤有效养分库的数据库。本研究所建议的框架并非是一个完美的系统，但是可以作为详细数据记录、过程建摸和野外校正的基准点。

conservative treatment of site resources that, if practiced over successive rotations, can form the basis for sustainable plantation management systems. Of particularly importance is the management of harvest residues, which can contain substantial quantities of plant available nutrients. Treatment of these materials following logging and subsequent site preparation for the new crop cycle is crucial to maintenance of site nutrient capital. Thus, a central theme of research in the network is the evaluation of alternative options for harvest residue management in relation to their effect on site nutrient stores and flux rates and tree productivity. A core set of treatments has been established across all sites with additional treatments tailored to each region depending on local management practices. While a prime consideration in all experiments is the provision of information required to assist with local management issues, this integrated experiment will ultimately allow the evaluation and generalization of results across a broad range of plantation sites in different countries.

Short rotation eucalypt plantations form a major component of the CIFOR network project, with 10 sites in 6 different countries. Collaborative research between CSIRO, Australia, and Kerala Forest Research Institute, India, funded by the Australian Centre for International Agricultural Research (ACIAR) contributes to this part of the CIFOR network project. In India, eucalypts are a major plantation species, covering a land area of 4.8 million ha (Davidson 1995). In Kerala State, two species of eucalypts (*Eucalyptus tereticornis* and *E. grandis*) are grown in short rotations (6 to 7 years), primarily to supply feedstock to the pulp and paper industry. The productivity of plantations in this region is generally low (5 to 10 m^3/ha/yr) compared to that achieved in many other parts of the world. In order to address this issue, in 1998 we established four inter-rotation site management experiments (two for each species), incorporating the CIFOR core treatments. The aim was to evaluate alternative options for management of harvest residues and to quantify their impact on soil fertility and tree productivity. We also incorporated a range of additional experiments in the study to investigate other silvicultural options such as fertilizer additions, weed control, use of cover crops and the effect of different tree spacing. Sankaran et al. (2000) have reported on outcomes from this research.

In this paper we focus on research results from southwestern Australia, where *E. globulus* plantations cover more than 200,000 ha. Most of these plantations are established on land previously used for various forms of agriculture, primarily grazing of cattle and sheep on improved pastures. In some areas trees are planted to reverse the effects of severe land and water degradation processes, such as water-logging and salinization, which have developed following clearing of the natural perennial vegetation. However, the main impetus for the establishment of most plantations has been the development of a commercial hardwood plantation industry. Although individual landholders have established some plantations, the majority of the plantation estate is owned and managed by industrial companies. The trees are grown in 10-year rotations to provide wood chips for export.

The landscape of southwestern Australia has developed over long time scales without major rejuvenation processes. As a consequence, the majority of soils are highly weathered and low in natural fertility. Following clearing for agriculture, large amounts of phosphate-based fertilizers have been applied to these soils to promote improved legume-based pastures. These management practices have improved soil fertility and increased production potential for the region as a whole. Consequently, where pastures are planted to eucalypts, soil nutrient supply rates support high growth rates of trees. However, recent research has indicated that this change in land use may, in the longer term, lead to a decline in availability of some plant nutrients, especially nitrogen and phosphorus (Grove et al. 2000; O'Connell et al. 2003). Therefore, implementing management strategies that aim to maintain and enhance soil fertility will be critical in the future for achieving sustainable levels of productivity.

In this region, the majority of plantations have been established since 1990. Harvesting of older stands has commenced and the management of second rotation crops will become an increasingly important aspect of stand management during the next few years. At harvest, large amounts of tree residues are deposited on the soil surface. One option for managing soil productivity is to develop judicious site management systems that are applied during the inter-rotation period of stand harvest and plantation re-establishment. Our research is focussing on options for treating harvest residues and the impact these strategies may have on soil organic matter, soil nutrient status, nutrient flux rates and tree growth. The aim is to develop guidelines to manage soil fertility in order to maintain high productivity of plantation eucalypts over successive rotations.

2 Materials and Methods

2.1 *Location and Site Description*

Detailed descriptions of the climate, soils and experimental sites and location of this research on *E. globulus* plantations have been reported elsewhere (O'Connell and Grove 1999; O'Connell et al. 2000) and are only briefly reiterated here to provide the context for this paper.

Southwestern Australia has a seasonal Mediterranean type climate with cool wet winters and warm, dry summers. The *E. globulus* plantation estate is limited to the southern part of this region where annual rainfall is in the range from 600 to 1500 mm. Rainfall at the experimental sites ranges from 800 to 1000 mm annually. Ancient Archaean crystalline rocks, primarily granites and granitic gneisses, dominate the geology of the region. These have weathered *in situ* resulting in deep soil profiles that are frequently mantled by laterite. Surface soils reflect the highly weathered nature of these parent materials and generally have low levels of plant-available nutrients (Turton et al. 1962).

In 1995, we established two inter-rotation site management experiments at Busselton and Manjimup following harvest of the first rotation *E. globulus* plantations (O'Connell et al. 2000). These sites are in similar rainfall zones but have contrasting soils. At Busselton, the soil is infertile grey sand (Podzol), total aboveground biomass of the first rotation crop was low (98 t/ha after 9 years) and about 31 t/ha of harvest residue was left on the soil surface after logging. At Manjimup the soils were heavily textured red earths (Ferralsol) with relatively high levels of plant available nutrients, total above-ground biomass of the first rotation plantation was high (275 t/ha after 9 years) and 81 t/ha of harvest residues (including 30 t/ha coarse woody debris) was left on site after logging.

2.2 *Experimental Methods*

At both sites, harvest residue management experiments were established using a randomised block design with four replicate blocks. The design incorporated the core CIFOR treatments BL_0 (no residues, all litter removed), BL_2 (single residues), and BL_3 (double residues) together with an additional treatment where all residues were burnt (B). At the same sites, separate randomised block experiments were conducted to evaluate responses to added N and P and to assess the role of inter-row agricultural legumes in early stages of the rotation.

The impacts of harvest management options on soil fertility were assessed through two sampling regimes. Firstly, soil samples were collected annually to evaluate changes in stores of organic carbon and nutrients over time. Results from seven sample collections are now available. Secondly, samples were collected monthly to examine the dynamics of soil mineral N (Raison et al. 1987) in relation to harvest residue management treatments. Details of the sampling protocols and analytical methods have been given by O'Connell et al. (2000). Tree growth (stem diameter and height) was measured annually. Aboveground biomass and N accumulation were estimated using allometric relationships with conical volume derived after harvesting some trees.

3 Results

3.1 *Amounts and Nutrient Content of Harvest Residues*

Markedly lower concentrations of total and labile P as well as exchangeable K in the coarse textured grey sand compared to the soil from the red earth site were the main features differentiating soil nutrient status of the two sites. This in turn was one of the factors that affected the relative productivity of each site and consequently the amounts of harvest residues present following logging of the first rotation stands (Table 1).

At the red earth site, which represents the upper range of plantation production in Western Australia, more than 50 t/ha of residues with end diameter <3 cm was deposited on the forest floor during harvesting. At the less fertile grey sand site 31 t/ha of residues remained following harvesting. There was a corresponding difference in the nutrient loads added in harvest residues (Table 1).

3.2 *Impact of Harvest Residues on Soil Nutrient Stores*

Changes in soil nutrients, organic carbon and pH have been monitored annually since December 1994 (red earth site) and June 1995 (grey sandsite). At both sites, measurements 5 years after stand establishment indicate that the pH of surface soil (0-5 cm) is higher by about 0.3 units on burn (B) and high residue treatments (BL_3) compared to zero and single residue treatments (BL_0 and BL_2). At this time, there were also significant effects of treatments ($P<0.05$) on surface soil (0-5 cm) organic

C at both sites (Figure 1), with high residue treatments resulting in higher organic C. Increased total organic C levels are also associated with correspondingly higher labile soil C concentrations at both sites.

Table 1. Dry weight and nutrients in harvest residues at the two experimental sites.

Site	Harvest residues	Dry weight (t/ha)	N (kg/ha)	P (kg/ha)	Ca (kg/ha)	Mg (kg/ha)	K (kg/ha)
Red earth	Leaves	21.5	259	15	315	30	96
	*Wood, bark & miscellaneous	29.6	88	8	282	29	86
	Total	51.1	347	23	597	59	182
Grey sand	Leaves	13.0	138	9	188	18	56
	Wood, bark & miscellaneous	18.3	81	7	200	18	55
	Total	31.3	219	16	388	36	111

* Wood residues <3cm diameter. At the red earth site an additional 30 t/ha of larger wood residues (>3cm diameter) was present on the soil surface following harvesting. This is a greater amount of wood residues than left under current harvest practice and it has not been included in calculating the nutrient stores in Table 3.

Changes in soil cations in response to residue treatments differed between sites. At both sites there were significant changes in exchangeable soil Ca levels. Surface soil Ca concentrations for the highest rates of residue retention (BL_3) were about double those where residue had been removed (BL_0 - Figure 2a, data only shown for red earth site). At the red earth site, these changes were also associated with changes in exchangeable Mg (Figure 2b) and again there was approximately a 2-fold difference in Mg concentrations between the extremes of treatments. For these cations the responses were due to both a decline over time in their levels in BL_0 and an increase in BL_3.

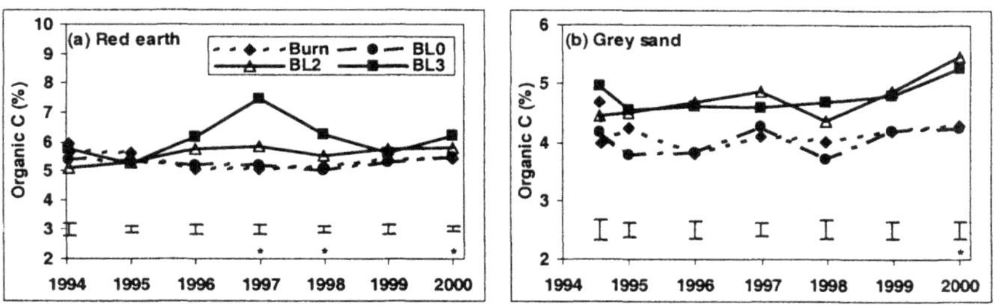

Figure 1. Impact of harvest residue treatments on concentrations of total organic C in surface 0-5 cm soil at (a) the red earth and (b) grey sand experimental site. Standard errors and ANOVA significance levels (*: P<0.05, **: P<0.01, ***: P<0.001) are shown for each year.

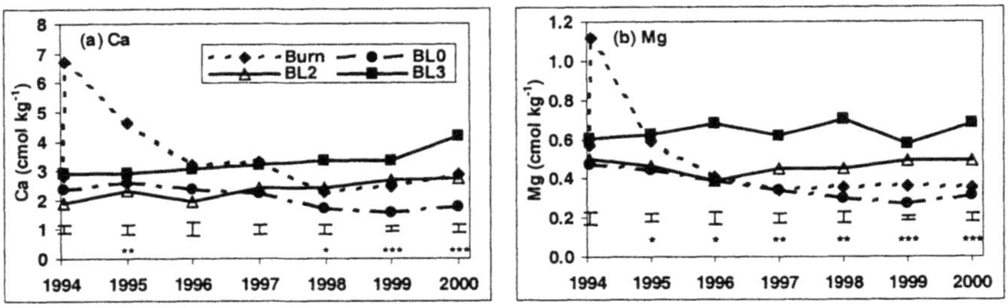

Figure 2. Effect of harvest residue treatments on concentrations of exchangeable (a) Ca and (b) Mg in surface soil (0-5 cm) at the red earth site. Measurements were conducted in December of each year. Standard errors and ANOVA significance levels (*: P<0.05, **: P<0.01, ***: P<0.001) are shown for each year.

Following harvest, exchangeable soil K declined for up to 3 years at both sites (O'Connell et al. 2000), but results from 1999 and 2000 indicate a reversal of this trend, with increasing exchangeable K at both sites. However, surface soil K concentrations at the 2000 sampling were still below the levels measured in soil following harvest by 15% to 55%.

3.3 *Impact of Harvest Residues on Soil Mineral N*

Harvest residue management had a significant impact on the concentrations of mineral N in soil and on the way this pool changed over time. Concentrations of mineral N were high immediately after harvest but declined over time through leaching and plant uptake. Two years after stand re-establishment (2nd rotation) the amounts of mineral N in soils were uniformly low in all treatments. Thereafter, at the red earth site, residue retention caused increased soil mineral N. At the grey sand site, residue management had no effect on soil mineral N concentrations after 2 years.

Temporal patterns of N mineralization were markedly influenced by the seasonal climate of southwestern Australia. Rates were high in late winter and spring (July-October) when soils were moist and decreased to low values as soil dried during summer and autumn (December-May). Patterns were similar at both experimental sites and for each of the harvest residue treatments. A model based on daily measures of soil temperature and monthly measures of soil moisture (O'Connell and Rance 1999) provided a good simulation of seasonal variation in rates of N mineralization over most of the measurement period (Figure 3). The model fit was poorest during the first 2 years after stand establishment. This is the period when large pools of mineral N are present in soil (O'Connell et al. 2000) and N mineralization measured against this high background is subject to larger experimental errors. When soil mineral N pools declined (1997 onwards), there was a better correspondence between observed and predicted N mineralization values.

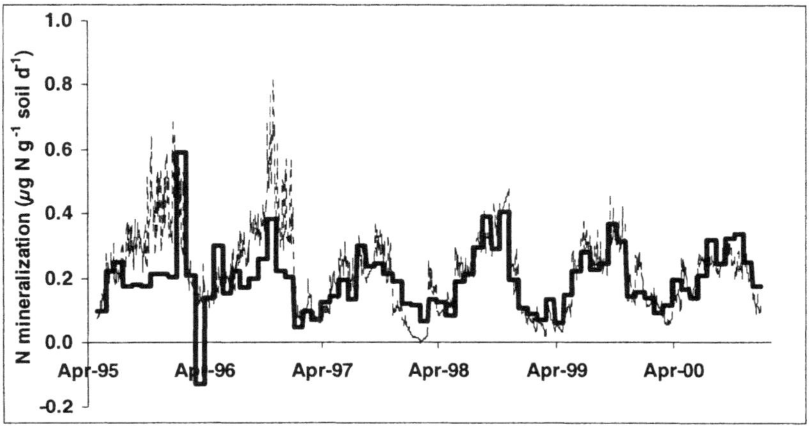

Figure 3. Seasonal pattern of N mineralization (red earth site, BL_0 treatment) in surface (0-10 cm) soil. Bold line shows monthly measured values; broken line the daily-simulated values.

At the red earth site, mean annual rates of N mineralization per unit soil weight (Figure 4), were significantly higher on residue-retained treatments compared to low residue treatments in the third to sixth years of measurement (1997 to 2000). At the grey sand site, annual N mineralization rates were significantly enhanced by residue retention, but only in the third and fourth years of measurement (1997 and 1998). Thereafter, mineralization rates were similar between treatments. Annual rates of mineralization per unit weight of soil were generally higher (up to about 2-fold) in the more heavily textured and fertile red earth soil than in the grey sand.

Amounts of N mineralized annually in surface 0-20 cm soil (Table 2) also differed between treatments. The significantly greater mineralization at the red earth site from 1997 to 1999 was associated with high residue treatments. At the grey sand site, annual amounts of mineralized N were only different in one measurement year (1997).

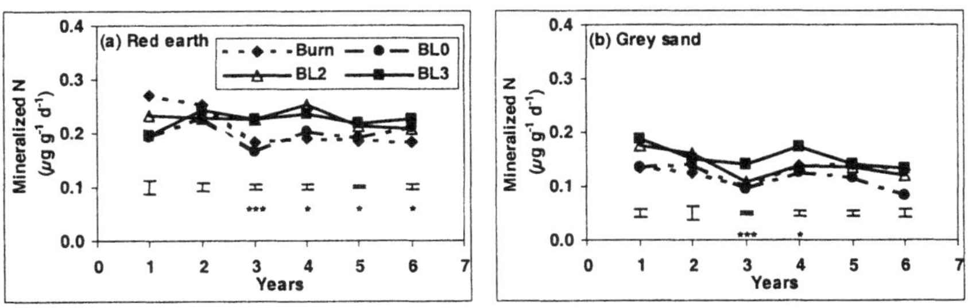

Figure 4. Annual mineralization rates in surface (0-10cm) soil at the two experimental sites. Standard errors and ANOVA significance levels (*: P<0.05, **: P<0.01, ***: P<0.001) are shown for each year.

Table 2. Net N mineralization (kg/ha) in 0-20 cm soil in relation to harvest residue treatments.

	1995[#]	1996	1997	1998	1999	2000
			Red earth			
B	42.7	70.5	75.1	85.1	85.6	82.4
BL_0	29.7	64.9	72.5	89.1	84.5	92.5
BL_2	44.8	81.2	88.7	109.5	95.6	89.5
BL_3	44.2	84.5	85.6	97.6	93.4	95.6
LSD (P<0.05)	28.3	25.8	11.1	15.5	9.0	16.2
ANOVA sig.	ns	ns	*	*	*	ns
			Grey sand			
B	54.3	84.4	56.5	70.3	71.3	64.9
BL_0	54.9	69.6	53.3	65.6	61.2	43.2
BL_2	61.7	78.4	54.4	68.8	66.8	57.5
BL_3	77.2	90.1	72.6	83.7	69.5	63.1
LSD (P<0.05)	22.5	48.9	12.5	19.3	14.1	17.3
ANOVA sig.	ns	ns	*	ns	ns	ns

[#] In 1995 N mineralization was measured only during the period May to December (252 days) at the red earth site and during July to December (196 days) at the grey sand site. Data shown for 1995 refer to net N mineralization measured during these periods.
ANOVA significance: ns not significant; * significant P < 0.05.

Response of potentially mineralizable N in soil to treatments reflected similar trends to those found for measured rates of *in situ* N mineralization. High residue treatments resulted in significantly higher average annual concentrations of potentially mineralizable N at both sites during each measurement period (Figure 5). As with *in situ* measured N mineralization rates, potential for N mineralization was systematically higher in the red earth soil compared to the grey sand. At both sites there was an initial decline in potentially mineralizable N, followed by a turning point at year 2-3 and then increasing potential N. This pattern may reflect an increasing contribution to potential mineral N supply from fine root turnover as the new stands develop. The lack of a clear trend in the residue-retained treatments at the red earth site may be due to the relatively high quantity of harvest residues masking the root turnover effect.

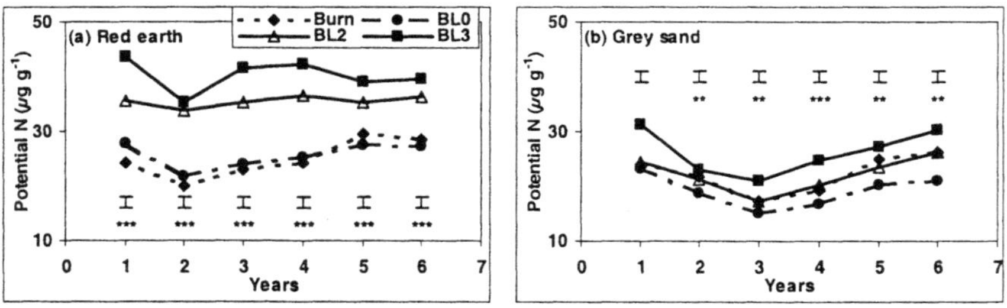

Figure 5. Potentially mineralizable N in surface (0-10 cm) soil. Standard errors and ANOVA significance levels are shown for each year.

3.4 *Impact of Harvest Residue Management on Tree Growth*

Initial responses in tree growth to harvest residue treatments differed between sites (O'Connell et al. 2000). Retention of residues enhanced growth during the first year after planting at the red earth site and initially depressed growth at the grey sand site (O'Connell et al. 2000). These effects declined over time. Five years after plantation re-establishment there were no significant effects of harvest residue management on tree growth at the more fertile red earth site (Figure 6). However, at the grey sand site, there was a significant ($P<0.01$) increase in growth at the high level of residues (BL_3) – about a 40% increase in growth over the zero residue treatment. Growth at this site was about one fourth of that at the red earth site, reflecting differences in the fertility and moisture holding capacity of soils at the two sites.

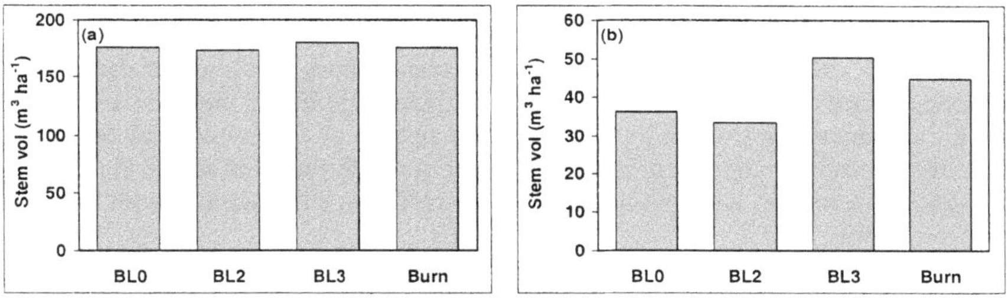

Figure 6. Stem volume of trees in 5-year old *E. globulus* at (a) the Manjimup and (b) the Busselton sites. LSD (5%), 22.3 at Manjimup, 7.8 at Busselton.

4 Discussion

Although large quantities of harvest residues were retained on the soil surface in some treatments in this study (up to 160 t/ha on BL_3 treatments at Manjimup), these quantities have made only minor impacts on the soil stores of organic C. Despite the statistically rigorous design and sampling procedures, changes in soil C were difficult to detect against the variable background levels. In the final sampling period (December 2000), significant treatment effects were observed in organic C concentrations in surface soil (0-5 cm) at both sites. However, in absolute terms, changes since harvesting in soil C storage to a depth of 20 cm on all treatments at both sites were small (within the range of about +/- 4 t/ha).

Leaf contained more than 70% of harvest residue-N and about 50% of the P. Much of these leaf nutrient stores were released through decomposition during the first year following harvest, probably exceeding the quantities of nutrients likely to be taken up by the newly replanted trees. This asynchrony between N supply and N uptake may lead to leaching losses from the site, especially where mineral N is present as nitrate. Immobilization of N in decomposing bark and woody residues may act as a buffer against N loss (Carlyle et al. 1998). Over longer time scales, larger woody residues become sources of N. Conservation of mineral N early in the rotation by immobilization of N in the soil by extractives leached from the eucalypt leaves (Aggangan et al. 1999) is also an important mechanism for limiting N loss during the period of stand re-establishment. At one of our sites, where nitrate was the predominant form of mineral N, retention of residues significantly reduced standing pools of mineral N and presumably also the loss of N through leaching (O'Connell et al. 2000).

Of the other nutrients in harvest residues, K is probably most at risk in regard to leaching loss. Potassium is very labile in decomposing residues, and a large proportion of K will be released with the first rains that follow harvest. In sandy soils or on sites with low cation exchange capacity, there is a strong possibility of leaching of K down the soil profile. Even where new seedlings have been planted soon after harvest, the limited spatial extent of root systems of the small plants will restrict the capture of much of this labile K. Annual soil sampling indicated a decline in surface soil exchangeable K in the first 3 to 4 years after logging (O'Connell et al. 2000), followed by a reversal of this trend in later samplings. At the coarse textured grey sand site in particular, where leaching of labile soil nutrients is most likely, trends in the data confirm a significant increase in surface soil K concentrations (Figure 7). This suggests uptake of K may now be occurring from deeper strata in the soil profile and contributing, through recycling in litterfall, throughfall, stemflow, and root turnover, to increases in soil exchangeable K levels in all treatments.

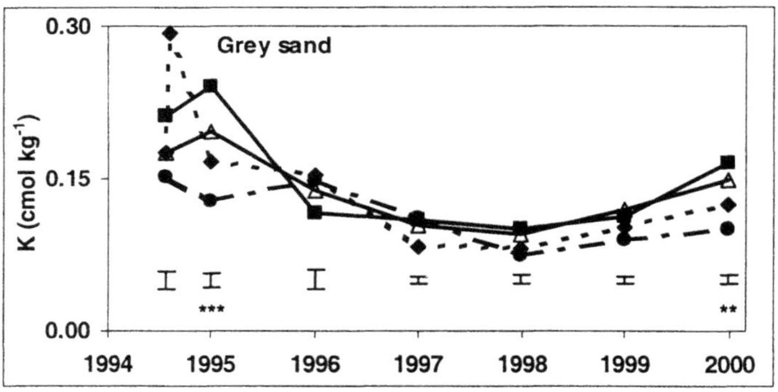

Figure 7. Changes in exchangeable K in surface soil (0-5 cm) since stand harvesting at the grey sand site. Standard errors and ANOVA significance levels are shown for each year. Symbols as for Figure 5.

Early in the rotation, before stand canopy closure, accumulation of soil mineral N occurs due to soil environmental conditions (moisture and temperature) favourable for mineralization and limited uptake of N by new plantations (O'Connell et al. 2000). This build up of the soil mineral N pool will, on many sites, result in leaching of N below the rooting zone. Leaching will be most likely where mineral N is present as nitrate or where soils are coarse textured (sandy) and have low cation exchange capacity. After canopy closure this situation is reversed, as tree roots occupy the site and take up a higher proportion of mineralized N. This will lead to lower concentrations of mineral N in soil and no leaching losses.

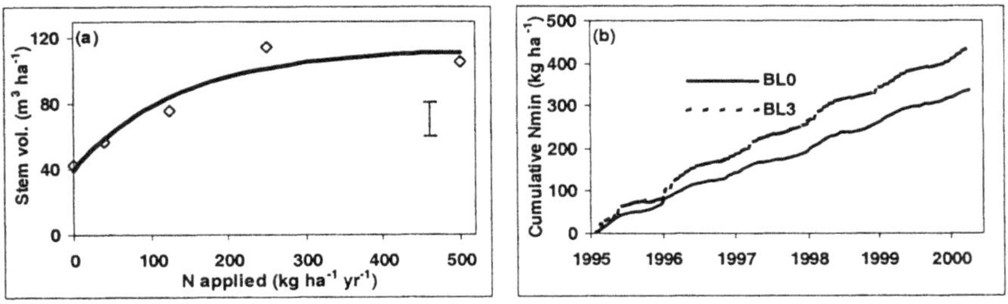

Figure 8. Tree growth response to applied N at age 5 (N applied during years 1 and 2) and cumulative soil N mineralization on BL_0 and BL_3 treatment at the grey sand site. Bar represents LSD (α=0.05).

At the Busselton (Grey sand) site, additional experiments conducted to provide information on the dynamics of tree nutrient uptake indicate significant tree growth responses to N added during the first two years (Bl_2 residue treatment common to all plots, Figure 8a). Part of this demand for N can be met through appropriate harvest residue management. For example, retention of residues may induce N immobilization early in the rotation resulting in reduced leaching losses and increased N mineralization later in the rotation when the root system is more fully developed. At Busselton, cumulative N mineralization over 5 years differed by 100 kg N/ha between BL_0 and BL_3 treatments (Figure 8b). However, because of the asynchrony between the time of supply of N through mineralization, and the ability of young plants to take up this N, even differences of this magnitude are insufficient to meet N demand by the developing plantation. A comparison of amounts of N released through mineralization of soil-stored N with aboveground accumulation of N in tree biomass at two levels of applied N is shown in Figure 9. Results illustrate the varying dynamics of N requirement and N supply. Early in the rotation, mineral N supply from soil and residues stores is high but is spatially dispersed in relation to the root system. After the first year, N demand increases and additional sources of N are required to achieve potential growth rates. Thus, even when harvest residues are retained so that soil N supply is optimised and N losses are minimised, application of N in fertilizers will still be necessary to maximize plantation productivity on many sites.

Our research was concentrated at two sites in southwestern Australia. Plantation productivity differs markedly at these sites, due principally to differences in their soil characteristics (soil texture, nutrient supplying capacity and moisture holding characteristics). Currently, experimental results extend over a period up to six years following stand harvesting and it is planned to continue monitoring the sites through the complete rotation to age approximately 10 years. Explaining and generalising research outcomes will be a key focus of future activities.

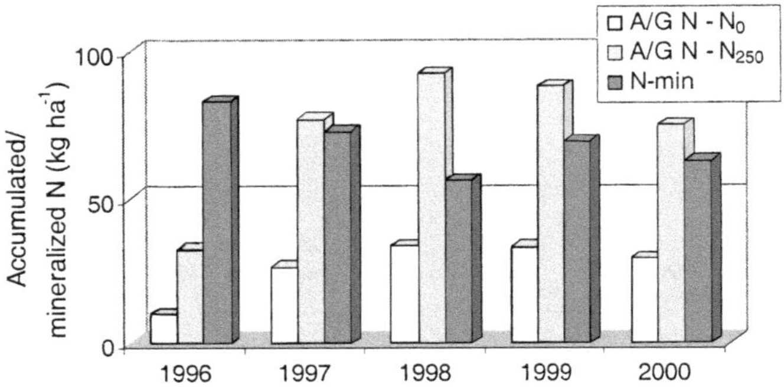

Figure 9. Annual above ground accumulation of N by trees at the grey sand site (Busselton) for 2 levels of applied N (0 and 250 kg/ha for 2 years), together with cumulative annual soil N mineralization on the BL_2 treatment during the first 5 years of the plantation rotation at the Busselton site.

Acknowledgments

Funding for this research was initially provided through Australian Government Industry Statement Funds to the CSIRO and subsequently by ACIAR as a collaborative research project with KFRI, India. Support was also provided by the Rural Industries Research and Development Corporation (RIRDC), Bunnings Treefarms and the Western Australian Department of Resources Development. Field assistance, in-kind support and land provided by Bunnings Treefarms Pty Ltd (now WA Plantation Resources Pty Ltd) and Mr Morris Cox are gratefully acknowledged. Many CSIRO colleagues, particularly G. Wan, P. Damon, T. Pham and S. Snelling provided technical assistance. CIFOR provided financial support for attending this meeting.

References

Aggangan, R.T., O'Connell, A.M., McGrath, J.F. and Dell, B. 1999. Impact of *Eucalyptus globulus* labill. leaf litter on C and N mineralization in soils from pasture and native forest. Soil Biol. Biochem. 31:1481-1487.

Carlyle, J.C., Bligh, M.W. and Nambiar, E.K.S. 1998. Woody residue management to reduce nitrogen and phosphorus leaching from sandy soil after clear-felling *Pinus radiata* plantations. Can. J. For. Res. 28:1222-1232.

Davidson, J. 1995. Ecological aspects of eucalypt plantations. In: Proc. of the Regional Expert Consultation on *Eucalyptus*, FAO Regional Office for Asia and the Pacific, Bangkok, Vol.I, pp35-72.

Grove, T.S., O'Connell, A.M, Mendham, D., Barrow, N.J. and Rance, S.J. 2000. Sustaining the productivity of tree crops on agricultural land in south-western Australia. Canberra, RIRDC/LWA/FWPRDC Joint Venture Report 01/09, 69p.

O'Connell, A.M. and Grove, T.S. 1999. Eucalypt plantations in south-western Australia. In: "Site management and productivity in tropical plantation forests", E.K.S. Nambiar, A. Tiarks, C. Cossalter and J. Ranger (eds), Workshop proc., Pietermaritzburg, South Africa, 16-20 February 1998, CIFOR, Bogor, Indonesia, pp53-59.

O'Connell, A.M. and Rance, S.J. 1999. Predicting nitrogen supply in plantation eucalypt forests. Soil Biol. Biochem. 31:1943-1951.

O'Connell, A.M., Grove, T.S., Mendham, D. and Rance, S.J. 2000. Eucalypt plantations in south-western Australia. In: "Site management and productivity in tropical plantation forests", E.K.S. Nambiar, A. Tiarks, C. Cossalter and J. Ranger (eds), CIFOR, Bogor, Indonesia, pp61-71.

O'Connell, A.M., Grove, T.S., Mendham, D. and Rance, S.J. 2003. Changes in soil nitrogen status and nitrogen supply rates when eucalypt plantations are established on agricultural land in south-western Australia. Soil Biol. Biochem. (in review).

Raison, R.J., Connell, M.J. and Khanna, P.K. 1987. Methodology for studying fluxes of soil mineral nitrogen *in situ*. Soil Biol. Biochem. 19:521-530.

Sankaran, K.V., Chacko, K.C., Pandalai, R.C., Kallarackal, J., Somen, C.K., Sharma, J.K., Balagopalan, M., Balasundaran, M., Kumaraswamy, S., Sankar, S., Gilkes, R.J., Grove, T.S., Mendham, D. and O'Connell, A.M. 2000. Effects of site management on *Eucalyptus* plantations in the monsoonal tropics - Kerala, India. In: "Site management and productivity in tropical plantation forests", E.K.S. Nambiar, A. Tiarks, C. Cossalter and J. Ranger (eds), A Progress Report, CIFOR, Bogor, Indonesia, pp51-60.

Tiarks, A., Nambiar, E.K.S. and Coossalter, C. 1998. Site management and productivity in tropical plantations. CIFOR Occasional Paper No.16, CIFOR, Bogor, Indonesia.

Turton, A.G., Marsh, N.L., McKenzie, R.M. and Mulcahy, M.J. 1962. The chemistry and mineralogy of lateritic soils in the south-west of Western Australia. CSIRO, Australia, Soil Publication No.20.

澳大利亚西南部桉树人工林轮伐间期立地管理：国际林业研究中心立地管理和生产力网络合作项目

ANTHONY M. O'CONNELL，TIM S. GROVE，DANIEL S. MENDHAM，STANLEY J. RANCE

CSIRO Forestry and Forest Products
Private Bag, No. 5, Wembley, Australia 6913

本文针对澳大利亚西南部不同土壤肥力条件下第二代蓝桉人工林采伐剩余物管理问题，就不同管理策略所产生的影响进行检测评估。本项研究是国际林业研究中心“热带人工林立地管理和生产力”网络研究项目的组成部分。研究项目中共同的核心处理是根据 8 个国家 16 种立地类型的试验进行评估。本项目也是由澳大利亚国家农业研究中心资助，与印度科瑞拉林业研究所合作研究的部分内容。在澳大利亚，随机区组试验设计中的各处理主要是采伐剩余物全部收获(BL_0)、全部保留(BL_2)、增加剩余物(BL_3)和焚烧剩余物(B)。我们研究(i)各处理对土壤碳和养分储量的影响，(ii)土壤氮矿化动态，(iii)养分吸收和树木生长。现报道第一代人工林砍伐迹地上重新造林林分 5.5 年后的研究结果。采伐剩余物的管理对碳、氮、磷、钙和镁的贮存和循环有显著的影响。可是，采伐剩余物的加倍（每公顷 160 吨）对土壤表面有机碳含量影响极小。土壤氮的变化比较敏感；大量剩余物的存留则导致较高矿化水平和提高潜在速效氮含量。然而，多数立地上，土壤有机质和剩余物矿化所提供的氮往往低于新建林分幼苗的实际需求量。由此可见，保留采伐剩余物只能部分地满足林木所需养分需求。在这种情况下，为提高林分生产力，通过施肥提高氮素的供应水平是十分必要的。

INTERACTION OF ASSOCIATIVE NITROGEN-FIXATION BACTERIA WITH *EUCALYPTUS*

LIHUA KANG AND HAIBIN MA

Research Institute of Tropical Forestry, Chinese Academy of Forestry
Guangzhou 510520, China

This study investigated the interaction of *Klebsiella oxytoca* NG13/pMC73A with *Eucalyptus*. *K. oxytoca* colonized rhizosphere, the root surface and even the inner cortex of eucalypt seedlings. The inoculation of the bacteria stimulated the excretion of Eucalypt root system and affected the components of amino acids, carbohydrates, phytohormones of root exudates. Compared to the control, inoculation of the bacteria significantly increased the height growth, dry matters, N contents etc of *Eucalyptus* seedlings.

1 Introduction

Biological N_2-fixation is very important for agriculture and forestry cultivation. It is reported that the total quantity of biological N_2-fixation on the earth is about 100 million tons each year, which is more than 3 times that of the industrial nitrogen fertilizer (You et al.1987). In the last few decades, increased attention has been paid to associative nitrogen-fixation, one of biological N_2-fixation methods. In particular, much of the efforts have been given to gramineous plants such as rice, sugar cane and pasture species in the last 20 years. Great progress has been made in study on the distribution of root surface sugar-based, phytolectin and bacteria exopolysaccharide (EPS), as well as the identification, cloning, regulation etc. of some relevant genes (Lin and You 1993; Brockwell 1963). Yet, many aspects of the mechanism of associative nitrogen-fixation are still not clear. In recent years attention has been focused on the interaction of associative nitrogen-fixation bacteria with plant species such as rice and pasture, but less efforts on the interaction with woody species.

Eucalyptus belongs to Myrtaceae, one of the three fastest growing tree species in the world. As one of the main planted tree species in South China, the total eucalypt plantation area is now more than 1 million hectares, and it increases by 50 to 100 thousand hectares every year. As there is a huge local and regional demand for pulp and other forest products, foreign investment now becomes one of the main forces in developing and managing eucalypt plantations in South China, particularly in the provinces of Guangdong, Guangxi and Hainan. To obtain high yield, application of chemical fertilizer is usually necessary. It is well accepted that consecutive application of chemical fertilizers may cause soil productivity depression, environmental pollution and other problems, thus restricting the sustainable development of eucalypt plantations in the region. By converting N_2 into ammonium nitrogen, the artificial inoculation of eucalypt with associative nitrogen-fixation bacteria can reduce the use of chemical fertilizer and mitigate the corresponding problems. This paper investigates the interaction of associative nitrogen-fixation bacteria with eucalypt.

2 Materials and Methods

2.1 *Source and Culture of Bacteria*

One associative nitrogen-fixation bacterium, *Klebsiella oxytoca* NG13/pMC73A with Ap^r and Km^r plasmid, provided by Guangdong Microbiology Institute was used for the present study. In liquid LB medium added with 0.1 g Ap and 0.05 g Km per liter, the bacteria were cultured for 2-3 days under 28-30°C.

2.2 *Seedlings and Inoculation*

Seeds of *E. urophylla* were surface sterilized in a 0.1% mercuric chloride solution for 5 minutes, and rinsed in sterile water. Then they were put in the sterile culture dishes with filter paper, and raised in the incubator under 28-30°C. When starting to germinate, seeds were transferred into containers filled with vermiculite and sand (1:1,v/v) sterilized under 100,000 Pa. When seedlings grew to 2-3 cm tall, 5 ml of bacteria solution per seedling was injected into the container around the roots with a sterile injector. Seedlings were observed with microscope, and the height, biomass and nitrogen were measured at 3-4 months old.

2.3 *Observation with SEM*

The seedling roots inoculated with bacteria were processed by washing away the surface impurities with phosphate buffer containing 0.88% NaCl. Then the roots were observed with JSM-T300 SEM.

2.4 *Counting of the Bacteria*

Three different types of roots were used to estimate *Klebsiella oxytoca* in the rhizosphere, on the root surface and inside the roots: a) the roots tightly stuck with the rhizosphere soil, b) the roots after washing away the rhizosphere soil with phosphate buffer containing 0.88% NaCl, and c) the roots treated in b) sterilized with 2% NaClO for 3 minutes and washed with sterile water. They were separately cut into small pieces of 0.3-0.5 cm, and grinded uniformly. After diluting to a proper concentration, the bacteria solution was dripped in the LB medium with Ap and Km, and cultured for 2-3 days under 28-30°C. The bacteria were counted with microscope.

2.5 *Collecting Roots Exudates*

River sand was soaked for 12 hours in 3% of HCl, washed with distilled water, and dried. The sand was then put into big test tubes full of Jesen nutrition solution, sterilized for 1 hours under 100,000 Pa. Then the seeds processed as in 2.2 were sowed, one germinating seed in each tube. While 1 ml bacteria solution was injected around each seedling, 1 ml sterile culture solution was injected around each control seedling. Both treatment and control had 4 seedlings, and repeated 3 times. The seedlings were watered with sterile water 1-2 times each week. Fresh weights were measured after 3-4 months. Thirty grams of roots were randomly selected and rinsed with 1-liter sterile water for each treatment, respectively. The rinsed solutions were separately collected and enriched. About 0.2 liter of enriched solution for each treatment was used for determining the root exudates.

2.6 *Determination of Root Exudates*

An Automatic Amino Acid Analyzer was used to determine the components of amino acids. A spectrophotometer was used to test the components of carbohydrates. The components of hormones were determined by using ELLISA (Wu et al. 1998).

2.7 *Culture of Bacteria in Medium added with Roots Exudates*

After filtered, and sterilized with Sartorius Minisart N 0.2 um, the root exudates were diluted to 5%, 10% and 20% of solutions, respectively. Then, they are added to the sterile LB liquid medium containing Ap and Km. The bacteria were inoculated, shake-cultured for 3 days and counted with microscope.

3 Results and Discussion

3.1 *Colony Location*

A large number of bacteria were observed with SEM on the root surface and in the cell air space of the seedlings inoculated with *Klebsiella oxytoca*. Moreover, *Klebsiella oxytoca* was observed in the thin walls of the cortical cells (Figure 1, left). Therefore, the locations of the associative nitrogen-fixation bacteria colonized in eucalypt are the same as those found in the gramineous plants (Old and Nicolson 1975; Martin 1977). However, the bacteria were not observed on the root surface and in the thin walls of the cortical cells of non-inoculated seedlings (control; Figure 1, right).

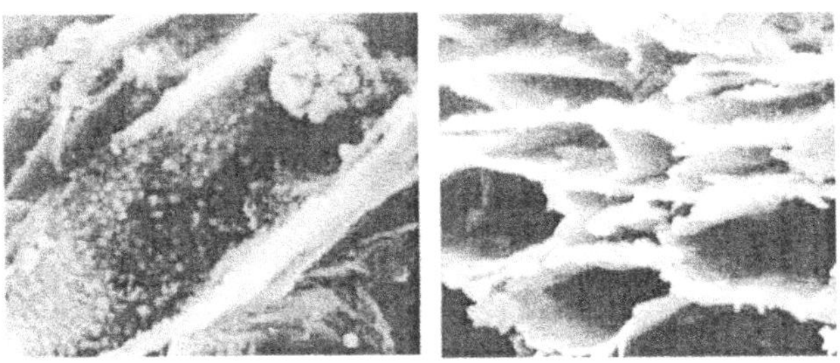

Figure 1. Left: *Klebsiella oxytoca* gathered at the root epidermis cell of the inoculated *Eucalyptus* seedling root epidermis cells (2,000X), and Right: root surface cells of non-inoculated seedling (2,000X).

3.2 *Distribution*

The density of *Klebsiella oxytoca* was estimated for different locations around the eucalypt root system (Table 1). The rhizosphere soil was most favorable for the growth of the bacteria, followed by the root surface and the inner roots. Similar distribution of the associative nitrogen-fixation bacteria was also found in the gramineous plants (You et al. 1987).

Table 1. Densities of *Klebsiella oxytoca* estimated for different locations around the *Eucalyptus* root system.

Treatment	Rhizosphere (No. of bacteria /g soil)	Root surface (No. of bacteria /g fresh root)	Inner root (No. of bacteria /g fresh root)
Inoculation	4.9×10^6	3.8×10^6	1.3×10^5
Non-inoculation	0.0	0.0	0.0

3.3 *Effect of Bacteria on Roots Exudates*

The interaction of associative nitrogen-fixation bacteria with gramineous plants showed that the carbohydrates in the root excretions could be the source of energy for microorganisms (Barber and Lynch 1977; You 1990).

Some exudates stimulate the growth of microorganisms. The organic acids and carbohydrates in these secretions were essential for the growth of microorganisms gathering around the rhizosphere, and their nitrogen-fixation function. Non-inoculated eucalypt roots can excrete many amino acids, carbohydrates and hormones, which would be fine carbon and nitrogen sources for nitrogen-fixation bacteria. Table 2 compared the contents of the different amino acids for inoculated and non-inoculated

Table 2. Amino acids and their contents of *Eucalyptus* root exudates.

Amino acids	Content (mg/g fresh roots)	
	Inoculation	Non-inoculation
Asp	0.033	0.014
Glu	0.047	0.034
Ser	0.026	0.020
Gly	0.014	0.013
Thr	0.013	0.009
His	0.086	0.052
Ala	0.020	0.021
Arg	0.041	0.025
Tyr	0.045	0.031
Val	0.037	0.021
Met	0.027	0.017
Phe	0.034	0.020
Ile	0.027	0.020
Leu	0.041	0.027
Lys	0.024	0.024
Total	0.466	0.299

seedlings. The contents of the different carbohydrates and hormones for the inoculated seedlings and control were presented in Table 3. The contents of 14 amino acids except alanine of root exudates were higher than those of the control. Compared with the control, the inoculated seedlings had higher amount of carbohydrates such as sucrose, fructose, polysaccharid except for glucose. In general, more hormones were generated in the inoculated seedling than in the control (Table 3). The results indicated that *Klebsiella oxytoca* in the rhizosphere selectively absorbed sucrose, fructose, polysaccharid and alanine for their metabolism with discharging amino acids and hormones.

Table 3. The contents of the carbohydrates and hormones in *Eucalyptus* root secretions.

Treatment	Content of carbohydrates (mg/g fresh root)				Content of hormones (Pmol/g fresh root)	
	Sucrose	Fructose	Glucose	Polysaccharid	GA	ABA
Inoculation	0.025	0.007	0.007	0.071	2713.33	260.0
Non-inoculation	0.032	0.008	0.002	0.129	1843.33	190.0

3.4 *Effect of Root Exudates on Bacteria*

Root exudates affected the growth of nitrogen-fixation bacteria (Table 4). It was shown in Table 4 that the density of bacteria significantly increased when the concentration of the root exudates was added. Compared with the control, the increment of bacteria density was relatively low (22.5%) at low concentration (5%) of root exudates added, highest at concentration from 5% to 10%, and slowed down again (177.8%) at concentration from 10% to 20%. This explained that the eucalypt rhizosphere was favorable for the growth of nitrogen-fixation bacteria.

Table 4. Effect of the root exudates concentration on the growth of nitrogen-fixation bacteria.

Concentration of root exudates (%)	Counts of bacteria ($\times 10^6$/ml bacteria solution)	Increment of density relative to the control (%)
Control	120.0	0.0
5	147.0	22.5
10	280.0	133.3
20	333.3	177.8

3.5 *Growth of Seedlings Inoculated with Bacteria*

Seedling growth was significantly stimulated by inoculation of the bacteria (Table 5). The height, dry weight of leaves, dry weight of stems, dry weight of roots and total biomass were all higher than the control without inoculation. Table 6 showed that the nitrogen contents of different parts (leave, stem, root) of the seedling inoculated with the bacteria were significantly higher than those of the control without inoculation.

This may be the consequence of *Klebsiella oxytoca* in the rhizosphere taking N_2 from air.

Table 5. Effect of inoculation with *Klebsiella oxytoca* on growth of *Eucalyptus*.

	Inoculation	Non-inoculation	Difference (%)	*F*-value
Height (cm)	46.6	35.9	29.81	107.2**
Leaves (pair/seedling)	26.5	19.3	37.31	6.191**
Dry weight of leaves (g/seedling)	0.745	0.552	35.01	5.249**
Dry weight of stems (g/seedling)	0.898	0.577	55.59	15.51**
Dry weight of roots (g/seedling)	0.901	0.504	78.93	14.85**
Total biomass (g/seedling)	2.545	1.632	55.89	13.78**

** Difference between inoculated and non-inoculated seedlings was significant at level of 0.01.

Table 6. Effect of inoculation with *Klebsiella oxytoca* on the content of nitrogen in different parts of seedlings (mg/seedling).

Part of seedling	Inoculation	Non-inoculation	Difference (%)
Leave	13.362	9.863	35.48
Stem	6.196	3.852	60.85
Root	6.984	3.485	100.40
Total	26.542	17.190	54.40

Interaction of associative nitrogen-fixation bacteria with eucalypt found in this study indicates that artificial inoculation of nitrogen-fixation bacteria is a promising approach to improving the nutrient conditions of eucalypt trees. It can increase the soil nutrition and productivity, stimulate the growth of tree, reduce the application of chemical fertilizer and alleviate the potential environmental issues in planting eucalypt. We believe that this method will widely be applied in managing eucalypt plantation.

References

Barber, D.A. and Lynch, J.M. 1977. Microbial growth in the rhizosphere. Soil Biol. Biochem. 9:305-308.

Brockwell, J. 1963. The release of organic substances by cereal roots into the soil. Appl. Microbiol. 11:3771-3883.

Martin, J.K. 1977. Role of wheat root exudates in associative nitrogen fixation. Soil Biol. Biochem. 9:1-7.

Lin, M. and You, C. 1993. Root Exudates of Rice and its interaction with *Alcaligenes faecalis*. Acta Agric. Nucl. Sini. 7:98-104.

Old, K.M. and Nicolson, T.H. 1975. Nonsymbiotic nitrogen fixation in the rhizosphere of rice. New Phytol. 74:57-58.

Wu, S.R., Chen, W.F. and Zhou, X. 1998. Determination of plant hormones by using ELLISA. Plant Physi. Comm. 5:53-57.

You, C., Jiang, Y. and Song, H. 1987. Biological nitrogen-fixation. Science Publishing House, pp322-347.

You, C. 1990. Associative nitrogen-fixation of rice rhizosphere. Agriculture Publishing House, pp294-297.

桉树与联合固氮菌相互作用的研究

康丽华, 马海宾

中国林业科学研究院热带林业研究所
510520 中国广州

本研究将联合固氮菌-催娩克氏菌(*Klebsiella oxytoca* G13/pMC73A)接种于木本植物桃金娘科（Myrtaceae）桉树属(*Eucalyptus*)的苗木，研究它们之间的相互作用关系。扫描电镜的观察结果表明：联合固氮菌不但可以在桉树根表定殖，而且还可进入根内定殖，并在接种桉树的根际、根表和根内均分离到该菌。接种联合固氮菌能刺激桉树根系的分泌作用，并对根系分泌物的氨基酸、糖及激素的含量有所影响。联合固氮菌对桉树生长有明显的促进作用。

INTER-ROTATION SITE MANAGEMENT OF EUCALYPT PLANTATIONS IN SOUTHWESTERN AUSTRALIA: A PARTNER PROJECT IN THE CIFOR NETWORK ON SITE MANAGEMENT AND PRODUCTIVITY

ANTHONY M. O'CONNELL, TIM S. GROVE, DANIEL S. MENDHAM AND STANLEY J. RANCE

CSIRO Forestry and Forest Products
Private Bag, No. 5, Wembley, Australia 6913

The impacts of alternative strategies for managing harvest residues in second rotation *Eucalyptus globulus* plantations are being examined at sites of contrasting soil fertility status in south-western Australia. This research is a component of the CIFOR network project, "Site management and productivity in tropical plantation forests", in which common core treatments are being evaluated at 16 sites in 8 different countries. Our project is part of a collaborative study with Kerala Forest Research Institute, India, supported by funding from Australian Centre for International Agricultural Research (ACIAR). In Australia, treatments being examined in randomized block experiments are complete harvest residue removal (BL0), residue retention (BL2), increased residue (BL3) and residue burning (B). We studied (i) impact of treatments on soil carbon and nutrient stores, (ii) dynamics of soil N mineralization and (iii) nutrient uptake and tree growth. Results up to 5.5 years after stand re-establishment following harvest of the first plantation rotation are reported. Significant impacts of harvest residue management on stores and fluxes of C, N, P, Ca and Mg have been found. However, application of harvest residues at high rates (up to 160 t/ha) has had a surprisingly small effect on organic C levels in the surface soil. The dynamics of soil N is particularly affected by residue treatments; higher rates of retention of residues leading to higher rates of mineralization and potentially available N. Nevertheless, the rate of supply of N through mineralization from soil organic matter and residues may be less than the requirement for N of newly established seedlings of the next tree crop on many sites. Hence conservative harvest residue management will only partially satisfy tree nutrient requirements at these sites. In such circumstances, N supply needs to be augmented with fertilizer to achieve optimum plantation productivity.

1 Introduction

The concepts and the rationale for initiating the CIFOR network project, "Site management and productivity in tropical plantation forests", have been outlined by Tiarks et al. (1998). From the initial planning meeting held at the Forest Research Institute of Malaysia in November 1995 has grown a substantial international collaboration between scientists in 8 countries, involving a network of 18-partner institutions that have established 16 experimental sites with common core treatments. The research focuses on the inter-rotation period of the plantation cycle. This is a period when sites are vulnerable to nutrient depletion through poor silvicultural practices. However, it is also a time when there are opportunities for

V. Plantation Management

USING A PROCESS-BASED FOREST MODEL TO ESTIMATE THE POTENTIAL PRODUCTIVITY OF *EUCALYPTUS* PLANTATIONS IN SOUTHERN CHINA

JIM D. MORRIS AND TOM BAKER

Forest Science Centre, P.O. Box 137, Heidelberg Victoria 3084, Australia

Process-based models are increasingly being applied as a planning tool by forest managers. The 3-PG forest model offers significant benefits for application to *Eucalyptus* plantation management in southern China. In addition to providing a rational basis for decisions on stand establishment, tending and harvesting, a well-constructed model may act as a framework for storage and retrieval of tree performance data and an indicator of priority areas for research.

Measurements of tree and stand growth, biomass relationships and water use from plantations on the Leizhou Peninsula of western Guangdong Province have been used to estimate model parameter values for major commercial eucalypt genotypes. The rapid early growth, high stand density and short rotations adopted by forest managers for woodchip production necessitate special forms of some parameters not included in other implementations of 3-PG.

Climate data inputs for modelling may be satisfactorily estimated from routine daily meteorological observations including temperature, rainfall and sunshine duration. Physiological parameters which are not amenable to direct measurement may be estimated by fitting the modelled growth predictions to one or more sets of growth observations, ideally for multiple species on sites with a range of climate and soil conditions or stand densities. The accuracy and reliability of modelled predictions will improve over time as more data sets become available to incrementally improve the precision of estimated parameters.

1 Introduction

Successful decision making to optimise the sustainable productivity of natural forests and plantations requires reliable, versatile and efficient methods for predicting the harvestable yield of wood products in relation to site conditions and stand management. Simple process-based models have attracted increasing interest as an integrated planning tool for this purpose, suited to adoption by forest managers in their own right or as a component of a sophisticated management information system (Battaglia and Sands 1998).

The 3-PG forest model (Landsberg and Waring 1997) has been widely applied and tested for both softwood and hardwood plantations in several countries (Dye 2001; Landsberg et al. 2001, 2003). The development and application of this model for *Eucalyptus* plantation management in southern China offers significant benefits. In addition to providing a rational basis for decisions on stand establishment, tending and harvesting, a well-constructed model may act as a framework for storage and retrieval of tree performance data and an indicator of priority areas for research.

The fast-growing plantations of southern China are managed for pulpwood production on rotations as short as 4 years. This leaves little time for reversal or correction of suboptimal establishment or tending decisions. In this situation a well-validated model may be particularly valuable as a predictor of the implications of planting season, spacing or species choice for growth on a given site. In addition, a model allows managers to look beyond the short rotation period to consider the possible consequences of thinning or wider initial spacing combined with longer rotations for solid wood or other products, without committing valuable productive land to exploratory research.

The structure of 3-PG has been well documented elsewhere (Landsberg and Waring 1997; Dye 2001; Morris and Collopy 2001; Landsberg et al. 2003), and the complete computer code is available as public domain software in multiple versions. This accessibility is a powerful advantage in that it allows the model to be extended or modified to provide additional information or cope with data sets beyond the scope of the original model. Multiple user interfaces have also been developed for 3-PG. A research-level interface which emphasises the versatility of the model by providing access to all its parameters and alternative functions is unlikely to be suitable as an operational growth prediction tool at a management level, where simplicity and clarity are valued more highly.

This paper focuses on the input data requirements for development of parameter files, application and testing of 3-PG for eucalypt plantations in China. Measurements of tree and stand growth, biomass relationships and water use from plantations of *E. urophylla* and eucalypt hybrid ABL12w5 on the Leizhou Peninsula of western Guangdong Province have been used to estimate model parameters. The methods adopted for deriving parameter values from these data sets may guide the collection of further data to extend the generality of the parameterised model to a wider range of sites and genotypes.

2 Input Data for Modelling

3-PG and similar models require three categories of information from which to predict stand growth, specifying the climate, site conditions and physiological characteristics of the species to be modelled. Site parameters in 3-PG include features of the soil environment (fertility, salinity, moisture availability) as well as stand management (initial stocking, thinning prescriptions, irrigation). The set of physiological parameters for a given species or genotype is, at least in concept, a universal specification which is independent of site and climate factors. The elucidation of these species parameters and their validation across a range of growing conditions is a task for research, so that application of the model at a management level can proceed without need to do more than select the species or genotype to be modelled. Much of the data required for the development and validation of species parameter sets is obtainable from growth plots or simple tree biomass studies.

3 Data Collection from Eucalypt Plantations on Leizhou Peninsula

Fifteen growth monitoring plots were established in stands of *E. urophylla* and the eucalypt hybrid ABL12 at Jijia forest farm in the western central Leizhou Peninsula of Guangdong Province. Plots were approximately 0.025 ha and contained up to 70 trees. Tree diameters and plot dominant height (mean of the 5 fattest trees) were recorded at 6-monthly intervals between September 1999 and May 2002. The measurement data covered an age range of 15 months to 71 months after planting, which is beyond the usual rotation age for these stands. Plot volumes were estimated from plot mean diameter (*D*) and dominant height (*H*) as

$$\text{Volume} = 0.33 \times D^2 \times H \times \text{stocking} \quad [1]$$

and used to calculate periodic volume increment. Figure 1 compares the current (CAI) and mean (MAI) annual increment for *E. urophylla* and ABL12, averaged over three plots of each species.

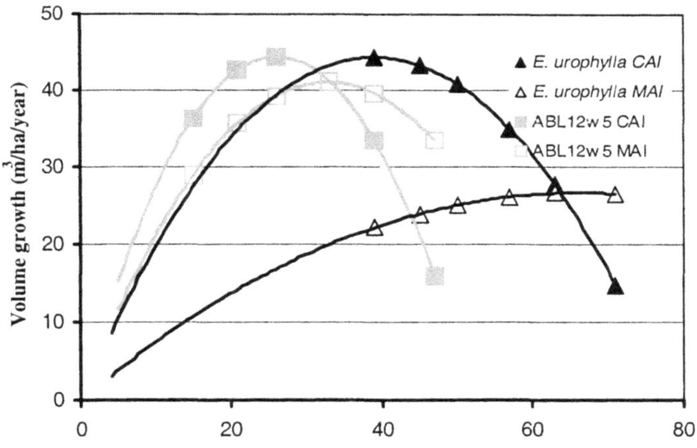

Figure 1. Current annual increment (CAI) and mean annual increment (MAI) of *E. urophylla* and ABL12 plantations at Jijia Forest Farm (means of 3 and 5 plots respectively).

Stand diameter distributions were also characterised from the plot measurement data. Although this information is not essential for prediction of total volume growth through time with 3-PG, it does allow a more precise estimate of harvestable wood product by size classes for the evaluation of thinning or harvesting strategies. When diameter distribution data are available to 3-PG they can also be used to refine the estimated volume of wood lost in tree mortality each year.

Measured diameter distributions that are not strongly bimodal can be readily described by a 3-parameter Weibull distribution, using a least squares approach to determine the parameters of best fit to the observed data. For use in 3-PG the Weibull parameters and the way they change with stand age are related to the mean stem biomass by regression, and the coefficients of these regressions are included in the species parameter set. Figure 2 shows examples of observed diameter distributions in 45 months old eucalypt plots together with their fitted Weibull distributions.

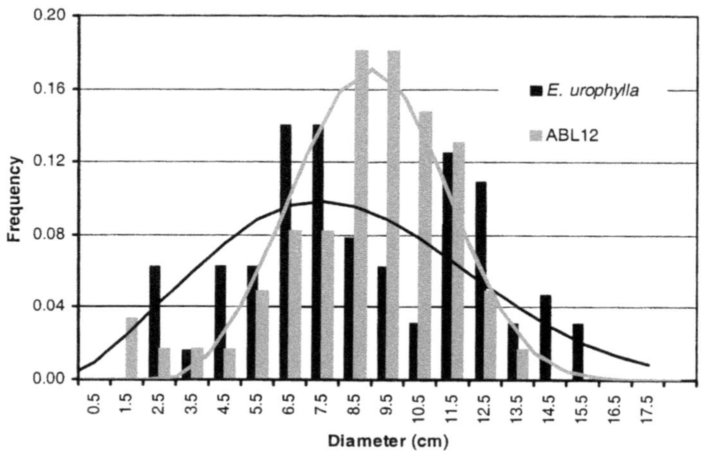

Figure 2. Observed diameter distributions in 45 month-old stands of *E. urophylla* and hybrid ABL12, and Weibull distributions fitted by a least squares procedure.

The above ground biomass of eight stands of *E. urophylla* seedlings, ABL12 clone w5 and *E. urophylla* clone U6 was assessed by harvesting 6 trees per stand, spanning the range of diameters present. Fresh and dry weights of leaves, twigs, live and dead branches, stem bark and stem wood were determined by a sampling procedure. Specific leaf area of a sample of 100 leaves per tree was assessed by scanning and image analysis, and used to estimate whole tree leaf area. Allometric relations between tree variables and diameter at breast height were determined, and applied to estimate stand volume and biomass from the growth plot measurements. The variation in specific leaf area and branch fraction with increasing tree diameter in *E. urophylla* and ABL12 is shown in Figure 3, displaying a clear difference between the species. Figure 4, by contrast, shows a single relationship between stem mass and diameter fitting the observations from all 8 stands. Quantifying these relations from biomass measurements ensures that the model parameters derived from them are well founded in actual growth data.

Root biomass was assessed by excavating 6 trees each of *E. urophylla* and ABL12 of a range of diameters, and by weighing fine roots sieved from soil cores of 60 cm depth collected from stands of both species.

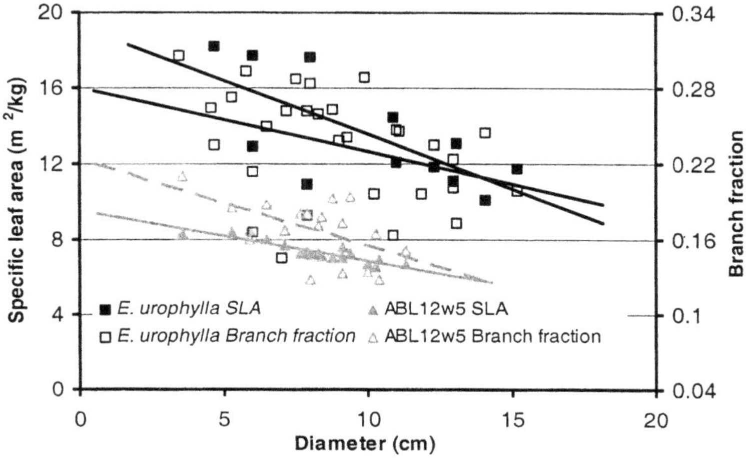

Figure 3. Variation in specific leaf area and branch fraction with increasing stem diameter in the Jijia plantations.

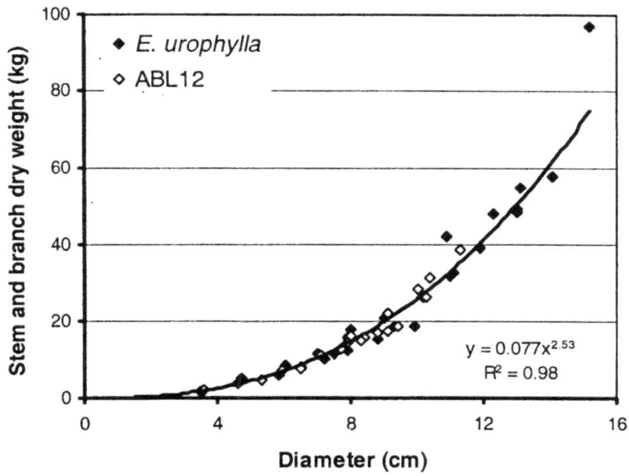

Figure 4. Allometric power curve relationships of stem and branch biomass with diameter in 8 stands of *E. urophylla* and ABL12 aged 15 to 71 months at Jijia.

4 Optimising 3-PG for Short Rotation Crops

The rapid early growth, high stand density and short rotations adopted by forest managers for woodchip production in western Guangdong necessitate special forms of some parameters which have not been included in other implementations of 3-PG.

The allocation of photosynthate to leaves and stems in 3-PG is determined by a pair of power functions for stem mass versus diameter (e.g. Figure 4) and foliage mass versus diameter. Following Landsberg and Waring (1997) it is common to assume that the parameters which define the foliage-diameter curve remain constant as the stand ages, as do the parameters of the stem mass curve. Biomass studies in an age series of stands at Jijia demonstrated that this is not the case, the curve in fact flattening and shifting to the right as the trees age. This has previously been ascribed to the effects of litter fall, with an assumption that the sum of current leaf mass and total leaf mass dropped as litter follows a constant relationship with stand mean diameter (J. Landsberg, pers. comm.). The rate of litter fall necessary to bring about the observed change in the leaf mass-diameter relationship at Jijia was therefore calculated, with results as in Figure 5. While it was possible to obtain a good fit to observed data for 21, 33 and 45 month old stands of *E. urophylla* clone U6 by this means, the implied rate of litter fall is around 40% of the canopy mass per month. Observed litter fall in these stands, from monthly collections, was approximately 5% to 8% per month.

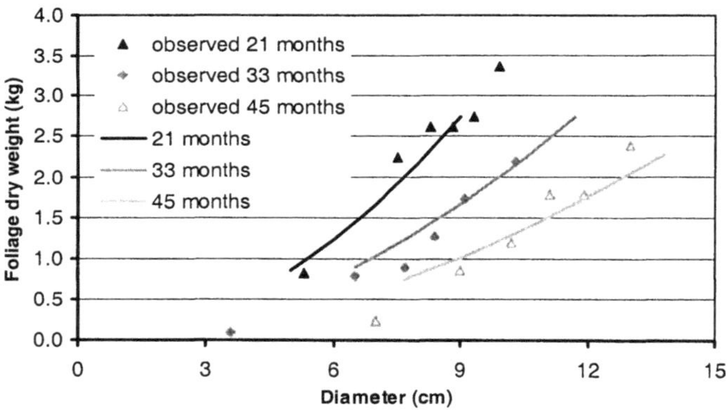

Figure 5. Observed foliage mass in relation to diameter for 3 stands of *E. urophylla* clone U6, and foliage mass predicted from a single allometric relation and monthly litter fall of 40% of the canopy mass.

Hence it must be assumed that the underlying relation between leaf mass and diameter has changed with time, in addition to canopy reduction by litter fall. The magnitude of this change and its critical role in 3-PG's allocation submodel make it essential to include it in the parameterisation of the model for short rotation plantations.

The monthly stem mass calculated by 3-PG includes the mass of branches and bark, and its conversion into stem wood volume uses a branch fraction which is usually parameterised to reduce with age from an initial high value to a stable fraction after several years. Variation in branching with stand density has not been included in the past, yet it is commonly observed that trees grown at low density have heavier branching and a smaller height to diameter ratio than trees at high density. Inverting the approximate relationship in equation [1] it is possible to estimate mean dominant height in 3-PG as

$$H = \frac{3(1\text{-branch fraction}) \times aS \times D^{(nS-2)}}{\text{wood density}} \qquad [2]$$

where aS and nS are the coefficient and power of the stem mass - diameter relationship (Figure 4). In the absence of measured branch fraction data from a range of stand density, equation [2] allows the assumed value of branch fraction to be checked by measurements of tree height and diameter. Stand density in the Jijia growth plots varies from 1735 to 2882 trees/ha, sufficient to impact significantly on stem wood volume as a result of branch fraction differences. The accuracy of modelling these plantations is therefore enhanced by incorporating a response to stand density in the parameterisation of branch fraction.

The 3-PG model was initially derived on the basis of data from cool temperate forests, where harvest rotation length is commonly several decades. In applying the model over such long time periods, the problem of estimating radiation absorption before canopy closure is often avoided by initialising the model at age 1 year or more, with an assumed mass of foliage, stems and roots. The approximation inherent in choosing these initial values is assumed to have little or no impact on growth at ages approaching rotation length. In fast-growing short rotation plantations this is clearly not so, and there is no option but to model stand growth from the seedling stage. Some additional parameters are then required to discount the early growth for the effects of uneven canopy distribution in young stands.

The emphasis on temperate species in data sets which have been used for developing and validating 3-PG also leaves some unresolved questions of appropriate parameter values for tropical species. In particular, the maximum value adopted for photosynthetic quantum efficiency (alphaC) has usually been no more than 3.9g carbon per MJ of absorbed radiation (J. Landsberg, pers. comm.). This leads to a credible parameterisation for temperate plantations of a number of species, but has been tested to only a limited extent in tropical climates. For the unquestionably high growth rates of young eucalypt plantations in southern China (e.g. Figure 2), there is some justification for exploring the consequences of allowing alphaC to assume higher values in deriving a set of species parameters.

5 Deriving Monthly Climate Data

Climate data inputs for modelling may be satisfactorily estimated from routine daily meteorological observations including temperature, rainfall and sunshine duration. Fitting or calibration of the model against historical growth measurements should ideally use actual monthly climate data for the appropriate period, but for predictions of future growth a single set of long term monthly mean values is often used. Where there is considerable annual variation in rainfall or other climate variables, it may be preferable to use an actual sequence of past observations, cyclically repeated as necessary for the period of modelling. Alternatively stochastic methods may be applied to generate an artificial climate sequence which mimics past weather patterns.

Daily maximum and minimum temperature and rainfall are commonly recorded by meteorological stations and can usually be obtained for one or more locations in the vicinity of the modelled site. Monthly values are readily calculated from these. Where rainfall and temperature gradients are known to exist in the area of interest, an interpolation procedure may be appropriate to estimate local conditions from observations at two or more stations. In locations with a cold winter climate, the number of frosts occurring each month is also required by 3-PG and may be satisfactorily estimated by regression on monthly mean minimum temperature.

Mean daily vapour pressure deficit (VPD) during daylight hours may be estimated from maximum and minimum temperature data (e.g. Dye 2001). If relative humidity observations are available they may be used to refine the estimated VPD.

Monthly mean wind speed is used to estimate aerodynamic conductance in the calculation of transpiration using a Penman-Monteith combination equation. In some versions of 3-PG this conductance is simply set to an assumed value, e.g. 0.2m/sec, reflecting an assumption that variation in mean wind speed is not great on a monthly timescale.

Monthly solar radiation is the starting point for calculating net photosynthesis, and hence is critically important to absorbed radiation models such as 3-PG. Unfortunately it has not been routinely measured at most meteorological stations, and hence must often be estimated from other observations. Zuo et al. (1996) derived a function for estimating radiation from daily sunshine duration data in China and southeast Asia, while Coops et al. (1998) described a method for estimating radiation from daily maximum and minimum temperature.

For modelling the growth of eucalypts on the Leizhou Peninsula, two years of meteorological data were available from intensive monitoring in plantations at Jijia and Hetou. A relation between the observed solar radiation data and daily sunshine duration from a meteorological station at Zhanjiang was derived by regression analysis, and used to estimate monthly solar radiation over the four year period from January 1998 to December 2001. The observed data sets for temperature, rainfall, wind speed and VPD were similarly extended by regressions on the corresponding Zhanjiang data. Table 1 shows the monthly mean value of each variable, averaged over four years.

6 Estimation of Site and Species Parameters

Parameters which are not amenable to direct measurement or which have not yet been reliably quantified may be estimated by fitting the modelled growth predictions to one or more sets of growth observations, ideally for multiple species on sites with a range of climate and soil conditions or stand densities. Parameter estimation may be systematic and objective, or an intuitive iterative procedure based on the refinement of an educated guess. The accuracy and reliability of modelled predictions will improve over time as more data sets become available to incrementally improve the precision of estimated parameters.

Table 1. Monthly values of 3-PG climate variables for western central Leizhou Peninsula, averaged over four years (1998-2001).

	Rainfall (mm)	Solar radiation (MJ/m)	Daytime VPD (kPa)	Wind speed (m/sec)	Maximum temperature (degree C)	Minimum temperature (degree C)
January	37	111	0.48	0.4	20.6	11.9
February	46	162	0.53	0.3	21.3	12.4
March	62	245	0.61	0.3	24.2	15.1
April	137	342	0.76	0.3	28.7	19.2
May	220	442	0.81	0.3	30.5	20.9
June	159	503	0.96	0.3	32.8	22.7
July	232	517	1.07	0.4	33.4	22.8
August	193	453	1.05	0.3	33.1	22.6
September	146	362	1.01	0.3	32.0	21.4
October	117	306	0.89	0.4	30.1	19.9
November	55	206	0.80	0.4	26.2	15.6
December	64	148	0.63	0.3	21.4	11.6

Most site and management parameters can be readily specified or measured, although it may sometimes be necessary to infer soil characteristics of the root zone including moisture holding capacity and rooting depth. The rainfall effectiveness parameter, which combines canopy interception, runoff and surface evaporation can be directly measured but will often be estimated from hydrological studies on similar sites, in the absence of a detailed water balance study in the area of interest. The site fertility parameter must be inferred from growth of vegetation on similar sites, since no satisfactory means of estimating it from objective measurements of nutrient concentrations or other soil factors have been found.

The 3-PG species parameters fall into four groups. The first of these contains parameters which determine the absorption of radiation by the canopy, including the canopy extinction coefficient, specific leaf area and incomplete canopy early growth parameters. The extinction coefficient is measurable with some difficulty from biomass studies in conjunction with above and below canopy radiation measurements. Specific leaf area and its variation with age may also be determined from biomass studies. The early growth parameters might be objectively determined from measurements of radiation and crown geometry, but are more likely to be inferred with the help of leaf area index measurements in young stands.

The second group of parameters determine net photosynthesis, comprising alphaC and the modifiers for limitations due to age, vapour pressure deficit (VPD), temperature, salinity and nutrition. Among these the parameterisation of temperature response may be estimated with caution from monthly mean daytime temperatures observed within the natural or planted range of the species; VPD response may be determined from water use measurements, and salinity response from performance on saline sites. AlphaC and the remaining modifier parameters will usually be inferred from growth data.

A third parameter group specify the carbohydrate allocation characteristics, litter fall and root turnover of the stand. Root allocation and turnover are likely to be inferred in most cases due to the difficulty of obtaining comprehensive root biomass data. The other parameters are amenable to measurement by biomass studies or regular litter fall monitoring, although leaf allocation parameters have also often been inferred due to lack of direct observational data.

The remaining parameters can all be derived from stand measurements, including those which determine tree mortality, initial biomass, diameter distribution, branch fraction and wood density. In summary, the principal parameters to be derived by fitting the model to sets of growth data are those related to quantum efficiency, fertility, leaf and root allocation and early growth.

7 Modelling Eucalypt Plantations on the Leizhou Peninsula

Table 2 lists the values of parameters which differ between *E. urophylla* and ABL12 in interim sets derived to fit 3-PG to the Jijia growth data. These parameters are sufficient to characterise the observed differences in diameter and volume growth of the two stand types, leading to the predicted values plotted with observed diameter and MAI for *E. urophylla* in Figure 6.

Although the model has been satisfactorily parameterised to fit the growth data from the Jijia plantations, this relatively straightforward exercise does not complete the formulation of a general parameter set for the species concerned. Validation and tuning of the parameters against growth data from sites with different climate and soil conditions are required before it can be assumed that the model's predictions are robust over the whole range of expected situations. The interim parameter estimates are however valid for predicting the response of plantations at Jijia to stand management options, and for quantifying the growth differences observed between measurement plots in terms of a fertility rating or site index.

The observed data in Figure 6 are from measurements averaged over several plots of each species, so the fitted parameters are a generalised set for the Jijia plantation area. An alternative approach would be to fit the model to the data of each plot individually, allowing only fertility to vary among plots of each species. The result would be a "fertility index" for each plot and its surrounding stand. This could be used in the same way as a traditional site index, integrating the soil and microclimate factors at each plot into a single value on the basis of observed growth data.

The parameterised model is also immediately able to be applied to explore options for stand management, such as the interaction of initial stocking, rotation length and maximum MAI. Potential yields from thinning at a given intensity and age are readily predicted together with the likely growth rate of the retained trees and their size distribution in relation to stocking, species and site fertility.

Table 2. Parameter values adopted for modelling growth of *E. urophylla* and ABL12 at Jijia. Parameters which did not differ between species are not listed.

Function	Parameter	*E. urophylla*	ABL12
Mortality	afact	-0.85	-0.2
	p32fac	4.5	12
Quantum efficiency	alphaC	2.7	3.4
Root allocation	rootco1	0.6	0.4
	rootco2	2	5
Stem allocation	nsfac	2.9	2.88
Branch fraction	initbf	0.6	0.3
	finbf	0.25	0.18
Specific leaf area	initsla	16	10
	finsla	11	7

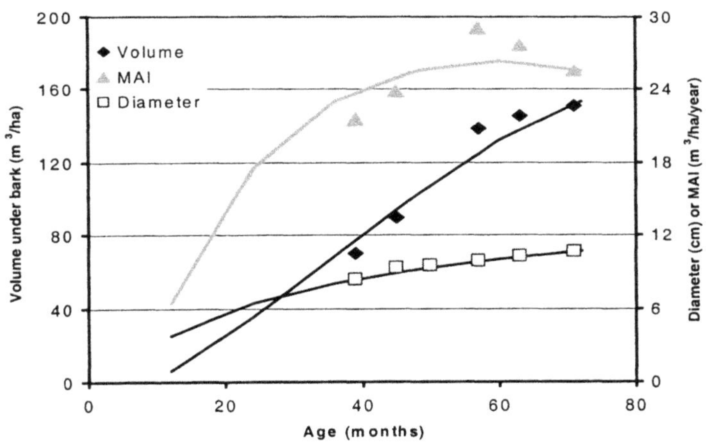

Figure 6. Diameter and MAI of *E. urophylla* observed at Jijia (mean of 3 plots), compared with 3-PG-predicted growth based on the interim parameter values in Table 2.

8 Further Research and Development

Although the model is ready for use now with the interim parameter sets derived in this study, it should be seen as the object of a continuous validation, extension and improvement process as opportunities arise to compare modelled and observed growth in various circumstances. A primary aim for ongoing research is then to collect stand growth data sets for contrasting climates and site conditions for the major species or genotypes in use for plantation establishment. Evaluation of the climate and site factors in terms of 3-PG parameters should accompany the measurements of tree diameter, height and stand density repeated at intervals through the rotation.

A more demanding but particularly valuable research objective towards better and more confident definition of the values of alphaC and other parameters is to conduct further biomass measurements for the key genotypes at a range of sites and ages. Similarly, measurements of leaf area index, water use and other stand level variables from a range of ages and site conditions would be equally useful. The encapsulation of the results of such research in the form of well-validated 3-PG parameter files represents a highly compact, accessible and lasting form of information storage. The need to check or confirm particular parameters which remain uncertain or in other ways unsatisfactory may in time become a driver for well-targeted new research.

Finally, the development of a DEM-based climate surface for the eucalypt-growing area of southern China at a grid resolution of 1 km or better is potentially of high value. This may be created as a component of a more general geographic information system, and would form a basis for spatial application of the growth model as a planning tool for plantation extension and to optimise the matching of genotypes to site conditions on the basis of local topographic data.

Acknowledgements

The data collection and model development reported here were funded by the Australian Centre for International Agricultural Research within Project FST97/77, "Eucalypts and Water". We are indebted to the management of Leizhou Forest Bureau for their generous hospitality, access to the plantations and support with field operations. Dr. Yan Hong of the Chinese Academy of Forestry, Beijing kindly provided collated meteorological data for derivation of a 3-PG climate file. Staff of the Research Institute of Tropical Forestry, South China Institute of Botany and China Eucalypt Research Centre have worked enthusiastically to collect plantation growth and other data, often in difficult conditions. We are grateful to them and to the management of their institutes for their friendly support and ongoing collaboration. We also thank Dr. Joe Landsberg for encouragement, advice and training associated with the introduction of 3-PG in China.

References

Battaglia, M. and Sands, P.J. 1998. Process-based productivity models and their application in forest management. For. Ecol. Mgt. 102:13-32.

Coops, N.C., Waring, R.H. and Landsberg, J.J. 1998. Assessing forest productivity in Australia and New Zealand using a physiologically-based model driven with averaged monthly weather data and satellite-derived estimates of canopy photosynthetic capacity. For. Ecol. Mgt. 104:113-127.

Dye, P.J. 2001. Modelling growth and water use in four *Pinus patula* stands with the 3-PG model. South African For. J. 191:53-64.

Landsberg, J.J. and Waring, R.H. 1997. A generalised model of forest productivity using simplified concepts of radiation-use efficiency, carbon balance and partitioning. For. Ecol. Mgt. 95:209-228.

Landsberg, J.J., Johnsen, K.H., Albaugh, T.J., Allen, H.L. and McKeand, S.E. 2001. Applying 3-PG, a simple process-based model designed to produce practical results, to data from loblolly pine experiments. For. Sci. 47:43-51.

Landsberg, J.J., Waring, R.H. and Coops, N. 2003. Performance of the forest productivity model 3-PG applied to a wide range of forest types. For. Ecol. Mgt. 172:199-214.

Morris, J.D. and Collopy, J.J. 2001. Validating plantation water use predictions from the 3PG forest growth model. In: F. Ghassemi, D. Post, M. Sivapalan and R. Vertessy (eds), Proc. of MODSIM 2001 - International Congress on "Modelling and simulation", Canberra, Australia, 10-13 December 2001, pp437-442.

Zuo, H., Hutchinson, M.F., McMahon, J.P. and Nix, H.A. 1996. Developing a mean monthly climatic database for China and southeast Asia. In: "Matching trees and sites", T.H. Booth (ed), Proc. of International Workshop, Bangkok, Thailand, 27-30 March 1995, ACIAR Proc. No.63, pp10-17.

利用林分生长过程模型预测华南地区桉树人工林生产潜力

JIM D. MORRIS，TOM BAKER

Forest Science Centre, P.O. Box 137, Heidelberg
Victoria 3084, Australia

生长过程模型作为一种规划手段日趋被林业经营管理者使用。3-PG 模型的应用对华南地区桉树人工林经营水平的提高有着显著的效果。完备的模型除为人工林营造、抚育和采伐提供合理的决策依据外，还可为树木的生长状况数据提供存储和修复平台，同时更能指示未来重点研究领域。

粤西雷州半岛桉树人工林林分生长，生物量分布关系和水分利用观测数据已经被用于几种主要桉树无性系（基因型）模型参数的估算。但经营者为了木片生产而采用的高种植密度、短轮发期和林分早期的快速生长需要一些模型参数的特例数据，而这些特例数据在其它 3-PG 模型的实际运作过程中还没有先例。

模拟所需的气候数据可以依据气象观测站日观测数据进行估计，如气温，降雨量和日照期。不能直接测定的生理参数可以通过拟合的方法进行估计，如用一次或多次林分生长观测数据拟合模型预测值，或更理想的是对生长于不同气候和土壤条件（或种植密度）的立地类型上多个树种的拟合。模拟结果的精度和可靠性随时间增长会不断改善，因为大量的数据资料会更有利于模型参数的精确估计。

THE RESPONSE OF A *EUCALYPTUS* HYBRID CLONE TO WEED CONTROL AND BURNING

KEITH M. LITTLE

Institute for Commercial Forestry Research
P.O. Box 100281, Scottsville 3209
Republic of South Africa

A field trial was initiated in 1995 to a *Eucalyptus* hybrid clone (GC540) to evaluate either the effects of various residue management options on weed and tree growth, or selective weed control on tree performance. This trial was situated in the coastal sub-tropical Zululand region of KwaZulu-Natal, South Africa. A combination of seven different treatments were used and included a weedy and weedfree treatment (these were imposed on plots where the plantation residues had been left *in situ* or alternatively burnt), a 2 m row weeding and the selective control of broadleaves or grasses only. Regardless of the treatment, tree growth suppression was detected where the vegetation was not controlled. However, the onset of competition occurred sooner and was much greater in the burnt + weedy (40 days after planting) treatment plots than in the unburnt + weedy plots (68 days after planting). Initial improvements in tree performance for those treatments where the vegetation was controlled was maintained over seven growing seasons, with significant differences recorded for the merchantable volume calculations. The best response, obtained with the unburnt + weedfree treatment, resulted in a 41% increase in merchantable volume when compared to the burnt + weedy treatment. No significant differences were detected between the residue management treatments when kept free of vegetation. Other benefits obtained from competition control include a reduced time to canopy closure and decreased stem variability.

1 Introduction

Initial research on site preparation in South Africa focused on methods of establishing eucalypt plantations in virgin grass veld, as well as on the re-establishment of eucalypt stands on sites where the post-harvest plantation residues had been burned. Recently there has been a shift away from the burning of these residues (Norris 1995) together with the realization that intensive soil preparation is not always necessary, especially on those sites where eucalypts are to be re-established (Smith et al. 2001). However, the retention of this plantation residue has meant that silvicultural operations, especially site preparation (where necessary), the preparation of a planting position, planting and vegetation control has become more difficult (Little et al. 1996). The manipulation of this plantation residue (burning or leaving the plantation residue *in situ*) would be expected to have a large impact on the presence and abundance of different weed species and their development, which in turn would have an impact on tree performance as has been demonstrated in *Acacia mearnsii* plantations (Little et al. 2001).

The beneficial effects of weed competition on initial tree growth in commercially planted eucalypt plantations have long been recognized (Schönau

1989), with the importance of competition control being stressed in terms of reduced seedling mortality (Dougherty and Lowery 1991), increased stem and stand uniformity (Mason and Kirongo 1998), reduced time to canopy closure (Endo and Wright 1993), increased yields (Little 1999) and reduced rotation times, as well as allowing for improved access into the compartment for various silvicultural operations.

The management of vegetation in the Zululand region of South Africa particularly critical due to the sub-tropical climate which favours an extended period during which vegetation is able to grow (Schulze 1997) and the sensitivity of the tree species grown (*Eucalyptus* hybrid clones) to competition from this vegetation means that vegetation control must be of a high standard (Little 1999). Several different management techniques exist that involve the manipulation of this competing vegetation to an acceptable level and include biological, physical, chemical and cultural weed control (Anderson 1996). Some of these practices include the manipulation of tree-row spacing, species choice and stand density, the maintenance of critical weed-free periods and distances (Little and Schumann 1996), selective control of vegetation, and using crop rotations and cover-crops (Little et al. 2002).

Although these practices have been demonstrated in pine plantations (e.g. Wagner 1989; Richardson 1993; Christie 1995; Zwolinski 1996), together with some research carried out on the control and impact of weeds on early eucalypt growth (e.g. Messina 1990; Wilkinson and Neilson 1990; Endo and Wright 1993; Little and Schumann 1996; Neilson and Ringrose 2001), little work has been reported on the longer term impacts of vegetation management on tree performance, especially when comparing different methods of plantation residue management. In 1995 a trial was initiated on a *Eucalyptus* hybrid to determine the long-term:

- impact of various residue management options on weed growth and tree performance, and
- influence of selective weed control on tree performance.

This report discusses the development of the trees from planting to seven years of age in relation to the above objectives.

2 Materials and Methods

2.1 *Description of Site*

The study was conducted near the coastal town of KwaMbonambi, KwaZulu-Natal (28°35'S and 32°05'E). The climate is classified as sub-tropical, with a mean annual rainfall and temperature of 1018 mm and 21.6°C respectively. The trial was located at an elevation of 87 m on a level site. Soil parent material is of aeolin origin and is classified as an arenic lixisol and arenic kandiustult respectively. The soil texture at the 0-20 cm depth interval is sandy with low organic carbon (0.35%) and clay (3%) contents. Historically the site on which the trial was situated supported indigenous coastal forest, before being cleared for the production of sugarcane (*Saccharum* L.

spp.). In the 1980's the site was converted to plantation forestry with the growth of *Eucalyptus grandis* on a short pulpwood rotation.

2.2 *Trial Design and Treatments*

On the 6th of July 1995 a trial was initiated on a stand of *Eucalyptus grandis* x *E. camaldulensis* clonal hybrid (GC540). The cuttings were planted at an inter-row espacement of 3 m and an intra-row spacing of 2.5 m that resulted in a stocking rate of 1333 stems/ha. Prior to planting a pre-plant spray with a non-selective herbicide (glyphosate at 4 l/ha) was undertaken to ensure weedfree conditions at the start of the trial.

Seven treatments were replicated four times (arranged in a RCBD) and imposed on the stand of hybrids. A description of the treatments and the abbreviations used to describe these treatments in the text and figures is given in Table 1. Each treatment plot consisted of 30 trees (5 rows x 6 trees in each row) with the inner net plot of twelve trees being measured (3 rows x 4 trees in each row). Each treatment plot covered an area of 225 m^2, with the total size of the trial being 6300 m^2.

Table 1. List of treatments in a vegetation management trial situated in Zululand, South Africa.

No.	Management of post-harvest plantation residues	Vegetation management	Abbreviation of treatment used in text
1	Post-harvest plantation residues burnt	No weed control	Burnt + weedy
2	Post-harvest plantation residues burnt	Complete vegetation control with glyphosate (4 l/ha) when required	Burnt + weedfree
3	Post-harvest plantation residues remain in situ	No weed control	Unburnt + weedy
4	Post-harvest plantation residues remain in situ	Complete vegetation control with glyphosate (4 l ha^{-1}) when required	Unburnt + weedfree
5	Post-harvest plantation residues remain in situ	Control of all vegetation for 1 m on either side of the tree row with glyphosate (4 l ha^{-1}) when required.	Unburnt + 2m row weeding
6	Post-harvest plantation residues remain in situ	Selective control of all broadleaved vegetation with MCPA (2.5 l ha^{-1}) when required such that only grasses remain	Unburnt + grass
7	Post-harvest plantation residues remain in situ	Selective control of all grasses with fluazifop-P-butyl (4 l ha^{-1}) when required such that only broadleaved vegetation remains	Unburnt + broadleaves

Before the cuttings were planted into those treatments where the plantation residues were retained, a 50 cm diameter was cleared such that the cuttings could be planted into mineral soil. After the initial pre-plant application of herbicide (glyphosate), no further weed control was carried out on the burnt + weedy and unburnt + weedy treatments. For the burnt + weedfree and unburnt + weedfree treatments, weed control was carried out on a regular basis. This consisted of the manual removal of weeds in the immediate vicinity of each tree, after which the rest of the area was sprayed with glyphosate at 4 l/ha. For the unburnt + 2 m row weeding treatment, the weeds in the 1 m area on either side of the tree rows were managed in a similar manner, with no further management of the weeds in the interrow. This treatment was included so as to test the long-term impact of implementing a 2 m row weeding, and was based upon results obtained from a study to determine optimum row weeding distances in eucalypt plantations (Little and Schumann 1996). In the unburnt + grass treatment all broadleaved vegetation was selectively controlled with MCPA (2.4 l/ha) when required such that only grasses remained. In a similar manner all grasses were selectively controlled with fluazifop-P-butyl (4 l/ha) such that only broadleaved vegetation remained in the unburnt + broadleaves treatment. These two treatments were included to determine if selective weed control could be considered as a viable option when compared to complete weed control in eucalypt plantations.

3 Measurements

3.1 *Vegetation Biomass, Density and Cover*

The presence of competing vegetation was quantified in two different manners. The first method involved the taking of above ground vegetation biomass samples 9 times. One biomass sample (0.5 m x 0.5 m) was collected from each of the treatment plots that contained vegetation on each measurement date. These biomass samples were taken from the border rows to avoid an impact on the tree performance of the measured trees. The samples were oven dried at 80°C to a constant mass and then weighed. The second method involved the determination of vegetation density on five occasions. The vegetation was assessed in one 0.5 m x 0.5 m area per treatment plot for either grasses or broadleaved plants. As the density remained fairly constant for the period over which the measurements were taken, only that at the end of the first growing season (six months) is discussed.

3.2 *Tree Growth and Variability*

Tree variates of either height (m), groundline diameter (cm) or diameter at breast height (cm) were measured at regular intervals during the development of the stand. These were used to determine the performance of the trees in the different treatments over time. The diameter at breast height (1.3 m above ground level) could only be determined when most of the trees from the different treatments were above 1.3 m in height. The suppression of trees in the treatments with vegetation meant that this measurement was delayed until 16 months after planting. From 16 months, the diameter at breast height was measured nine times until clear felling at 7 years, and the basal area per hectare (m^2/ha) was calculated using the following equation:

$$Basal\ area\ per\ hectare = \pi\left(\frac{Dbh_{ob}}{2}\right)^2 \cdot Stocking \qquad [1]$$

where Dbh_{ob} is the over bark diameter at breast height and the stocking is obtained from the respective treatments plots. Due to the lack of a suitable volume equation for the clone planted, the Dbh_{ob} to merchantable volume relationship from a second vegetation management trial planted to *E. grandis* x *E. camaldulensis* (GC340) was used (Little 1999). In this trial the merchantable underbark volume (V_m in m^3) on an individual tree basis was determined at seven years of age. The V_m equates to an underbark volume up to the minimum overbark diameter (0.07 m) that can be utilized economically. The Dbh_{ob} and V_m data received a $\log_{10}$ transformation, for the purpose of equalising the variances (Rayner 1967), and an exponential regression was fitted.

$$\log_{10}(V_m) = 0.593 - 8.87 \cdot 0.215^{\log_{10}(Dbh_{ob})} \qquad [2]$$

This regression (Eqn [2]) accounted for 98.4 % of the variation within the data set and was used to calculate the estimated V_m from the final Dbh_{ob} measurements in this trial. From this the total merchantable volume per hectare (V_{mha} in m^3/ha) was then calculated using the stocking obtained from the respective treatments plots.

In order to quantify the effect of the treatments on tree variability, the coefficients of variation for groundline diameter and diameter at breast height were calculated on a treatment pot basis in a similar manner as used by Weiner (1985 and 1986), Knox et al. (1989) and Bouvet (1997).

3.3 *Analysis*

Before comparisons between individual treatment means were made, an overall *F*-test was carried out. The *F*-test is an overall test of the significance of the differences that have been observed between the means of all the treatments in the experiment (Mead and Curnow 1983). Only if the *F*-value was significant were treatment differences further investigated using least significant differences (*lsd's*). Unless otherwise stipulated, comparisons between the tree variates were analyzed as a randomized complete block design with the use of Genstat® for Windows™ (Lane and Payne 1996). In addition, Bartlett's test (Snedcor and Cochran 1956) was used to test the assumption of homogeneity of variance in order for a valid analysis of variance to be performed.

4 Results and Discussion

4.1 *Vegetation Biomass, Density and Cover*

The development of the vegetation biomass and percentage cover was initially very rapid, this despite the planting of the cuttings into a site that was free of vegetation.

The development of this biomass was the greatest in the burnt + weedy and unburnt + weedy plots, with 2.05 and 1.40 t/ha recorded only 40 days after planting (Figure 1). The biomass in all treatment plots continued to increase through the first growing season and only started to decline from 8 months (246 days) onwards. This could be due to a combination of the senescence of the vegetation in autumn due to a decrease in available moisture (March-April) and the shading effect of the tree crowns preventing the continued, vigorous growth of the vegetation. Although the vegetation did continue to survive and grow once the trees had become established, the biomass never reached the same levels that were attained in the first growing season. Although the rapid growth of the vegetation had resulted in an almost complete ground cover 3 months (95 days) after planting, differences were detected in the density of this vegetation (Figure 2). When compared to the unburnt treatment plots, almost double the amount of stems/m^2 were recorded in the burnt + weedy plots. The retention of the plantation residues possibly acted as a physical and chemical barrier to the germination and growth of existing and incoming seeds (Schumann et al. 1995). In addition to the lack of this barrier in the burnt + weedy plots, seed germination may have been stimulated in those weed species requiring fire to break dormancy.

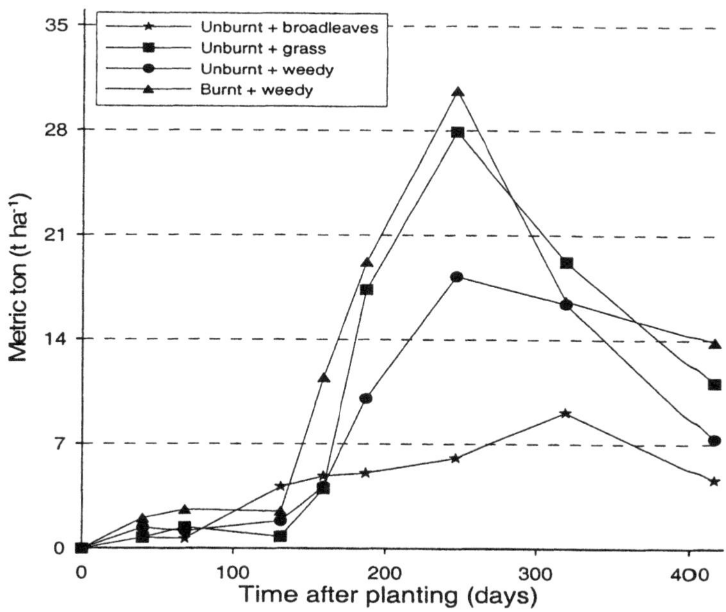

Figure 1. Development of above-ground vegetation biomass for the first 14 months growth in a vegetation management trial situated in Zululand, South Africa.

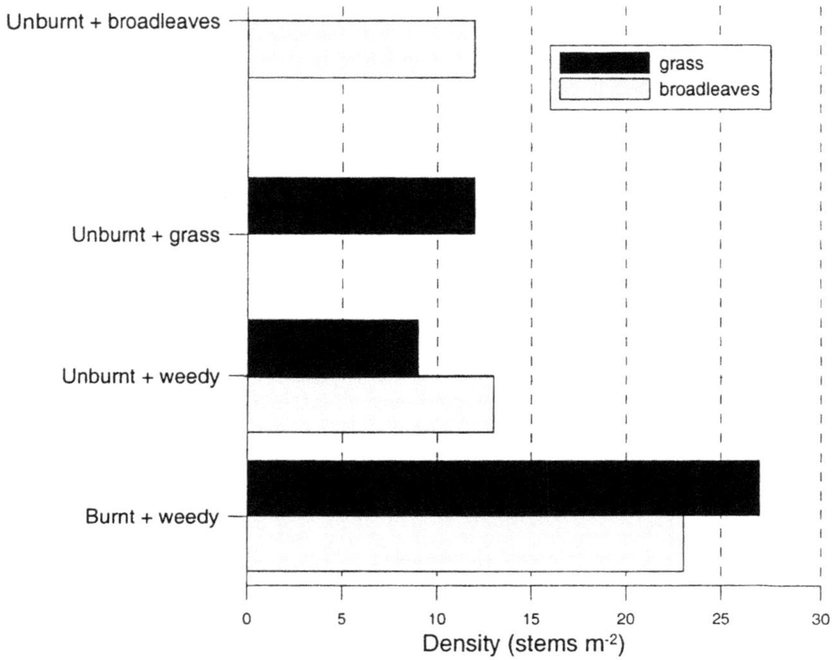

Figure 2. Density of grasses and broadleaves after 6 months in a vegetation management trial situated in Zululand, South Africa.

4.2 *Early Tree Growth and Variability*

The early effects of competition (up to 16 months after planting) of those treatments containing vegetation are illustrated with the use of the groundline diameter (Figure 3). Depending on the presence and initial development of competitive vegetation, three groups of treatments emerge. When competing vegetation was kept away from the trees, either by complete weeding (burnt + weedfree and unburnt + weedfree) or by carrying out a 2 m row weeding (unburnt + 2 m row weeding), no significant differences were detected between these three treatments (Figure 3). The rapid growth and development of the competitive vegetation meant that divergence in terms of groundline diameter was detected between 40 and 68 days after planting in the burnt + weedy treatment and between 68 and 131 days after planting in the unburnt + weedy, unburnt + broadleaves and unburnt + grass treatment plots. Little and Rolando (2002)[#] found in a combined analysis of 22 weed control trials on eucalypts in South Africa that the onset of weed induced tree growth suppression occurred on average 67 days after planting for sites that had been burnt, compared to

[#] Data obtained from the Institute for Commercial Forestry Research field trial data base

157 days after planting for those site where the plantation residues had not been burnt. The rapid initial growth of the vegetation in the burnt + weedy treatment meant that tree growth was even more suppressed than in the unburnt treatments that contained vegetation. Similar results were found in a residue and weed control management trial conducted on *Acacia mearnsii* in South Africa (Little et al. 2001).

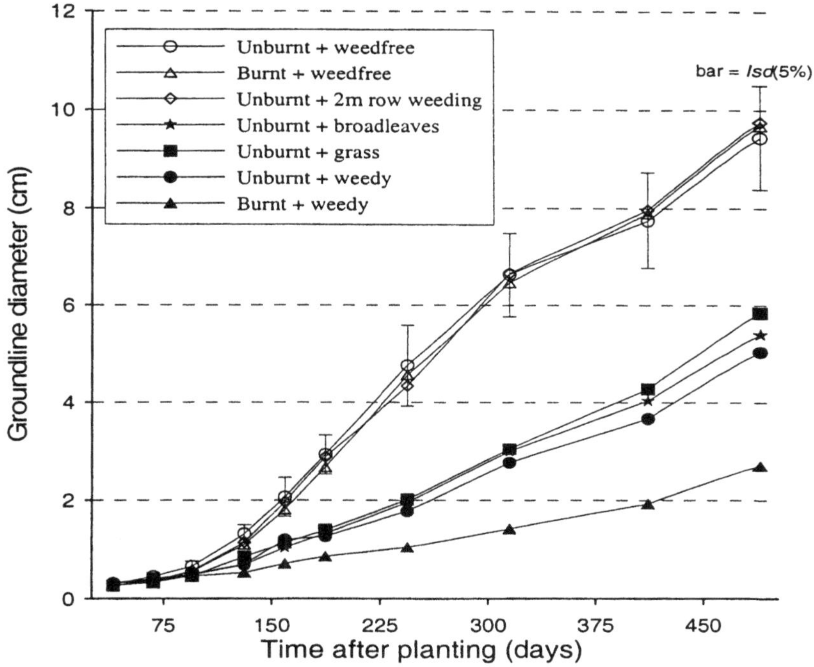

Figure 3. Development of *Eucalyptus grandis* x *E. camaldulensis* groundline diameter with time for the first 16 months growth in a vegetation management trial situated in Zululand, South Africa.

Two of the most important factors affecting variability in tree performance include the use of genetically variable seedlings and possible variability associated with differences in the environment in which these seedlings are grown. Another manner in which the environment in which the cuttings are grown can be altered is through the use of different silvicultural practices. To determine the impact of the residue and weed management treatments on the development of cutting variability with time, the coefficient of variation was determined on a treatment plot basis and plotted for groundline diameter (Figure 4). At the time that the first measurement was taken, all the treatments exhibited a high degree of variability, this despite the use of genetically identical cuttings in this trial. This high degree of variability could have been the result of one or a combination of the following factors: variability introduced whilst at the nursery, variability introduced when the cuttings were planted into the field (Morris 1994), variability associated with small differences that occur in the topsoil, microclimate and topography, and variability associated with the replanting of dead trees. From the first measurement date there was a decline in the coefficient of variation for those treatments in which the vegetation was either controlled or kept away from the trees. Although Little et al. (2001) detected differences in *Acacia mearnsii* seedling variability between treatments where the residues had been burnt or left *in situ*, no such differences were detected in this trial. This could in part be attributed to the removal of the residue from around the pit into which the cuttings were planted, resulting in the planting of the cutting into mineral soil. Inter-specific competition between the cuttings and the vegetation resulted in an initial increase in the coefficient of variation for the rest of the treatments before declining. In comparison to the unburnt + weedy, unburnt + grass and unburnt + broadleaves where the coefficient of variation declined from 8 months after planting, a reduction in variability in the burnt + weedy treatment only occurred from 14 months onwards.

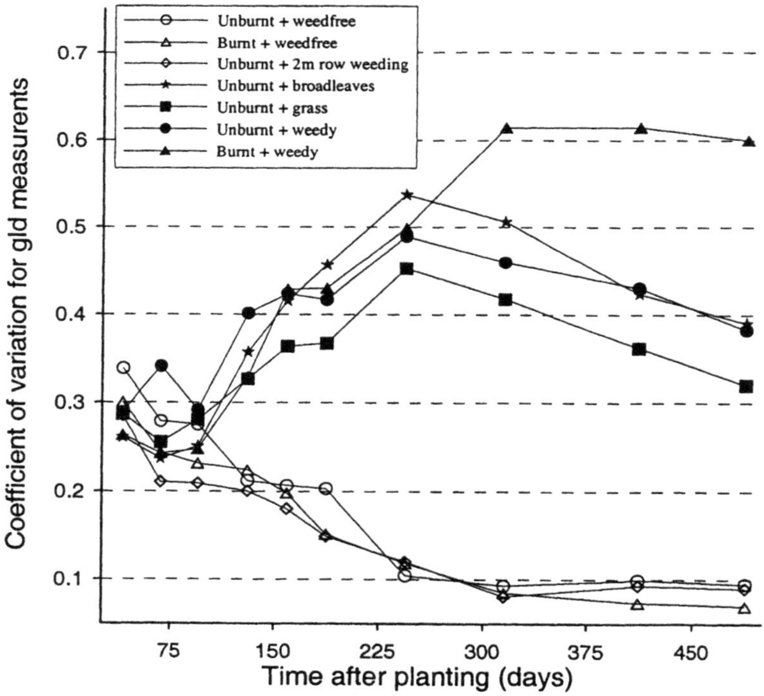

Figure 4. Changes in the coefficient of variation for *Eucalyptus grandis* x *E. camaldulensis* groundline diameters over the first 16 months in a vegetation management trial situated in Zululand, South Africa.

Table 2. Summary of analyses of variance and data for final *Eucalyptus grandis* x *E. camaldulensis* growth measurements in a vegetation management trial situated in Zululand, South Africa.

Source of variation	DF	Stocking (stems/ha)	Height (m)	Dbh_{ob} (cm)	Basal area per hectare (m^2/ha)	Merchantable volume per hectare (m^3/ha)
Summary of analysis of variance (sum of squares)						
Replications	3	24691^{ns}	1.97^{ns}	2.45^{ns}	43.21^{ns}	3444.4^{ns}
Treatments	6	29982^{ns}	76.45^{*}	31.66^{**}	281.19^{**}	24307.9^{**}
Trees (residual)	18	333333	83.30	11.97	114.54	7584.1
Total	27					
Summary of data (treatment means)						
Unburnt+weedfree		1222	26.73^{a}	18.93^{a}	34.84^{a}	274.5^{a}
Burnt+weedfree		1194	25.81^{ab}	19.04^{a}	34.16^{a}	269.4^{a}
Unburnt+2m row		1197	25.78^{ab}	18.69^{a}	32.89^{ab}	266.1^{a}
Unburnt+broadleaves		1250	25.84^{ab}	17.02^{b}	29.67^{b}	227.2^{b}
Unburnt+grass		1250	23.62^{abc}	16.78^{b}	28.33^{b}	220.3^{bc}
Unburnt+weedy		1167	23.39^{bc}	17.13^{b}	27.70^{b}	212.7^{bc}
Burnt+weedy		1167	21.80^{c}	16.29^{b}	26.13^{b}	194.5^{c}
Grand mean		1206	24.71	17.70	30.53	237.80
Standard error		96.2	1.52	0.58	1.78	14.51
Coeff. of var. (units)		11.3	8.7	4.6	8.3	8.6
Bartlett's test for homogeneity of variance. Chi-square for 7 df.		6.70^{ns}	11.12^{ns}	7.41^{ns}	4.92^{ns}	6.35^{ns}

Note: * signifies significance at $p<0.05$, ** significance at $p<0.01$ and ^{ns} non-significance.
Within each column, values followed by the same letter are not significantly different; $p<0.05$ according to Students *t*-test.

4.3 *Treatment Differences at Seven Years of Age*

No significant treatment differences were detected for tree survival for the duration of the trial (Table 2). Benefits of early weed control were maintained through to seven years as indicated by the basal area growth curves from 16 months after planting onwards (Table 2 and Figure 5). This effect is highlighted in the final height, diameter at breast height and volume calculation as shown in Table 2. No significant differences were found between those treatments with complete weed control and the unburnt + 2 m row weeding. This confirms the results obtained by Little and Schumann (1996) which showed that provided the weeds were kept 1 m away from the trees there would be no potential loss in yield. Whilst not adversely affecting tree performance, the retention of vegetation in the inter-row will have the dual benefit of soil stabilization and reducing weed control costs by up to one third. A significant reduction in growth occurred where the vegetation was retained on the site (burnt + weedy, unburnt + weedy, unburnt + grass and unburnt + broadleaves).

In this respect the benefits of establishment weed control are justified in eucalypt plantations grown on a short pulpwood rotation under the prevailing climatic conditions. The selective management of vegetation (unburnt + broadleaves and unburnt + grass) is not a viable option as these two treatments were not significantly different from weedy treatments. A greater difference in merchantable volume was observed between the unburnt + weedy treatment and the unburnt + weedfree treatment (29%) compared to the 38% for the burnt + weedy treatment and the burnt + weedfree treatment (Table 2). The implications of this are that if one were to burn and did no weed control, the onset of competition would be earlier and the degree of suppression greater than if one did not burn the plantation residues.

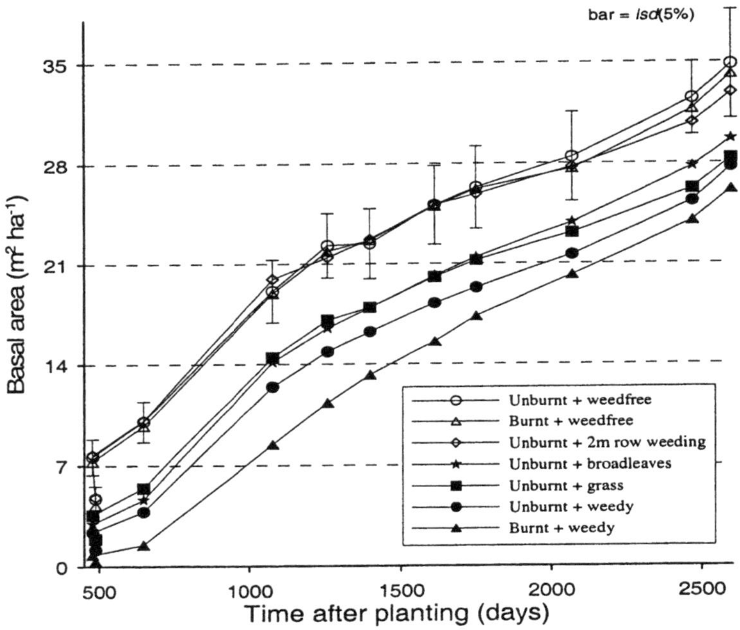

Figure 5. The development of *Eucalyptus grandis* x *E. camaldulensis* basal area per hectare until felled at seven years in a vegetation management trial situated in Zululand, South Africa.

Acknowledgements

The author would like to express thanks to staff of Mondi Forests (Zululand), Mike Maxfield, Denis Oscroft and Gert van den Berg for their assistance during the implementation and maintenance of the trial. Thanks must also go to Carol Rolando for her comments regarding this manuscript.

References

Anderson, W.P. 1996. Weed science principles and applications. West Publishing Company, New York, pp1-26.

Bouvet, J.-M. 1997. Effect of spacing on juvenile growth and variability of *Eucalyptus* clones. Can. J. For. Res. 27:174-179.

Christie, S.I. 1995. The effect of the grass *Setaria megaphylla* on the growth of *Pinus patula*. PhD Thesis, University of Witwatersrand.

Dougherty, P.M. and Lowery, R.F. 1991. Spot size of herbaceous control impacts loblolly pine seedling survival and growth. Southern J. Appl. For. 15:193-199.

Endo, M. and Wright, J.A. 1993. Growth of *Eucalyptus grandis* plantation under different levels of competing vegetation control. In: H. Gjerstad (ed), Proc. of the International Conference on "Forest vegetation management: ecology, practice and policy", Auburn University, Alabama, USA, School of Forestry Report 1, pp168-175.

Knox, R.G., Peet, R.K. and Christensen, N.L. 1989. Population dynamics in loblolly pine stands: changes in skewness and size inequality. Ecology 70:1153-1166.

Lane, P.W. and Payne, R.W. 1996. Genstat® for Windows™ an Introductory Course. The Numerical Algoriths Group Limited, UK.

Little, K.M. 1999. The influence of vegetation control on the growth and pulping properties of a *Eucalyptus grandis x camaldulensis* hybrid clone. PhD Thesis, University of Natal, Pietermaritzburg.

Little, K.M. and Schumann, A.W. 1996. A new systematic trial design for the optimization of interspecific weed control. In: R.C.H. Sheperd (ed), Proc. of the Eleventh Australian Weeds Conference, Melbourne, Weed Science Society of Victoria Inc., Victoria, Australia, pp440-444.

Little, K.M., Smith, C.W. and Norris, C.H. 2001. The influence of various methods of plantation residue management on replanted *Acacia mearnsii* growth. Aust. For. 63:226-234.

Little, K.M., Schumann, A.W. and Noble, A.D. 2002. The impact of a cowpea planted as a cover-crop on the performance of a *Eucalyptus grandis x camaldulensis* hybrid clone. For. Ecol. Mgt. 168:43-52.

Little, K.M., Eccles, N., Oscroft, D., Viero, P. and Eastwood, T. 1996. Growth of eucalypts on slash managed and burned sites in Zululand. ICFR, Pietermaritzburg, South Africa, ICFR Bulletin Series 3/96.

Mason, E.G. and Kirongo, B. 1998. Responses of radiata pine clones to varying levels of pasture competition in a semiarid environment. In: R.G. Wagner and D.G. Thompson (eds), Proc. of Third International Conference on "Forest vegetation management: popular summaries", Ontario Forest Research Institute, Forest Research Information Paper No.141:187-189.

Mead, R. and Curnow, R.N. 1983. Statistical methods in agricultural and experimental biology. Chapman and Hall, London, pp33-46.

Messina, M.G. 1990. Herbicides increase growth responses to fertiliser in a 5-year-old *Eucalyptus regnans* plantations. N. Z. J. For. Sci. 20:168-175.

Morris, A. 1994. Trial series R176 final report: factors influencing post-planting survival in late season re-establishment of *Pinus patula*. Forest Research Document 16/94, Usutu Pulp Company, Swaziland.

Neilson, W.A. and Ringrose, C. 2001. Effect of initial herbicide treatment and planting material on woody weed development and the growth of *Eucalyptus nitens* and *Eucalyptus regnans*. Weed Res. 41:301-309.

Norris, C.H. 1995. Site preparation options for South African soils. ICFR, Pietermaritzburg, South Africa, ICFR Bulletin Series 6/95.

Richardson, B. 1993. Vegetation management practices in plantation forests of Australia and New Zealand. Can. J. For. Res. 23:1989-2005.

Rayner, A.A. 1967. A first course in biometry for agriculture students. University of Natal Press, Pietermaritzburg, pp242-268.

Schönau, A.P.G. 1989. Requirements for intensive silviculture. South African For. J. 150:40-49.

Schulze, R.E. 1997. South African atlas of agrohydrology and climatology. Water Research Commission, Pretoria, Report TT82/96.

Schumann, A.W., Little, K.M. and Eccles, N.S. 1995. Suppression of seed germination and early seedling growth by plantation harvest residues. South African J. Plant and Soil 12:170-172.

Smith, C.W., Little, K.M. and Norris, C.H. 2001. The effect of land preparation at re-establishment on the productivity of fast growing hardwoods in South Africa. Aust. For. 64:165-174.

Snedcor, G.W. and Cochran, W.G. 1956. Statistical methods applied to experiments in agriculture and biology. The Iowa State College Press, Iowa, pp285-268.

Wagner, R.G. 1989. Competition thresholds for the survival and growth of ponderosa pine seedlings associated with woody and herbaceous vegetation. New For. 3:151-170.

Wilkinson, G.R. and Neilson, W.A. 1990. Effect of herbicides on woody weed control and growth of plantation eucalypt seedlings. Aust. For. 53:69-78.

Weiner, J. 1985. Size hierarchies in experimental populations of annual plants. Ecology 66:743-752.

Weiner, J. 1986. How competition for light and nutrients affects size variability in *Ipomoea tricolor*. Ecology 67:1425-1427.

Zwolinski, J.B. 1996. Weed control and large bare root stock improve early growth of *Pinus radiata* in South Africa. N. Z. J. For. Sci. 26:63-172.

一个桉树杂种无性系对杂草控制和采伐剩余物焚烧的反应

KEITH M. LITTLE

Institute for Commercial Forestry Research
P.O. Box 100281, Scottsville 3209
Republic of South Africa

1995 年对一个桉树杂种无性系（GC304）开始田间试验。试验点位于南非纳塔耳和克瓦祖鲁的祖鲁兰区域的亚热带沿海地区。为了研究不同采伐剩余物处理方式对杂草和林木生长的影响，以及林下植被有选择地控制与林木生长的关系，试验设有 8 个不同处理，分别为采伐剩余物焚烧+林下植被不清除（对照）、采伐剩余物焚烧+林下植被完全清除（对照）、采伐剩余物不焚烧+林下植被不清除（对照）、采伐剩余物不焚烧+林下植被完全清除（对照）、采伐剩余物不焚烧+每行保留 2m 宽林下植被、采伐剩余物不焚烧+林下阔叶植被完全清除和采伐剩余物不焚烧+林下杂草完全清除。试验结果表明，不管处理与否，在不控制林下植被的地方林木生长均受到抑制。但焚烧+林下植被不清除处理的竞争开始的时间（植后 40 天）要比不焚烧+林下植被不清除的时间（植后 60 天）早而且影响要大。林下植被受控制的地方，林木生长初期增益可以维持 6 个生长季节以上，这种增益特别是在材积生长上显著差异。与焚烧+林下植被不清除的对照相比，不焚烧+林下植被完全清除的反应最好，其材积增加 52%。在没有林下植被的情况下，林分对不同采伐剩余物处理方式的反应没有太大差异。竞争控制的其它益处包括林冠的提早郁闭和树干变异的缩小。

A STUDY ON EFFECT OF DENSITY/SPACING ON TREE GROWTH AND STAND VOLUME OF *EUCALYPTUS UROPHYLLA* PLANTATION FOR PULP WOOD

GUANGYOU LI [1], JIANMIN XU [1], ZHAOHUA LU [1] AND JIANBO CHEN [2]

[1]*Research Institute of Tropical Forestry, Chinese Academy of Forestry Guangzhou 510520, China*
[2]*Guangxi Research Institute of Forestry Nanning 530001, China*

In this paper, the dynamic change of tree height, *DBH* and stand volume in different density/spacing treatments, the relationship between *DBH* and crown width, the basic density management for 5.9-year old *E. urophylla* plantation in Laibin County of Guangxi, China, were studied. The results indicated that plantation density had a negative correlation with height, *DBH* and standing volume growth. The superiority order of the density treatments in stand volume was 1667>2222>3333 trees/ha. After 3.7 years, spacing 2 m × 3 m performed better in height growth than 1 m × 6 m. The best treatment was 1667 trees/ha with stand volume 2.04 times of that of 3333 trees/ha. Three spacing treatments ranked in stand volume in order: 1667>2222>3333 trees/ha at 4.7-year old but 2222>1667>3333 trees/ha at 5.9-year old. The treatment of density 2222 trees/ha yielded 127.6 m^3/ha, 1.17 times as much as that of 3333 trees/ha. The relationship between crown width and *DBH* could be described as $Cw=0.4724\times DBH^{0.6715}$. By means of the regression analysis, the basic density for *E. urophylla* plantations was developed, and the desirable planting spacing for *E. urophylla* pulp wood production was also suggested.

1 Introduction

An effective way to solve timber shortage is to develop short-rotation plantations in southern China. *Eucalyptus urophylla* S. T. Blake is an excellent species to meet this objective. Density and spacing management is important for *E. urophylla* plantations. In many countries such as Australia, Brazil, India and American, a lot of studies on plantation density and spacing have been conducted (e.g. Grewal et al. 1994). The species studied include *E. citrildora*, *E. exserta*, *E. grandis* × *E. urophylla* et al., but not *E. urophylla* (Qi 1989; Wu 1992). This study aims to explore the feasibility of a better density and spacing design with higher growth than the existing ones used by local farmers.

2 Materials and Methods

2.1 *Experimental Site*

The site is located at Weidu Forest Farm, Laibin County in Guangxi, China (23°50′N, 109°08′E). It has tropical monsoon climate with annual mean temperature of 20.7°C, the lowest temperature of -3.3°C, and annual mean rainfall of 1376 mm, mostly from May to August. The soil was latosol developed from sandy rock with 10-15 g/kg organic matter, 1.20-1.29 g/kg hydrolytic nitrogen, 3.15-4.23 mg/kg available P and pH of 4.5-5.5. The vegetation mainly consisted of *Echinochloa hainanensis* Chia., *Baeckea frutescens* L., *Melastoma candidum* D. Don, *Rhodomyrtus tomentosa* Hassk., *Dicranopteris dichotoma* Burm. F. and other thorn shrubs.

2.2 *Design and Planting*

Randomized complete block design was used with 4 blocks. The seedlings with similar height (about 20 cm) were selected before planting. Planting hole size was 40 cm × 40 cm × 40 cm, and 358.2 kg/ha superphosphate as basal fertilizer was applied into the holes after the site was prepared. The experiment involved four density or spacing treatments (see Table 1). It was planted in May 1992. Tree height, *DBH* and crown width were measured every year, until 1998. The survival rates of the four treatments ware 81.0%, 82.0%, 78.0% and 78.0%, respectively at 5.9-year old.

Table 1. Density and/or spacing schemes in the trial.

Code	Density (trees/ha)	Spacing (m × m)
A	1666	1.0 × 6.0
B	1666	2.0 × 3.0
C	2222	1.5 × 3.0
D	3333	1.0 × 3.0

2.3 *Analysis*

Individual and stand volume were calculated with the following formulas:

$V_I = H \times \mathrm{DBH}^2 / 30{,}000$, (Mckenney et al. 1991), and

$V_S = V_I \times D \times S$,

where V_I=individual volume (m^3), H=height (m), DBH=diameter at breast-height (cm), V_S=stand volume (m^3/ha), D=density (trees/ha), and survival (%). Height, *DBH*, crown width and stand volume at ages 1.7-5.9 years were analyzed by using multiple comparison of variance and correlation of SAS software (SAS Institute 1990; Chen and Xu 1997).

3 Results and Discussion

3.1 *Relationship between Density/Spacing and Growth*

Variance analysis and *t*-test (Tables 2, 3 and 4) indicated that there were significant differences between density/spacing treatments in height, *DBH* or individual volume at 5.9-year old. Among the four treatments, B had the highest height and *DBH* growth, but D had the lowest, for ages from 1.7 to 5.9 years. The mean height and *DBH* of B were 1.14 and 1.9 times, respectively of those of D at age of 5.9 years (Tables 2 and 3).

Table 2. Multi-range F-test of difference in height among spacing at 5.9- year old.

Treatment code	Mean height (m)	Multi-range *t*-test
B	16.81	a
C	15.97	ab
A	15.04	ab
D	14.69	b

Note: significant at 5% level.

Table 3. Multi-range *t*-test of difference in DBH among spacing at 5.9 years.

Treatment code	Mean DBH (cm)	Multi-range *t*-test
B	12.34	a
C	11.81	a
A	11.76	a
D	9.25	b

Note: significant at 5% level.

Table 4. Individual volumes of and difference among spacing treatments.

Age (year)	Individual volume (m^3)				Duncan *t*
	A	B	C	D	
4.7	0.05588	0.06890	0.05262	0.03137	10.78*
5.9	0.06992	0.08533	0.07362	0.04190	7.67*

* significant at 5% level.

The superiority order of three planting density in wood growth (height, *DBH* and individual volume) was 1667>2222>3333 trees/ha. After 3.7 years, the growth of spacing 2 m × 3 m (B) was higher than that of 1 m × 6 m (A) (e.g. Tables 1, 2 and 3).

DBH was a function of the density (*D*): $DBH=311.469\times D^{-0.4316}$, $r=-0.8528$, closing to be significant. Table 4 showed that density/spacing treatments significantly differed in individual volume. At ages of 4.7 and 5.9 years, the treatment B had the highest individual volumes, 2.2 and 2.04 times, respectively, of those of treatment D. In addition, the differences between treatment D and the other 3 treatments were also significant.

Stand volume ranked from high to low with three planting densities as 1667>2222>3333 trees/ha at 4.7-year old but 2222>1667>3333 trees/ha at 5.9-year old. The stand volume of treatment B was 94.185 m^3/ha at 5.9-year old, higher than A. The best treatment in stand volume at 5.9-year old was C with 127.6 m^3/ha, 1.17 times that of D. It was concluded that the best density was

1666-2222 trees/ha. The growth of B was better than A, although both had the same density.

3.2 *Relationship between Crown Width and DBH*

To obtain the best prediction of crown width (cm) by *DBH* (cm), 5 regression functions were compared:

- $C_W = a + lnDBH$,
- $C_W = a + b \times DBH$,
- $C_W = a \times e^{(b \times DBH)}$,
- $C_W = a \times DBH^b$, and
- $lnC_W = a + b \times lnDBH$,

where $a(b)$=regression parameters. We found that the best function to regress crown width against *DBH*: $C_W = a \times DBH^b$ at age of 5.9 years. The regression results for three treatments (A, B and C) were presented in Table 5.

Table 5. Regression of crown width against *DBH* with function: $Cw = a \times DBH^b$.

Parameter	Spacing treatment		
	A	B	C
a	0.2559	0.5905	0.5264
b	0.9087	0.5559	0.6726
r^2	0.5922**	0.5207**	0.6518**

** significant at1% level.

3.3 *Basic Plantation Density*

A general regression of C_W with *DBH* was:

$C_W = 0.4724 \times DBH^{0.6715}$, $r = 0.678$,

significant at 1% level. Further validation showed no significant difference between predicted and observed crown width. With the predicted crown width (C_W), the basic density of plantation was estimated with the following formulas,

$N_0 = 10,000 / C_{WSD}$, and $C_{WSD} = 3.1416 \times C_W^2 / 4$.

The basic density estimated for plantation was tabulated for a wide range of *DBH* classes in Table 6. By using this table, a desirable planting density can be obtained for given *DBH*.

Acknowledgements

The authors thank Weidu Forest Farm of Guangxi of China for supporting the experiment and providing assistances, and Dr. Rongsheng Li for valuable discussion and review.

Table 6. Predicted basic density of *E. urophylla* plantation for different *DBH* classes.

DBH class (cm)	C_W (m)	C_{WSD} (m)	Density (trees/ha)	DBH class (cm)	C_W (m)	C_{WSD} (m)	Density (trees/ha)
6	1.57	1.94	5143	17	3.17	7.87	1270
7	1.74	2.39	4181	18	3.29	8.50	1176
8	1.91	2.86	3495	19	3.41	9.14	1094
9	2.07	3.35	2984	20	3.53	9.80	1021
10	2.22	3.86	2590	21	3.65	10.46	956
11	2.36	4.39	2279	22	3.76	11.13	898
12	2.51	4.93	2027	23	3.88	11.82	846
13	2.64	5.49	1821	24	3.99	12.51	799
14	2.78	6.07	1648	25	4.10	13.22	757
15	2.91	6.66	1502	26	4.21	13.93	718
16	3.04	7.26	1378	27	4.32	14.66	682

Note: C_{WSD}=theoretical crown area and C_W=predicted crown width.

References

Chen, Z. and Xu, X. 1997. Key to exercises in SAS program of biostatistics (Supplement). Computer and Agricultural Sciences Publishing House, Beijing, pp24-34.

Grewal, S.S., Juneja, M.L. and Kehar-Singh, K. 1994. Influence of vegetation changes on forest hydrology and erosion in small mountain watersheds. J. Trop. For. Sci. 6:408-421.

Mckenney, D.W., Davis, J.S., Turnbull, J.W. and Searle, S.D. 1991. The impact of Australian tree species research in China. Canberra: ACIAR Economic Assessment Series, Canberra, 1991(12):6-7.

Qi, S. 1989. *Eucalyptus* in China. China Forestry Publishing House, Beijing, pp164-167.

SAS Institute Inc. 1990. SAS/STAT user's guide, version 6, Vol.1 and Vol.2, fourth edition. SAS Institute Inc, Cary, NC, pp891-996 and 1661-1673.

Wu, C. 1992. A study on optimal planting of *E. grandis* × *E. urophylla*. *Eucalyptus* Sci. Tech. 1992(2):50-53.

尾叶桉纸浆林造林密度控制技术的研究

李光友[1]，徐建民[1]，陆钊华[1]，陈健波[2]

[1]中国林业科学研究院热带林业研究所
510520 中国广州
[2]广西林业科学研究院
530001 中国南宁

对 5.9 年生尾叶桉纸浆林生长的分析表明：（1）密度与树高、胸径呈反比，三种密度的树高、胸径大小排序均为：B（密度 1667 株/公顷）>C（密度 2222 株/公顷）>D（密度 3333 株/公顷），密度相同株行距不同的树高和胸径排序为：B（2 m × 3 m，密度 1667 株/公顷）>A（1 m × 6 m，密度 1667 株/公顷）；（2）密度对单株材积的影响与对胸径影响的规律一致，5.9 年生时单株材积生长量以密度 B 最高，达到 0.0853 m^3/株，是最低单株材积的密度处理 D 的 2.04 倍，株行距处理的单株材积排序为 B>A；（3）4.7 年生时不同密度的单位面积蓄积量大小排序为 B>C>D，密度 B 达到 94.2 m^3/ha，是最低蓄积量的 D 处理的 1.15 倍；5.9 年生时蓄积量最大者为密度 C，达到 127.60 m^3/ha，是最低蓄积量的 D 处理的 1.17 倍。随着时间的变化，株行距控制尤为重要，4.7、5.9 年生时单位面积蓄积量均是 B>A，株行距 B 分别比 A 高 24.82%和 23.53%；（4）对胸径 *DBH* 和冠幅 C_W 值用方程进行回归拟合，结果以乘幂 $Cw=a\times DBH^b$ 拟合最好。方程表达式为 $Cw=0.4724\times DBH^{0.6715}$；（5）编制的经营密度表反映出，随着胸径 *DBH* 的递增，基本经营密度 N_0 逐渐下降，当 $DBH\geq18$cm 时，N_0 的下降趋势渐趋平缓。

POTENTIAL THREATS OF PLANT PATHOGENS TO *EUCALYPTUS* PLANTATIONS IN CHINA

GILES E.ST.J. HARDY, TREENA BURGESS AND BERNIE DELL

School of Biological Sciences and Biotechnology
Division of Science and Engineering
Murdoch University, Murdoch
Western Australia 6150

The eucalypt plantation forestry industry in China has expanded rapidly in the last decade. With this expansion, it is expected that pest and disease problems will increase. Devastating eucalypt pathogens already present in Asia include *Phytophthora cinnamomi*, *Cryphonectria cubensis*, *Coniothyrium* spp., *Botryosphaeria* spp., *Mycosphaerella* spp. and *Cylindrocladium* spp. There is an urgent need to train Chinese foresters and scientists to recognise eucalypt diseases and to establish a database for easy and rapid identification of disease problems. Good identification procedures in association with quarantine and breeding programs should help to reduce the impact of eucalypt diseases in Chinese plantations. This article gives an overview of the potential major threats to the Chinese eucalypt plantation industry and some strategies to manage disease incursions and their spread.

1 Importance of Eucalypts in China

In China, *Eucalyptus* spp. are being established as plantation crops for timber, pulp and other purposes. Currently there are over 1,000,000 ha of eucalypt plantations and this area will increase as demand for wood products increases over the next few years. These demands include the establishment of plantations for medium-density to low-density short fiber pulp for paper, saw-log timber, firewood, power poles, furniture, shelter belts (Zheng 1988). Other potential uses include high quality charcoal for iron and steel production and environmental purposes such as erosion control. China has an active and reforestation program and could potentially have the most extensive eucalypt plantation stands worldwide. In addition to the plantations, there are over one billion individual *Eucalyptus* trees dispersed throughout southern China (Wang et al. 1994).

Species included in recent plantation establishment in China have consisted mainly of *E. grandis* and *E. urophylla* and hybrids of the two (Turnbull 2000). However, in the past the most widely planted species were *E. exserta*, *Corymbia citriodora*, *E. globulus* and a hybrid of unknown origin known as *E. leizhou* No.1. There is substantial potential to breed and select for superior genotypes and obtain superior material from countries such as South Africa, Brazil and others that have had a long history of eucalypt plantation forestry.

2 Potential of Disease Outbreaks in Eucalypt Plantations

In the majority of countries where *Eucalyptus* spp. have been established as exotics, plantations have initially been successful and relatively pest and pathogen-free. This is mainly due to the absence of pests and pathogens naturally associated with them in their regions of origin (Wingfield 1999). However, as the planted areas have increased in these countries and more plantations established 'off-site', disease problems have also increased (Wingfield 1999; Wingfield et al. 2001). Therefore, it is to be expected that in China the incidence and severity of diseases in plantations will increase in the near future and start to place serious management constraints to the industry.

It has also long been recognized that tree breeding and intensive management practices can result in substantial yield increases. In countries such as Brazil and South Africa, clonal forestry has made huge economic gains due to large improvements in yields. There is no doubt that China will follow the route taken by these countries and others such as Morocco, Chile and Spain where improved provenance selections and clones are routinely used in plantations. Improved selection together with improved silvicultural techniques, such as better land selection, fertilizer use, weed control and use of mycorrhizal fungi have also resulted in large yield increases. China has already started to benefit from such approaches to plantation establishment and will continue to do so.

The potential benefits gained by improved breeding and silvicultural practices will need to be balanced by the increased risks associated with nutrient imbalances, pests and diseases. This is particularly true in plantations, which are usually of a uniform age, densely planted, and of a narrow genetic range. Consequently, it must be expected that fungal and bacterial diseases together with pests will, with time, become important economic constraints to long-term plantation health management. The majority of pathogens present on exotic eucalypts have been introduced with the germplasm, either on seed or chaff or as endophytes of vegetative material (Burley 1987). These pathogens cause a range of stem and leaf diseases. Since soil has rarely been introduced along with plant stock, it is unlikely root pathogens have been introduced in this way. However, endemic soil borne pathogens have the potential to become problems in plantations.

China can obtain a substantial knowledge base about the potential threats of plant pathogens to their rapidly expanding plantation industry, from countries with years of experience in eucalypt plantation forestry and from Australia, where the majority of eucalypts are endemic. This article examines a range of soil borne foliar and canker pathogens that have the potential to be major threat to *Eucalyptus* plantations in China.

3 Soil Borne Pathogens

A number of *Phytophthora* spp. have been recorded on eucalypts. These include *P. boehmeriae*, *P. cactorum*, *P. cinnamomi*, *P. citricola*, *P. drechsleri*, *P. heveae*, *P. nicotianae* and *P. parasitica*. *P. cinnamomi* is the best known of these and has a very wide host range and is distributed throughout the world. Although this group of pathogens is commonly referred to as 'fungi' it now belongs to the Kingdom Chromista (Irwin and Ribeiro 1996). Members of this group produce a range of spore types that are adapted for dispersal (motile zoospores produced from sporangia) and survival in soil and plant debris as encysted zoospores, chlamydospores and oospores. Essentially any region with warm wet environments with rainfall above 600 mm has the potential to support *Phytophthora* spp. and the potential for disease outbreaks.

Phytophthora spp. infect fine roots, lateral roots and tree collars producing basal cankers with stem girdling and tree death. They attack the cortex, phloem and cambial tissues in roots and stems causing necrosis. Plants will either die or exhibit dieback symptoms. More recently, *Phytophthora* spp. have been found as aerial pathogens of plantation eucalypts (Dick, pers. comm.) in New Zealand.

Phytophthora spp. spread readily in water, soil and plant material, especially plant nursery stock treated with fungicides. Many fungicides are fungistatic and not fungicidal to this group of pathogens (Hardy and Sivasithamparam 2002) and apparently healthy plants, harboring *Phytophthora* spp. can be sold and planted into disease-free regions.

4 Stem Pathogens

Cryphonectria cubensis is widespread and the cause of serious and destructive stem canker disease in tropical and sub-tropical regions of the world (Burgess and Wingfield 2002), although it has also been isolated from eucalypts in the temperate regions of South Africa (Wingfield et al. 1989) and the southwest of Western Australia (Davison and Coates 1991). In South-East Asia, *C. cubensis* has been found on *E. grandis* in Sumatra and *E. camaldulensis* in Vietnam (Wingfield, pers. comm.) and in South Kalimantan and Thailand (Floyd et al. 2002).

In affected trees it can cause large basal cankers that can kill trees in the first 2-3 years of growth. It can also cause extensive perennial cankers on the trunks of older trees several meters in length. Under favorable climatic conditions, up to 50% mortality in susceptible species or clones in plantations has been recorded (Alfenas et al. 1983). Reduced growth rates cause a reduction in the yield of surviving trees. *C. cubensis* can also reduce coppicing as the pathogen can grow from cut stumps to attack developing coppice stems (Barnard et al. 1987).

The cankers kill the phloem, cambium and sapwood, and can girdle trees. In young trees the first symptoms are usually sunken areas in the bark near the base of the tree, whilst in older trees, cankers are frequently associated with swelling, bark cracks and abundant kino seepage. *Cryphonectria* canker is favored by temperatures above 23°C and by high rainfall (1500-2400 mm) (Hodges et al. 1979). Removal of the outer bark reveals defined demarcation lines between healthy and diseased tissues.

Pycnidia and perithecia are produced either together or separately on dead bark near the cankers are clearly visible to the naked eye. Perithecia form under the bark surface through which the necks protrude especially in humid weather. Whilst pycnida tend to be more superficially borne on the bark and these ooze masses of yellow conidia.

There is considerable intraspecific and interspecific variation in susceptibility to *Cryphonectria* (Conradie et al. 1992). Hodges et al. (1979) considered *E. saligna* and *Corymbia maculata* as highly susceptible; *E. grandis*, *E. propinqua* and *E. tereticornis* as moderately susceptible; *E. microcorys*, *E. paniculata* and *E. robusta* as moderately resistant; *C. citriodora* and *E. urophylla* as highly resistant. In South Africa, breeding and selection for resistant clones of *E. grandis* has been successful. In summary, this pathogen poses a substantial threat to plantations in China.

A number of *Coniothyrium* spp. have been described from eucalypts, most of them causing leaf diseases of little economic importance. In contrast, *Coniothyrium zuluense* has recently emerged as a serious primary stem pathogen in the sub-tropical regions of South Africa (Wingfield et al. 1997). The disease was first observed in 1988 on a single clone of *E. grandis* in Zululand. Since its discovery, it has become widespread in South Africa and affects many different eucalypt species, clones and hybrids. Subsequently, *C. zuluense* has been reported in Uruguay, Argentina, Thailand and Mexico where it is also causing significant disease (Wingfield, unpubl.).

This disease is typified by small dark lesions on young green stem tissue during the early part of growing season, hence its common name of 'measles disease'. Small lesions then coalesce to form large necrotic patches and swellings develop and crack and exude large amounts of kino.

In susceptible clones, the stem cankers coalesce and girdle the tree. As with *C. cubensis*, the most effective means to manage this disease is through the planting of resistant clones.

5 Opportunistic Stem Pathogens

There are a number of eucalypt pathogens that have an endophytic habit and become detrimental opportunistic pathogens when trees are stressed. Since exotic eucalypts are often exposed to a wide variety of stresses these opportunistic pathogens can become important economic pathogens in plantations (Burgess and Wingfield 2002). Environmental stresses include drought, hot or cold winds, aerial pollutants, nutritional imbalances, excess or insufficient water, hail damage, insect damage and damage from other pathogens. Endophytic pathogens can pose greater threats to plantation forestry than non-endophytic plant pathogens as they can cause disease in response to stress throughout the life of their host (Burgess and Wingfield 2002). Since plantation eucalypts in China are planted outside their natural range the threat from such endophytic opportunistic pathogens is very real. It is critical that if clonal forestry is being undertaken that careful screening of these fungi is undertaken.

Botryosphaeria spp. are the most important endophytic pathogens of eucalypts (Burgess and Wingfield 2002; Smith et al. 1996). Symptoms of *Botryosphaeria* disease include twig dieback and stem cankers. The cankers are commonly sunken and the infected tissue can be darkly coloured. Bark cracking and kino leakage are often present. They have a wide host range amongst woody plants and are consequently likely to be present in China. Management is difficult as the disease manifests itself when associated with any of the stresses listed above. However, there is good evidence that clones of *E. grandis* can differ in their susceptibility to *B. parva* and consequently, the opportunity exists to select clones from this and other *Eucalyptus* species with a high degree of tolerance (Wingfield, pers. comm.). *Endothia gyrosa* and *Colletotrichum gloeosporiodes* are other opportunistic stress pathogens often associated with *Botryosphaeria* (Smith et al. 1998; van der Westhuizen et al. 1993).

6 Foliar Pathogens

Mycosphaerella spp. and their anamorphs are extremely common pathogens of eucalypts throughout the world and are the major cause leaf blights in exotic eucalypt plantations worldwide (Crous 1998). These fungi form necrotic lesions on leaves and young shoots of eucalypts causing a disease commonly referred to as *Mycosphaerella* leaf blotch disease.

Some eucalypt species are more susceptible to others to *Mycosphaerella* leaf blotch and different species of *Mycosphaerella* vary in their host range. *Mycosphaerella* spp. found on eucalypt are specific to eucalypts and appear to be spread by seed. *E. globulus*, the plantation species of choice in many parts of Australia, is particularly susceptible to *Mycosphaerella* leaf blotch disease (Crous 1998). Recent work in Western Australia has shown that *E. globulus* yields are reduced by this disease and up to 13 different *Mycosphaerella* species and their anamorphs are associated with the disease (Maxwell et al., pers. comm.).

Mycosphaerella spp. restrict the growth of infected trees by reducing the photosynthetic area of the leaves. In the most severe cases, they can reduce the photosynthetic capacity of trees to such an extent that death eventually occurs. Lesions formed by different species vary in colour and shape and their preference for the upper or lower leaf surface. Taxonomy is difficult and relies mainly on the germination pattern of the ascospores. Molecular techniques for identification are being developed.

The disease is most serious on the juvenile leaves of susceptible trees. In general, once the adult leaves develop the disease disappears. Provenances and clones differ in their susceptibility. In addition, chemical control during the first year may be feasible and economically viable in high impact regions.

Eucalyptus rust (*Puccinia psidii*), was first recorded on eucalypts in Brazil in 1944, where it apparently adapted from native members of the *Myrtaceae* (the guava family). In Brazil, it is severely limiting the early growth of highly susceptible species and provenances. Today, it is a serious problem in Central and South America and the Caribbean. There are also unconfirmed reports of its occurrence in Taiwan and India (Wingfield, pers. comm.).

Puccinia psidii has a very wide host range, and infects eight genera and 25 species of plants belonging to the *Myrtaceae*. Seed sources of *E. grandis* from South Africa have been found to be highly susceptible to this rust. The following species of *Eucalyptus* are also known to be susceptible to *P. psidii*, *E. camaldulensis*, *E. citriodora*, *E. phaeotricha*, *E. saligna*, and *E. urophylla*.

P. psidii readily attacks foliage, inflorescences and young, succulent twigs of eucalypts younger than two years of age. First signs of infection are the appearance of tiny, raised spots or pustules on the infected tissue. After a few days, these pustules become a distinctive egg-yolk yellow colour. Infected areas often coalesce and parts of the plant can be completely covered with pustules. The spores can be both air and water dispersed. Secondary infections occur a few days after the first pustules appear. Following invasion, infected plants shrivel and leaves are deformed. Leaf deformation is possibly due to a host response to fungal infection.

The rust is seasonal in its transmission and severity does depend on the susceptibility of genetic material and weather conditions. Presence of free water at any time for more than 3 hours and temperatures of 18-23°C are favorable to the uredial cycles. Spores are spread by wind and rain splash (Coutinho et al. 1998).

Cylindrocladium spp. are known to cause severe leaf spots and often shoot blight of eucalypts. *Cylindrocladium* spp. have been found wherever eucalypts are grown, but cause the most severe damage in the tropics and sub-tropics (Crous et al. 1991; Crous and Wingfield 1994). The most devastating of these species in South-East Asia is *C. quiqueseptatum* (Floyd et al. 2002; Su-See 1999).

Lesions cover the foliage and new shoots of susceptible young trees and cause the leaves to be shed and the branches to die-off. Unlike *Mycosphaerella* spp., species of *Cylindrocladium* found on eucalypts usually have a wide host range (Crous et al. 1991). They cause a range of serious diseases from damping-off of seedlings, seedling and shoot blight, leaf spot, stem cankers and in some countries, even root diseases.

7 Nursery Diseases

Disease control must start at the seedling nursery, as there are many opportunities for diseases to have devastating affects on the production of eucalypts during seedling production. The control of diseases in seedling nurseries must take into account the overall management of the nursery and consider the role of the disease triangle in the management regime. If close attention to management practices during seedling production are maintained it is possible to minimise if not eliminate, nursery diseases.

The appropriate growing medium for the seedlings is a critical component in nursery management. Soils or artificial media must be well drained, have good water-holding capacity and air-filled porosity and be slightly acid (pH5-6). Poorly draining media are extremely conducive to soil borne plant pathogens and to poor plant growth. It is critical not to have too high a seedling density as this reduces airflow and as seedlings grow, reduced airflow provides an ideal microenvironment for many foliar pathogens, including *Botrytis cinerea*.

Strict quarantine and hygiene measures are vital during nursery production. New diseases must be prevented from entering China. Consequently, it is critical that quarantine measures are strictly adhered to and all attempts are made to prevent new pathogens entering China in seed or vegetative plant material. It is also important to adhere to strict hygiene when moving plants between nurseries within the country.

Growing media should be sterilized by physical (steam, solarization) or chemical (fumigants, such as methyl-bromide, chloropicrin, formaldehyde, or metam sodium) treatments. Seedling trays must also be sterilized between crops.

Post-sterilization hygiene is also critical to maintain healthy stock. Once seed has been sown, it is critical to place seedling trays on raised benches where possible to avoid contact with soil and water splash. Water sources should be treated by filtration, irradiation with ultra-violet light or by chlorination. The use of fungicides will usually differ between the types of diseases being controlled. However, fungicide usage should be aimed at controlling more than one disease. It is critical to alternate fungicides on a very regular basis to ensure fungicide resistant pathogens do not develop.

8 Control Measures

As mentioned previously, large-scale plantation forestry is relatively recent in China and to date there have been few disease problems. However, experience indicates that disease problems will emerge and that some of these problems will be major. In order to reduce the impact of plantation diseases, a number of strategies are available.

- Maintain strict quarantine and hygiene in the nursery during seedling production.
- Implement regular plantation health surveys on a scientific basis. This will require the training of foresters to recognize pests and diseases and how to collect suitable material and how to process it. Associated with this is the training of scientists as pathologists.
- Vitally important is that the pathogens are correctly identified. To facilitate this reference culture collections and herbarium material should be maintained. In addition electronic and paper records of all the major *Eucalyptus* pathogens to facilitate rapid identification. Also require links with scientists in other countries with more experience in forest pathology. Fungal taxonomy is currently undergoing extensive review and classic taxonomic techniques are no longer sufficient for identification purposes. Molecular taxonomy is being very important, and reference herbarium material should be matched with relevant molecular records.
- The best way to control pathogens is to keep them out. This can be managed if strict hygiene and quarantine regulations are maintained, especially when importing plant material both for plantations and breeding purposes. Good nursery sanitation will also help prevent the spread of disease.

- Once a pathogen is present, the best method of control is to implement screening procedures in tree breeding programs to ensure that susceptible clones are not selected.

Acknowledgements

We thank Mike Wingfield for sharing his experience and up to date information on *Eucalyptus* diseases.

References

Alfenas, A.C., Jeng, R.S. and Hubbes, M. 1983. Virulence of *Cryphonectria cubensis* on *Eucalyptus* species differing in resistance. European J. For. Path. 13:197-205.

Barnard, E.L., Geary, T., English, J.T. and Gilly, S.P. 1987. Basal cankers and coppice failure of *Eucalyptus grandis* in Florida. Plant Disease 71:358-361.

Burgess, T. and Wingfield, M.J. 2002. Impact of fungi in natural forest ecosystems: a focus on *Eucalyptus.* In: "Microorganisms in plant conservation and biodiversity", K. Sivasithamparam, K.W. Dixon and R.L. Barrett (eds), Kluwer Academic Publishers, Dordrecht, pp291-311.

Burley, J. 1987. Problems of tree seed certification in developing countries. Comm. For. Rev. 66:151-159.

Conradie, E., Swart, W.J. and Wingfield, M.J. 1992. Susceptibility of *Eucalyptus grandis* to *Cryphonectria cubensis.* European J. For. Path. 22:312-315.

Coutinho, T.A., Wingfield, M.J., Alfenas, A.C. and Crous, P.W. 1998. *Eucalyptus* rust: a disease with the potential for serious international implications. Plant Disease 82:819-825.

Crous, P.W. 1998. *Mycosphaerella* spp. and their Anamorphs associated with leaf spot diseases of *Eucalyptus.* American Phytopathological Society, St Paul, Minnesota, Mycologia Memoir No.21.

Crous, P.W. and Wingfield, M.J. 1994. A monograph of *Cylindrocladium*, including anamorphs of *Calonectria*. Mycotaxon LI:341-435.

Crous, P.W., Phillips, A.J.L. and Wingfield, M.J. 1991. The genera *Cylindrocladium* and *Cylindrocladiella* in South Africa, with special reference to forest nurseries. South African For. J. 157:69-85.

Davison, E.M. and Coates, D.J. 1991. Identification of *Cryphonectria cubensis* and *Endothia gyrosa* from eucalypts in Western Australia using isozyme analysis. Aust. Plant Path. 20:157-160.

Floyd, R., Wylie, R., Old, K.M., Dudzinski, M.J. and Kile, G.A. 2002. Pest risk analysis of *Eucalyptus* spp. at risk from incursions of plant pests and pathogens through Australia's northern border. Consultancy Report to Biosecurity Australia, 90p.

Hardy, G.E.St.J. and Sivasithamparam, K. 2002. Phytosanitary considerations in species recovery programs. In: "Microorganisms in plant conservation and biodiversity", K. Sivasithamparam, K.W. Dixon and R.L. Barrett (eds), Kluwer Academic Publishers, Dordrecht, Chapter 13, pp335-365.

Hodges, C.S., Geary, T.F. and Cordell, C.E. 1979. The occurrence of *Diaporthe cubensis* in Florida, Hawaii, and Puerto Rico. Plant Disease Rep. 63:216-220.

Irwin, D.C. and Ribeiro, O.K. 1996. *Phytophthora Diseases Worldwide.* American Phytopathological Society, St Paul, MN, 562p.

Smith, H., Wingfield, M.J. and Petrini, O. 1996. *Botryosphaeria dothidea* endophytic in *Eucalyptus grandis* and *Eucalyptus nitens* in South Africa. For. Ecol. Mgt 89:189-195.

Smith, H., Wingfield, M.J. and Coutinho, T.A. 1998. *Eucalyptus* die-back in South Afica associated with *Colletotrichum gloeosporiodes.* South African J. Bot. 64:226-228.

Su-See, L. 1999. Forest health in plantation forests in South-East Asia. Aust. Plant Path. 28:283-291.

Turnbull, J.W. 2000. Economic and social importance of eucalypts. In: "Diseases and pathogens of eucalypts", P.J. Keane, G.A. Kile, F.D. Podger and B.N. Brown (eds), CSIRO Publishing, Melbourne, Australia, pp1-10.

van der Westhuizen, I.P., Wingfield, M.J., Kemp, G.H.J. and Swart, W.J. 1993. First report of the canker pathogen *Endothia gyrosa* on *Eucalyptus* in South Africa. Plant Path. 42:661-663.

Wang, H., Jiang, Z. and Yan, H. 1994. Australian trees grown in China. In: "Australian tree species research in China", A.G. Brown (ed), Australian Centre for International Agricultural Research, Canberra, ACIAR Proc. No.48, pp19-25.

Wingfield, M.J. 1999. Pathogens in exotic plantation forestry. Intl. For. Rev. 1:163-168.

Wingfield, M.J., Swart, W.J. and Abear, B.J. 1989. First record of *Cryphonectria* canker of *Eucalyptus* in South Africa. Phytophylactica 21:311-313.

Wingfield, M.J., Crous, P.W. and Coutinho, T.A. 1997. A serious canker disease of *Eucalyptus* in South Africa caused by a new species of *Coniothyrium.* Mycopathologia 136:139-145.

Wingfield, M.J., Slippers, B., Roux, J. and Wingfield, B.D. 2001. Worldwide movement of exotic forest fungi especially in the tropics and Southern Hemisphere. Bioscience 51:134-140.

Zheng, H. 1988. The role of *Eucalyptus* plantations in southern China. In: "Multipurpose tree species for small-farm use", D. Withington, K.D. MacDicken, C.B. Sastry and N.R. Adams (eds), Winrock International and IDRC, Canada, pp97-85.

植物病原菌对中国桉树人工林的潜在威胁

GILES E.ST.J. HARDY，TREENA BURGESS，BERNIE DELL

School of Biological Sciences and Biotechnology
Division of Science and Engineering
Murdoch University, Murdoch
Western Australia 6150

桉树人工林产业不仅在中国，而且在世界其他各地也都在迅速发展。然而，长期以来，由于单一树种反复，以及不能适地适树的栽培模式造成了桉树人工林树种的遗传变异大大降低，人工林所面临的病虫害风险日益骤增，而且，不同的树种和种源试验、育种程序和人工杂交都倾向于选择最好的种植材料进行无性繁殖，致而产生了大量的无性系人工林。这样就导致根、茎、叶部病原菌成为影响这一产业经济收入的重要因素。这些病原菌包括 *Aulographina eucalypti*, *Botryosphaeria* spp., *Ceratocystis fimbriata*, *Cryphonectria cubensis*, *Coniothyrium* spp., *Cylindrocladium* spp., *Cryphonectria eucalypti* ss. nov. (syn: *Endothia gyrosa*), *Cytospora eucalypticola*, *Erythricium salmonicolor*, *Kirramyces destructans*, *Mycosphaerella* spp., *Phytophthora* spp. 和 *Puccinia psidii*。在本文中，我们主要讨论一些在中国需花费很大成本防治的病害。讨论内容包括许多细节问题以及这些病原菌的分离和鉴定。此外，还讨论了有利于病害传播的环境因素如营养缺乏、霜冻和其他恶劣的生长环境以及病害的防治措施。最后将探讨良好的卫生和检疫措施对防止外来病原物侵入中国的重要意义。

EFFECT OF TREE SPACING ON STORM RESISTANCE OF YOUNG *EUCALYPTUS UROPHYLLA* PLANTATION

SHAOXIONG CHEN

China Eucalypt Research Center
Zhanjiang 524022, Guangdong, China

In June 1994, a typhoon (No.3) struck Leizhou Peninsula of Guangdong Province with a maximum wind speed of 32.6 m/s. This typhoon caused an average of 10-50% windfall and windbreak in young eucalypt plantation in Zhanjiang. Two spacing trials were investigated at three days and one month after the typhoon. Trees were scored according to 4 classes, based on the angle between stem and ground:

A class=0 or stem break,
B class=0-60, need to be recovered by hand,
C class=60-90, most trees naturally recovered in 3 months, but growth was affected, and
D class=no damage.

The results showed that the greater the planting density, the stronger the storm resistance. The percentage of trees scored A class in trial No.1 were 0.44%, 1.49% and 3.42%, and in B class were 2.20%, 5.33% and 12.14% for the spacing 3 m × 1 m, 3 m × 2 m and 3 m × 3 m, respectively. For the C class, more than 80% of trees had recovered within one month after the typhoon. The percentages of trees scored C class were 5.7%, 4.9% and 4.0%, respectively, one month after the typhoon. A two-year old eucalypt plantation was damaged by the typhoon more severely than the one-year old plantation. The percentages of trees scored A class in trial No.2 were 3.3% and 12.5%, and B class were 23.9% and 33.2% for 3 m × 1 m and 3 m × 2.5 m, respectively. Both spacing treatments of 3 m × 3 m and 3 m × 2.5 m were considered unsuitable for use in the coastal areas because of their heavy loss caused by typhoon.

1 Introduction

It is estimated that total area of eucalypt plantation in China is nearly 1.5 million ha, half of which is distributed along the south-east coast where the climate is tropical to subtropical with an average number of 2 to 5 typhoons (wind-speed at 17 to 33 m/s) each year, resulting in serious damage to eucalypt plantations, especially when they are young. In June 1994, a typhoon (No.3) struck Leizhou Peninsula of Guangdong Province with a maximum wind speed of 32.6 m/s. This typhoon caused an average of 10-50% windfall and windbreak in young eucalypt plantations, and 30-80% in agricultural crops. Rubber plantations suffered a damage of 60-90%, and acacia plantations 80-100% in the same typhoon. Storm damage has become one of the main obstacles for the development of eucalypt plantations in southern China. This paper deals with the effect of planting spacing on the degree of damage caused by heavy storms.

2 Materials and Methods

Both trials No.1 and No.2 were located in Leizhou Peninsula, at the center of damaged area by typhoon No.3. In trial No.1, the clonal stand of *Eucalyptus urophylla*, including 3 spacing treatments of 3 m × 1 m (S1), 3 m × 2 m (S2) and 3 m × 3 m (S3), was established in April 1993. In trial No.2, the hybrid clonal stand of *E. urophylla* × *E. grandis*, including 2 spacing treatments of 3 m × 1 m (S4) and 3 m × 2.5 m (S5), was established in April 1992. Randomized complete block design was employed, with 3 and 2 replications for trials No.1 and No.2, respectively. The area of each plot was 0.169 ha in Trial No.1, but not even in trial No.2 with treatment S4 having 143- and 115-tree, and S5 106- and 48-tree plots in the respective replications. All areas were ploughed with a Shearer Majestic disc plough followed by a single run with Shearer offset discs. At three days and one month after the typhoon, trees were scored according to 4 classes, based on the angle between stem and ground:

- A class=0 or stem break,
- B class=0-60, need to be recovered by hand,
- C class=60-90, most trees will recover naturally in three months, but growth is affected, and,
- D class=no damage.

3 Results

3.1 *Trial No.1*

At 14 months after planting, tree growth did not differ significantly among the 3 spacing treatments in trial No.1 (Table 1).

Table 1. Tree growth at age of 14 months in trial No.1.

Treatment	Spacing (trees/ha)	Survival before typhoon (%)	Mean DBH (cm)	Mean height (m)
S1	3m×1m (3333)	97.6	6.1	7.6
		97.1	5.5	7.9
		96.5	5.8	8.0
S2	3m×2m (1666)	96.8	6.5	7.9
		96.1	6.4	7.5
		95.0	6.5	7.8
S3	3m×3m (1111)	95.1	6.4	5.9
		98.5	6.2	7.5
		96.2	6.6	7.2

Significant difference occurred between spacing treatments in their resistance to wind damage assessed three days after the typhoon. The mean percentages of trees scored A class over three blocks were 0.44%, 1.49% and 3.42% (totaling 7, 12 and 18 trees), for S1, S2 and S3, respectively (Figure 1). The percentages of trees scored B class were 2.20%, 5.33% and 12.14% for S1, S2 and S3, respectively (Figure 2). Thus, the wider the planting density, the more trees suffered A or B class damage, and spacing 3 m × 3 m had the lowest capacity of resistance to storm damage among the 3 treatments.

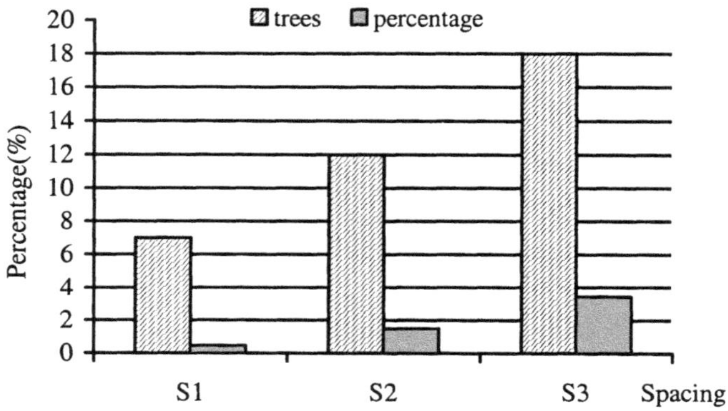

Figure 1. Trees and percentage scored as A class.

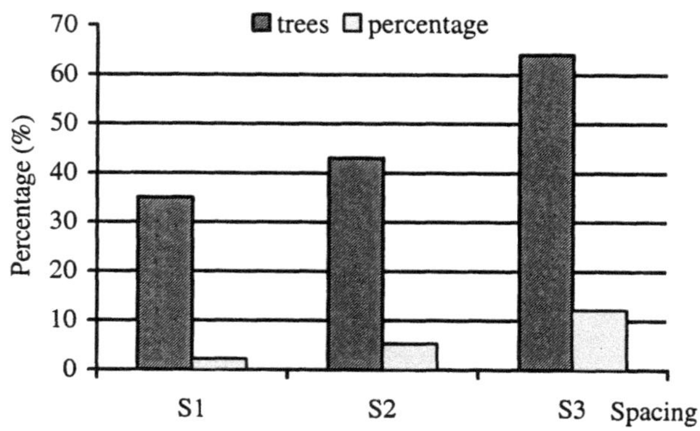

Figure 2. Trees and percentage scored as B class.

One month after the typhoon, the trial was assessed again. More than 80% of trees previously assessed as C class had recovered (Figure 3). The mean percentages of trees scored C class over three blocks were 5.7%, 4.9% and 4.0% for S1, S2 and S3, respectively, one month after typhoon, which was considered the normal frequency of leaning trees for plantations in the region. No significant difference was found among three spacing treatments.

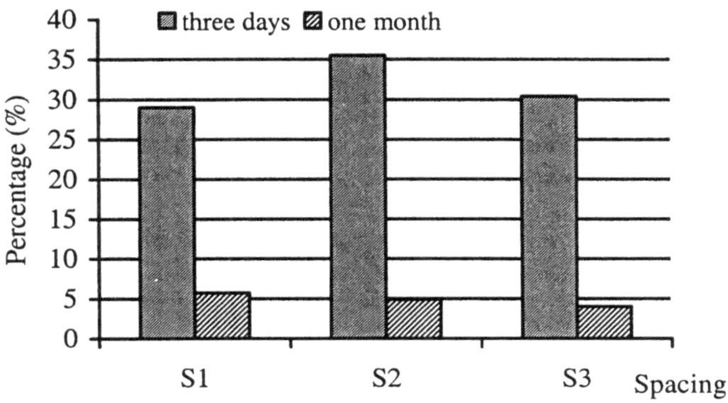

Figure 3. Percentages of trees scored as C class three days and one month after typhoon.

3.2 *Trial No.2*

At two-year old, Trial No.2 eucalypt stand was heavily damaged by typhoon No.3 (Table 2). The mean percentage of trees scored as A class over two replications was 3.3% and 12.5%, and as B class was 23.9% and 33.2% for treatments S4 and S5, respectively. The percentages of A, B and C class trees in S4 were much lower than in S5, indicating that trees at 3 m × 1 m spacing were more resistant to storm damage than those at 3 m × 2.5 m (Figure 4).

Table 2. Wind-damaged trees scored as A, B and C class one month after typhoon.

Treatment	Spacing	Total trees		A class		B class		C class	
				Trees	%	Trees	%	Trees	%
S4	3m×1m	Rep1	143	7	4.9	38	26.2	8	5.6
		Rep 2	115	2	1.7	25	21.7	15	13.0
S5	3m×2.5m	Rep 1	106	11	10.4	35	33.0	17	16.0
		Rep 2	48	7	14.6	16	33.3	5	10.4

4 Conclusions

The higher the planting density, the stronger the storm resistance. Older eucalypt plantations are more susceptible to typhoon damage. Spacings of 3 m × 3 m and 3 m × 2.5 m are not recommended for plantations in the coastal areas considering heavy typhoon damage. Further research is required to explore the influence of root distribution and crown characteristics on resistance to storm damage.

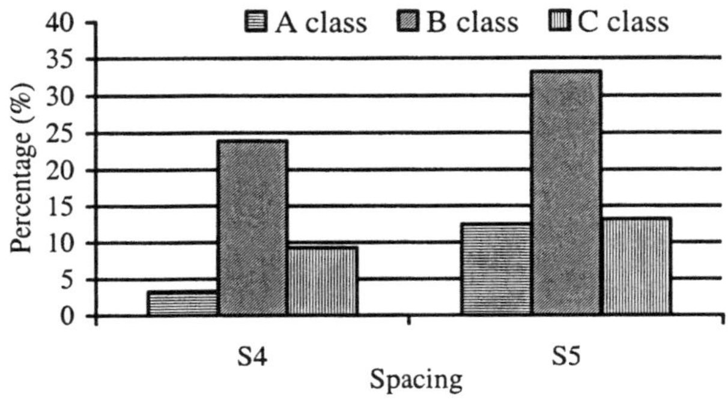

Figure 4. Average percentages of trees scored as A, B and C class one month after typhoon.

林木株行距对幼年桉树人工林抗风能力的影响

陈少雄

国家林业局桉树研究中心
524022 中国广东湛江

1994 年 6 月，三号台风侵袭了广东雷州半岛，记录的最高风速达 32.6 m/s。这次台风造成了湛江地区幼年桉树人工林平均 10-50%的树木被刮断或吹倒。在台风过后 3 天和一个月分别进行两次林分株行距与风害关系的调查。根据树干同地面的夹角，将桉树分成 4 级：

A 级=0 或茎干折断；
B 级=0-60，需要人工扶直；
C 级=60-90，大多数树木在 3 个月内自然恢复，但生长已受到影响；
D 级=无损伤。

调查结果表明，种植密度越大，抗风能力越强。1 号试验中对于 3 m × 1 m，3 m × 2 m，3 m × 3 m 三种株行距，A 级树占的百分比分别为 0.44%、1.49%和 4.42%，B 级树占的百分比分别为 2.20%、5.33%和 12.14%。80%多的 C 级树在台风过后 1 个月内已经恢复。台风过后 1 个月 C 级树占的百分比分别为 5.7%、4.9%和 4.0%。一块两年生桉树人工林地比另一块一年生林地受台风毁坏更严重。2 号试验中对 3 m × 1 m 和 3 m × 2.5 m 二种间距，A 级树分别为 3.3%、12.5%，B 级树分别为 23.9%、33.2%。由于台风过后损失严重，本调查研究认为 3 m × 3 m 和 3 m × 2.5 m 的间距不适合沿海地区造林。

VI. Wood Processing and Utilization

UTILIZATION OF PLANTATION-GROWN *EUCALYPTUS*: NEW RESOURCES ... NEW APPROACHES

EVAN D. SHIELD

Forestry/Forest Industry Consultant
3/115 Chester Road, Annerley
Q., 4103, Australia

Plantation-grown *Eucalyptus*, typically harvested when young, have wood properties distinctly different to those of trees of the same species harvested from Australia's native forests. Characterisation of these wood properties is important. However, it can also lead to definition of utilization difficulties, occasionally a long list if higher-value utilization for lumber or plywood is intended. This paper does not dwell on the detail of such problems or their solution. Rather, it seeks to demonstrate the economic logic of overcoming them through the application of modern technology to produce new generation products. Thus, it is concerned with the utilization of plantation-grown *Eucalyptus* in the manufacture of such products as medium density fibreboard (MDF), oriented strand board (OSB) and laminate strand lumber (LSL). These are considered "sunrise" products, technically well suited to substitution for hardboard, particleboard, lumber and plywood in a wide variety of applications. Moreover, it is possible that they can be more cost-competitive than some of these "sunset" products.

1 Introduction

On a global scale, there is increasing interest and activity in utilization of logs of plantation-grown *Eucalyptus* species for higher-value products, these being defined for present convenience as those other than pulpwood (for paper-making) and fuelwood. To a substantial extent, the current and expanding global surplus of hardwood pulpwood supply over demand, reflected in more or less continuously falling prices for hardwood pulpwood since 1995, is a motivation for this.

This paper has focus on utilization of plantation-grown *Eucalyptus* for higher-value products. In light of both technological developments in wood products manufacture and economics, it questions whether conventional approaches in the management of *Eucalyptus* plantations are now the most appropriate to an objective to manufacture of higher-value products, typically lumber and plywood.

2 Forest Management for Conventional Higher-Value Utilization

It is, perhaps, now universally recognized that the production of lumber and rotary veneer (for plywood or laminated veneer lumber (LVL)) from plantation-grown *Eucalyptus* species demands that plantations are both thinned and pruned. This is certainly true if:

a) the products are intended for higher-value, open markets (which tend to discriminate heavily against lower grade products);
b) higher grade lumber or veneer yield is to be maximized;
c) and, if, as sound economics demand, short rotations are to be obtained.

On a global scale, several *Eucalyptus* taxa have been demonstrated to provide a short-rotation potential for higher-value products. Perhaps, this is best demonstrated at a large, operational scale in South America. However, it is also probable that the best experimental evidence of what can be achieved with plantation-grown *Eucalyptus* is to be found in South Africa (albeit with a single species, *E. grandis*). This evidence is based on long-standing and thorough research.

All the evidence suggests that these are the fundamental requirements for short-rotation production of higher-value lumber or rotary veneer products:

a) selection of superior sites for plantation establishment;
b) good matching of the plantation taxa to site (and in this, the genus *Eucalyptus* has much to offer. Indeed, it may have a unique potential.);
c) intensive early age management, with specific emphasis on pre-establishment site preparation, fertilizer application and competition control;
d) adequate control of pests and diseases;
e) and, genetic improvement, with emphasis on clonal propagation. (Alone, the derived uniformity of individual trees within the stand may justify its adoption).

Of course, these requirements are also fundamental to low-cost pulpwood production. However, for the conventional approach to production of higher-value products, all of these matters are of little real significance without thinning and pruning.

Thinning has the express purpose of producing logs of larger diameter and many economic arguments can be mounted in favour of larger diameter logs. Indeed, it is not unreasonable to claim that the only good plantation-grown *Eucalyptus* logs are those of larger diameter. For most species, radial wood property differentiation...in a multi-faceted sense...would justify this claim.

On superior sites, this author favours radical thinning regimes for *Eucalyptus* plantations intended for conventional higher-value utilization. On such sites, a radical regime begins with a heavy non-commercial thinning at an age as young as 18 months and includes a single commercial thinning at age 10 years with removal of 50% of tree numbers, leaving as few as 150 - 200 (essentially open-grown) stems per hectare for rotation-length harvesting.

Experimental and observational evidence strongly supports pruning for *Eucalyptus* plantations as an imperative if they are intended for conventional higher-value utilization. Many species of this genus are well suited to repeated heavy pruning with removal of a substantial proportion of the green crown proven feasible.

It is critically important to ultimate wood quality that pruning commences at an early age and that it is done well. However, doing this task well is expensive and, unfortunately, this does lead to application of inferior, cheaper techniques. Some of these are a threat to the health of the stand. Patently, other techniques are no more than tokens.

The differential between the costs of plantation production of *Eucalyptus* pulpwood and larger, pruned logs best suited to conventional higher-value utilization is substantial. Immediately, it should be noted that land value for superior sites and the longer rotation required (perhaps a factor of two) in providing logs of larger diameter will contribute to this differential. So too will the costs of non-commercial thinning and, in particular, high standard pruning.

3 Inherent Defects of *Eucalyptus* Logs

Younger, fast-grown *Eucalyptus* logs possess a wide range of inherent defects representing barriers to their easy, efficient and low-cost conversion to lumber and rotary veneer. To a greater or lesser extent in both individual species and situations, these defects include:

a) spiral grain;
b) growth stresses;
c) knots;
d) higher rates of normal shrinkage and collapse, or abnormal shrinkage;
e) limited heartwood durability;
f) kino (gum) veins and pockets;
g) brittle heart and, perhaps, core decay;
h) tension wood;
i) wet-wood, (bacterial) discolouration and (fungal) decay;
j) saturated moisture content;
k) higher extractives content in heartwood; and,
l) significant radial differentials in a range of physical and mechanical wood properties.

Such defects can and do provide vast opportunities for research. They also provide some skeptical commentators with excuses to deny both the possibility and the profitability of managing *Eucalyptus* plantations for higher-value utilization.

This author's allegiance lies neither with those who promote postponement of proceeding without the (occasionally dubious) benefits of research programs, nor with those who would deny feasibility. Both seem to exhibit an unwarranted arrogance, denying the reality of the achievements of practitioners in forest management and in plantation-grown log conversion and product processing.

A successful *Eucalyptus* sawmiller friend in Brazil has a framed notice in his office which can be translated from the Portuguese to read: "*No one told me it was impossible...so I went ahead and did it!*" In the present context, this author applauds the sentiment.

However, it is almost undoubtedly true that the inherent defects of young, fast-grown *Eucalyptus* logs will continue to deny to the manufacture of lumber and rotary veneer products the efficiency and economy of conversion and processing which attach to logs of plantation conifers such as radiata pine, the Southern pines and the Caribbean pines. The higher potential growth rates of well-selected

Eucalyptus taxa planted on superior sites are required to offset this apparently permanent disadvantage.

4 Some Economic Fundamentals

4.1 *The Cost of Time*

Time is, perhaps, the most expensive factor in plantation management. This is well understood by all those who have analysed plantation economics. They recognise that real (inflation adjusted) long-term interest rates must be applied to both anticipated costs and revenues. With costs crowded into the early years of a rotation and revenues predominantly at the end, the rotation length assumes substantial importance.

Less well recognized is the importance of an upward risk adjustment to the long-term real interest rate as the rotation length extends. This premium is required to protect the investment against technological changes with the potential to nullify the intended utilization of the plantation crop.

Consider this example: in Australia, there are substantial efforts being made by some public agencies to demonstrate the technical and financial feasibility of establishing *Eucalyptus* plantations for lumber products on sites which can be described, generously, as only marginal. Some have annual rainfalls averages possibly as low as 400 mm. Inevitably, the species suited to these sites are those of higher density, such as *E. occidentalis*, *E. cladocalyx*, *E. astringens*, *E. leucoxylon* and *E. argophloia*. Perhaps as a result of recognition of the higher wood density, suggested applications for the anticipated lumber include flooring and top-end wooden furniture (to which applications there are, as yet, no firm assurances of lumber suitability). Rotation lengths at long as 40 years have been mooted, this being the time estimated (optimistically) by one agency to produce a stem with breast-height diameter of 40 cm.

Currently, substantial research and development endeavours are underway to augment the surface hardness of, and to dye with permanent colours, the wood of radiata pine. It is understood the key focus of these endeavours now is to reduce the costs of these modifying treatments. Within a 40 years period, there would seem to be reasonable prospects that a modified and cheaper radiata pine lumber product could have come to dominate the intended market for higher density *Eucalyptus* wood grown on marginal sites under very long rotations. Thus, this author would argue that a substantial risk-adjustment premium should be applied to the real interest rate used to determine feasibility of such a project.

This example may be at one extreme of the range of rotation years considered for *Eucalyptus* plantations for higher-value utilization. However, it is not the only example illustrating the cost of time.

Additionally, it is worth recording the comment made to the author by the former chief executive of Argentina's largest higher-value *Eucalyptus* processing enterprise (with annual production of 36,000 cubic metres of plywood and 78,000 cubic metres of lumber). It was to the effect that the sole economic advantage of *E.*

grandis as a raw material for this production was that, under an intensive thinning and pruning regime, plantation logs for conversion could be produced in a rotation length of 15 years. In terms of ease of conversion and processing, wood quality obtained and market acceptance of the products, he did not believe any advantages were derived from this raw material. (Possibly, this prejudice derived from the man's background in *Pinus radiata*).

On the basis of relative competitiveness, this author believes that *Eucalyptus* should not be considered for plantations for higher-value utilization unless the rotation length is short and that perhaps, a maximum of 20 years is appropriate. However, there is a caveat to this opinion and it is that, if a rotation length as long as 20 years is under consideration, the species involved should be an intrinsically more valuable one. Based on experience in Galicia in North West Spain, *E. globulus ssp. globulus* ... a white wood with an attractive quarter-sawn figure ... is considered to be such a species.

The objective of minimizing rotation lengths gains greater validity when the following matters are also brought into consideration.

4.2 *Global Competitiveness*

Increasing globalisation of wood products' markets is an undeniable reality. There is also increasing globalisation of investment in forest management (for example, in both New Zealand and Australia) and in the institutional funding of major forest products' manufacturing facilities. All lead to the perceived need for any plantation project to achieve global competitiveness and to this author's opinion that, without that, it is better to abandon ambitions for any such project. This comment is particularly relevant to those projects designed to serve international markets.

Without a globally competitive production cost position, investment in the establishment of the plantations can scarcely be warranted, even by a public agency which, typically, can bring a wider range of justifications to their investment decisions. For example, these may include creation of employment opportunities, substitution for native forest wood resources lost to conservation tenures, regional development and combat of land degradation agencies such as erosion and salination.

Further, without a globally competitive cost position, investment in conversion of the plantation's log products and the processing of conversion products is severely inhibited. These activities are closer to the market place and the imperative for competitiveness is probably more keenly understood.

This author believes that early achievement of full scale-economies in closely integrated conversion and processing facilities is an important contribution to global competitiveness. Previously, for South American green-field projects involving *Eucalyptus* plantations for conventional higher-value utilization, his recommendations were for an intensively managed, compact plantation estate of at least 25,000 hectares with annual harvest (strongly dominated by saw/peeler logs) exceeding 650,000 cubic metres. In this concept, the scale of sawmilling and rotary peeling undertakings were very large, providing adequate appropriate mill residues for a scale-economic facility for MDF manufacture. However, as this paper reveals, this concept should now be reviewed.

4.3 *Perhaps All the Products are Commodities*

This author believes that, to an increasing degree, the conventional higher-value products of *Eucalyptus* plantations...such as lumber and the plywood and LVL products of rotary veneer...are commodities. In simplest terms, this means that they have many competitors in open markets and there are many producers of these competitive materials. In all probability, their long-term prices in open markets are falling in real terms.

It merits note that claims...occasionally outrageous ones, considering the species nominated...are often made of ambitions to produce "high-value" lumber from plantation-grown *Eucalyptus* logs. Great caution should be attached to such claims. The reality is that no *Eucalyptus* lumber can be regarded as of heritage or heirloom quality. It simply is not the equivalent of those hardwood species...such as rosewood, black walnut, cherry, teak, mahogany or oak...having long-standing international reputations and, therefore, demand levels which combine with restricted supply to allow them to transcend the barrier of commodity status. For more utilitarian purposes, *Eucalyptus* species can and do provide lumber that is functionally effective and aesthetically pleasing. Nevertheless, it is considered unfortunate, but prudent, not to expect that the lumber of even the best of them can be positioned in a status beyond that of a commodity.

In economic and commercial terms, the appropriate strategic response to the reality of the commodity status of *Eucalyptus* lumber and rotary veneer is to ensure that...on a continual basis...the costs of production fall in real (inflation adjusted) terms.

For a seller of log raw materials from plantations, meeting this objective is a daunting challenge. However, it is not an impossible one. Perhaps it is best approached through the corollary of continual improvements in log quality providing real improvements in conversion and processing efficiencies. These would include production of pruned logs of larger diameter (improving conversion costs and recoveries) with reduced juvenile wood/knotty core diameters (improving grade yields). (In improving log diameter, there may be a place for late-age fertilizer applications, perhaps coincident with commercial thinning). They would also include improvements in the physical and mechanical properties of the wood, most probably through a well-targeted program of genetic improvement.

In conversion and processing, continual improvements in manufacturing productivity are the keys to reductions in costs of production. With technological improvements in conversion and processing being made on an almost continuous basis, regular capital investments to exploit the opportunities they offer for improvements in manufacturing productivity seem to be essential.

Such investment may require additional log raw material supply. For example, this would be the case if the investment was made in equipment with faster sawing/peeling speeds. This transfer of responsibility for the realization of manufacturing productivity gains back to the forest demands of plantation managers an unrelenting pursuit of improvements in plantation productivity. The alternatives of augmenting supply by sourcing logs from more distant plantations (perhaps in other ownerships) or by choosing those of lower quality can quickly deny to the enterprise the anticipated benefits of the capital investment made.

However, as the following sub-section indicates, there may be limitations to what can be achieved through even superior endeavours at continual cost reduction in real terms.

4.4 *Manufacturing Processes*

The production of lumber and rotary veneer has undergone a dramatic transformation in recent years through technological improvements in manufacturing processes.

When this author's career commenced, a lathe for rotary veneer production was considered at the cutting edge of technology if it was capable of peeling at the rate of one block per minute. Today, the same company that manufactured these lathes can boast of a spindle-less lathe capable of peeling 20 (smaller diameter) blocks per minute and a triple-spindle lathe capable of 15 blocks per minute with blocks to a maximum diameter of 700 mm. In like fashion, sawing speeds have risen to commonly advertised maxima of 200 metres per minute for small diameter logs of lower density. Today, small-log sawing lines for conifer stud lumber production boast capacities of 20 logs per minute. Additionally, technology advances have permitted saw kerfs and sizing variations to be reduced.

However, none of these welcomed improvements has provided lumber or the plywood products of rotary veneer with immunity from competition. In the world of wood, lumber and plywood have long been substitution targets for a wide variety of reconstituted products. Initially, it was hardboard, then particleboard and more recently MDF, OSB and LSL. Synthetic film laminates, providing appearance equivalence and service improvements, have aided some of this substitution.

This substitution has been such that the term "*sunset*" products might well now apply to lumber and plywood. Even a relative newcomer such as LVL is not immune from competition.

Initially, larger cross-sections of structural lumber were the targets for substitution by LVL. Subsequently, however, this product found its most substantial demand for application in flanges for fabricated I-Beams (typically with a web of structural plywood or, now more commonly, OSB). In North America, it is now estimated that 60% of LVL production is consumed in this application.

However, LVL itself is now under increasing competitive pressure. Cheaper machine stress graded structural lumber is substituting for LVL in I-Beam flanges. So too is a more recent product, laminated strand lumber (LSL). This product is also beginning to erode the market share of (more expensive) LVL in conventional beam applications.

What are the characteristics of the reconstituted products now appearing to hold ... or to be on the verge of gaining and, ultimately, holding ... competitive advantages over the longer established products of lumber and the plywood and LVL products of rotary veneer?

Clearly, at the most immediate level, there is the matter of cost competitiveness and, importantly, cost competitiveness evaluated at the level of the finished product. Additionally, there is the matter of ease and efficacy in application.

How then do these products achieve their combination of cost competitiveness and ease and efficacy in application? In this author's opinion, there are many components in the answer to this question. Thus, these products:

a) are dependent on lower cost raw materials, viz., pulpwood logs or chips rather than larger, more defect-free logs, which are more expensive to produce in plantation forestry;
b) have more *systemic* manufacturing processes;
c) employ manufacturing processes providing substantially higher recoveries of saleable products as a proportion of raw materials consumption;
d) employ manufacturing processes having the flexibility in set-up to manufacture products with a range of properties. Thus, there is opportunity to vary product dimensions, product density and mechanical properties to underwrite efficacy in individual applications; and,
e) in a single set-up, their manufacturing processes provide a single grade of product.

In addition, the manufacturing processes for reconstituted products are themselves undergoing transformations through technological innovation. For example, the continuous press has provided a quantum improvement in the productivity of MDF and OSB mills. Importantly, for the former product, this improvement is most marked in the production of thin panels. One very important and recent development* has been the capacity to produce both OSB (a structural plywood substitute) and LSL (a structural lumber substitute) on a common production line equipped with a continuous press.

Items a), c) and d) above are self-explanatory. With regard to items b) and e), some further explanation is warranted.

Item b): A *systemic* manufacturing process is one involving bulk feed of raw materials and a continuous flow through the entire manufacturing process. It is highly efficient, but capital intensive and the individual components of the process are intimately interrelated. A *systemic* manufacturing process involves high degrees of standardization, mechanisation and automation. Variable costs of production are low and, typically, individual equipment components in such a process are used at 100% of their capacity. Significantly, *systemic* manufacturing processes are judged to be most appropriate ones for the production of commodities.

In contrast, conventional sawmilling and plywood/LVL manufacturing are best described as *disconnected line flow processes*. They entail single piece raw material feeds to primary conversion and out-turn of multiple intermediate products (sizes

---

* Siempelkamp, the German manufacturer of plant and equipment for the production of reconstituted wood products, has provided an excellent introduction to this potential in two papers presented to a symposium held in Beijing in October 1998. The symposium theme was "*New building materials*" and the two relevant papers, by Paul Niedermaier, were entitled "*CSL + LSL: An engineered wood product for structural lumber application*" and "*OSB: An engineered wood product building material with high strength and stability*".

and grades, for example). These intermediate products must be first segregated, then aggregated, for example, for the typically separate process of drying. For both lumber and rotary veneer, there is often the additional task of up-grading some intermediate products. For example, this may be achieved by defect docking/clipping, followed by finger jointing (lumber) or composing (veneer). Intermediate product storage is inevitable in this manufacturing system, as is under-utilization of the capacity of some plant items. This manufacturing process has greater flexibility than a *systemic* process. It is also more labour intensive and less amenable to automation. Most importantly, it is less appropriate for the production of commodities than a *systemic* manufacturing process.

Item 2): Regardless of the efficiency with which it is undertaken, the conversion of logs by sawing and rotary peeling results in products with physical and mechanical property differentials. To large extent, these differentials are the result of radial wood property variations and the occurrence of defects such as knots. Thus, the products of these conversion processes must be graded and, in open market situations, it is rare that less than four grades of lumber or rotary veneer are involved. Profitable sale of the lower grades ... especially those of lumber from plantation-grown logs ... is often a formidable challenge. Indeed, it may be a perennial impossibility.

For rotary veneer, because of the nature of its utilization, there are better prospects for lower grades. For example, when graded for structural adequacy, they can be incorporated into plywood as core stock or into the center of LVL panels. However, if the proportion of lower grades is excessive as a result of inadequate log quality, these options may not utilize all the lower grade veneer available.

In a single processing set-up and assuming no technical faults occur, reconstituted products ... hardboard, particleboard, MDF, OSB and LSL ... have no such grade differentials. Finely comminuted raw materials ... as fibres, flakes or strands ... are blended to provide a high degree of product homogeneity within a single manufacturing set-up.

Moreover, it is often possible to blend different raw materials to permit beneficial use of one that, alone, might otherwise be rejected as unsuitable. Within limits, this capability might be applied to optimize product properties also.

All the indications in this sub-section point to continuation of market place substitution of lumber and the plywood and LVL products of rotary veneer by reconstituted wood products. It is expected that substitution pressure will be more intense in those applications:

a) where there is a high labour component in manufacturing products which are price sensitive. For example, it is anticipated that MDF will continue to substitute for lumber in lower-end cabinet goods (for example, kitchen and bathroom fixtures) and shop-fitting. It will also continue throughout the price range of domestic and commercial furniture. For high-end furniture, this substitution will be aided by sliced veneer lamination;
b) where large sections of structural lumber are required for domestic construction ... as in beams. Here, it is believed that LSL will substitute for both lumber and LVL;

c) where use of LVL is apparently unwarranted, as in flanges of I-Beams. Machine stress graded structural lumber and LSL ... both cheaper materials ... have potential to effect this substitution. Indeed, the process of substitution is now a reality;

d) where structural plywood is used in dry applications such as sheathing, flooring and bracing. Here, the substitution will be primarily by OSB. (However, MDF is also used as sheathing in Japan and HDF is commonly used as a base for new generation flooring products). In contrast, it is not anticipated that OSB will enjoy substantial success in substituting for plywood in wet applications such as concrete formwork. In such applications, swelling properties of reconstituted products may render them unsuitable. (Such appears to have been the experience in North America. However, in Western Europe, use of (possibly higher quality) OSB in concrete formwork does appear to be more commonly practised).

It is most interesting to note that one of the majors amongst U.S. forest products companies, Louisiana Pacific, has announced a strategy to sell off its large forest estate and its substantial sawmilling and plywood businesses. It will retain only its OSB and engineered wood products businesses. Could that be a sign that there is widening realization of where the more competitive business opportunities lie?

4.5 *Market Perceptions of Eucalyptus Products*

It is unfortunate, but true, that ... on a global scale ... markets have poor perceptions of higher-value *Eucalyptus* products. Typically, species of this genus are widely perceived to be able to provide high quality pulpwood and good fuel wood. However, for higher-value applications, the inherent defects are judged by many to be formidable obstacles to acceptance of *Eucalyptus* lumber or plywood.

Indeed, this author has found that perhaps only a single species ... jarrah or *E. marginata* ... produced exclusively from Western Australian native forests (and now, because of public land tenure conversion, subject to a heavily curtailed supply) has wider global recognition.

Thus, at a global scale ... and, indeed, perhaps anywhere beyond Australia's shores ...there is no *"demand pull"* (and it might be noted that *Eucalyptus* lumber's share in a more or less static annual consumption in Australia has been in decline for the past 30 years). It follows that marketing endeavours must be based on *"supply push"* and, in this, cost competitiveness is central to success.

4.6 *Global Supply and Demand*

Wood Resources International (1998 and 2000) has indicated that, on a global scale, supply of industrial hardwood is likely to exceed demand for a period of at least 30 years. Moreover, they have indicated that the same over-supply will apply to softwood (conifer) industrial wood for at least the next 20 years. The world's softwood plantations ... for example, in the Southern States of the U.S.A., Brazil, Chile, New Zealand and Australia ... are more mature than those of *Eucalyptus* managed for higher-value utilization. Moreover, there is demonstrated potential for softwood products to substitute for those of hardwood in some applications. Thus, it is reasonably concluded that success in marketing the conventional higher-value products of *Eucalyptus* plantations is dependent on being able to substitute them for those now established in the world's market places.

Primarily, such substitution marketing demands cost competitiveness. In this, short rotations are essential. In turn, so too are site selections and silvicultural management strategies and techniques able to contribute to achievement of that objective. Compared to that of most plantation conifer species, the faster growth rate potential of many *Eucalyptus* species represents a sound fundamental on which to build through application of such strategies and techniques.

Substitution marketing may be assisted by the fact that the products are plantation-grown. Even greater assistance may be derived from some form of recognized certification that the products are sourced from plantations meeting established criteria...such as those of the Forest Stewardship Council...for ecological and social sustainability. However, in general terms, such "green" labeling of products is of little purpose if the fundamental of cost competitiveness is lacking. Claims that certification provides opportunities to gain a price premium appear...at worst...to be a myth. At best, realization of the opportunities is very rare.

The most significant substitution opportunity for *Eucalyptus* lumber and plywood is believed to be in the substantial global trade in tropical hardwood. Indeed, it is possible to demonstrate how effective *E. grandis* can be...in technical terms...as a potential substitute for the lumber and plywood of red meranti from the tropical forests of South East Asia.

5 *Eucalyptus* for Reconstituted Products

5.1 *Hardboard*

The wood of *Eucalyptus* species is used in the commercial production of the reconstituted products hardboard, particleboard and MDF. Indeed, for the first of these applications, it is probably a superior raw material and has been so for more than 60 years.

North American hardboard siding producers have paid out millions of dollars in compensation for failures of their products. These failures have, perhaps, irredeemably tarnished the reputation of hardboard for siding application there. In contrast, the Australian producer of *Weathertex*™ wet-process hardboard siding offers a 25-year product guarantee. In manufacture of this hardboard, no added adhesive is required and *Eucalyptus* raw materials are used exclusively.

However, in global markets, it merits acknowledgement that hardboard is a product at or near the end of its life cycle. Moreover, its manufacture may be associated with what are now unacceptable levels of pollution. The costs of "*green-field*" mills with pollution abatement equipment installed and the life-cycle position of the hardboard product combine to deny the possibility of secure investments.

5.2 *Particleboard and MDF*

Perhaps, the first significant commercial utilization of *Eucalyptus* for particleboard and MDF manufacture was in South Africa at a mill in White River, near Nelspruit. The small mill was built in 1990 at the time of South Africa's economic isolation. It has the unusual configuration of two independent fibre preparation lines ... one for particleboard, the other for MDF ... feeding a common press line. Raw material used for both products is plantation-grown *E. grandis*.

The author is also aware of other mills producing particleboard from log and mill residue raw materials of *Eucalyptus*. These are the mills of Eucatex (Sao Paulo state, Brazil), Masisa (Concordia, Entre Rios province, Argentina) and Sino-Forest (Zhaoqing, Guangdong, China).

Production of *Eucalyptus* MDF manufacture is better documented. In their 2001 MDF Industry Up-Date, Metso Panelboard reported no less than ten mills dependent on plantation-grown *Eucalyptus* raw materials. These mills ... and the author's assumption of the species used ... are found in Spain (*E. globulus ssp. globulus*), Argentina (*E. grandis*), India (*E. tereticornis*), Thailand (*E. camaldulensis*), Vietnam (*E. camaldulensis*), Tunisia (*E. camaldulensis*), Australia (*E. nitens*) and South Africa (*E. grandis*). Other mills listed in the Metso Panelboard directory reported unidentified hardwood raw materials. This opens the possibility that the number of mills using *Eucalyptus* raw materials is greater than ten. Certainly, in China, *Eucalyptus* wood is also included in furnish for MDF production. There, it is understood it is blended with other fibres.

A research report from the U.S. Forest Products Laboratory (Madison, Wisconsin) indicates the conduct of trials there to determine the suitability of plantation-grown *E. saligna* from Brazil for MDF manufacture. The results showed "*that nearly all mechanical properties of the panels at all thickness levels were above minimum requirements for MDF as specified in the ANSI-AHA and Euro MDF standards*". However, it was also noted "*additional work is needed to ascertain the performance of MDF panels from this species through pilot and production scale trials*".

The Australian MDF manufacturer using *E. nitens* is Starwood Australia. Their mill is located at Bell Bay in Northern Tasmania and the *Eucalyptus*-based product is made exclusively from wood of this plantation-grown species. A subsidiary of Japan's pioneer MDF manufacturer, Hokushin Co. Ltd., claims this product has excellent machinability and high moisture resistance (stability). It is recommended specifically for applications in areas with high ambient humidity and extreme temperature variation. Kitchens, bathrooms, laundries, flooring and window frames and reveals are nominated as sites and/or applications.

5.3 *OSB*

It is more difficult to determine if *Eucalyptus* wood is used in the commercial manufacture of OSB. It is probable that it is not. However, several years ago, Siempelkamp provided advice that they had tested the plantation-grown wood of four species provided from South America. The species were *E. grandis*, *E. cloeziana*, *E. nitens* and *E. globulus ssp. globulus*.

Strands produced from the higher density wood of *E. cloeziana* were found to be brittle and the derived OSB panel had low internal bonding. However, Siempelkamp provided the opinion that wood of this species could be used in a mixture with those of other *Eucalyptus* species to the extent of 30%. The other three species were found to provide wood considered to be suitable, the derived OSB panels testing at CEN (European Standard) Grades 3 and 4, the higher of the four grades recognized. Panel densities exceeded 680 kg/m^3, and were more likely to fall in the range 700-720 kg/m^3. (OSB produced in North America has a typical density of 640-660 kg/m^3).

Siempelkamp stressed that their results were from limited, laboratory-level research and that further development under operational settings might well reduce the density levels.

Tasmanian native forest regrowth *Eucalyptus* wood has also been tested in Canada by Durand Raute to determine suitability for OSB production. This product had a density of 720 kg/m^3.

5.4 *LSL*

It is understood that a sole manufacturer, the Weyerhaeuser subsidiary Trus Joist, produces LSL. The company has three mills, the oldest one in Minnesota, the second in Kentucky and the third now being commissioned in Ontario, Canada. The raw materials used are Aspen and Yellow Poplar, pioneer native forest species providing low-cost wood with basic densities close to 400 kg/m^3. Each mill has a Siempelkamp single-daylight steam press, that at the most recent mill sized at 8 feet x 64 feet and rated as capable of an annual production of 500,000 m^3.

However, Lignor, a start-up Australian company, has recently begun promotion of a proposal to manufacture of ESL (or *Eucalyptus* Strand Lumber, a registered trade-name). Essentially, this proposal appears to be a search for equity participants. It is understood that Lignor has had a number of Australian *Eucalyptus* species tested by Siempelkamp to determine their suitability, with a focus on those available in the South West corner of Western Australia. If the promoters are to be believed, their product has outstanding characteristics.

5.5 *Scrimber*

More than 20 years ago, Australia's CSIRO developed a novel reconstituted structural product which was given the name *Scrimber*. In many respects, it was similar to LSL, the notable difference being that, rather than bonding dried, machined strands, the *Scrimber* product relied on *"scrim"* resulting from longitudinal crushing of small diameter, debarked logs. Although the product was originally conceived as a means of beneficially utilizing low-value *Eucalyptus* raw materials, the concept was applied to thinnings from radiata pine plantations in South Australia. Both a pilot plant and a small commercial plant (capacity 50,000 m^3/yr) were built before technical and commercial difficulties resulted in the project being abandoned by its public-sector owners more than 10 years ago.

However, it is known that immediately prior to its closure, a trial was conducted in the commercial plant by CSIRO using 90 m^3 of thinnings from a plantation of *E. grandis*. It is understood that an excellent structural product (density 700 kg/m^3) was manufactured.

Late last year, plant and equipment from the South Australian mill was sold. Currently, there are endeavours being made by the owner of a private-sector plywood mill at Wagga Wagga in Southern New South Wales to resurrect the commercial manufacture of *Scrimber*. It is understood this company has interest in testing the suitability of both radiata pine and plantation-grown *Eucalyptus* raw materials.

While renewed interest in *Scrimber* is to be welcomed, it merits reporting that there are several knowledgeable people who are not convinced of prospects for achievement of technical and commercial feasibility.

5.6 *Summary*

It is reasonable to conclude that there are sufficient indications of the suitability of *Eucalyptus* raw materials for the manufacture of MDF, OSB and LSL (or, perhaps, its *Scrimber* equivalent) to justify further research, including pilot-scale trials. Equally clear is that higher density taxa and those with high extractives contents may not prove satisfactory. It is suggested that applied research should have focus on rapid screening of candidate species at laboratory level. In addition, there is need for work on optimization of production set-up parameters for those, such as lower density *E. grandis*, which now appear suitable and require the next stage of experimentation intended for product property verification based on trials at pilot mill scale.

6 Conclusion

This paper is concluded with expression of the author's opinions concerning preferred options for the higher-value utilization of wood from *Eucalyptus* plantations. These opinions derive from the considerations examined above and his wide experience of the expectations of plantation owners, even the irrational ones (all too commonly encountered). The irrational expectations lead to one important caveat that must be mentioned. The following opinions are based on the assumption that the *Eucalyptus* taxa for the plantation project are selected so that utilization potential is maximized.

Globally, there are innumerable examples of where *Eucalyptus* plantation projects are undertaken with what this author terms "*silvicultural supremacy*". This is translated to mean that the project is based on taxa which survive and grow well, leaving the utilization challenges ... occasionally insurmountable ones in open market situations ... to another day, another responsibility.

In this regard, the genus *Eucalyptus* provides abundant opportunities for serious, even permanent, mistakes to be made. Many of these mistakes are based on the production of wood of high density and, less frequently, on taxa exhibiting inferior form. This author believes the more serious errors can be avoided if taxa with lower mature-age basic densities (more commonly published than are data for plantation-grown taxa) are chosen for plantation establishment. These taxa are likely to offer greater flexibility in utilization and greater ease in conversion and processing.

6.1 *Lumber*

It is believed that higher-grade appearance quality lumber has superior prospects for avoiding substitution by reconstituted products. Accordingly, it is recommended that efforts should be directed at the production of appearance grade lumber, with the concurrent production of structural glu-laminated lumber from lower grades. From the larger logs, the possibility of production of radially sliced veneer should not be overlooked. However, this will require wood of premium quality.

For these purposes, only premium sites should be selected for production of logs, with the intentions of minimizing rotation lengths and maximizing log diameters. Such sites are at the higher end of the spectrum of superior sites.

Furniture, flooring and joinery (builders' woodwork) should be the prime application markets targeted for the appearance grade lumber produced. The selection of *Eucalyptus* taxa for production for these applications should be undertaken with great care. Perceptions of an attractive appearance are critical to success and they deserve to be tested in individual market places.

Other than for the suggested production of structural glu-laminated lumber, *Eucalyptus* species should not be considered for the production of structural lumber. The time required to season thicker sections of lumber...and the entailed risks of induced seasoning defects...is simply not competitive with that commonly achieved for structural conifer lumber.

6.2 *Rotary Veneer Products*

Perhaps because of the even more critical dependence for economic production of rotary veneer on *Eucalyptus* logs of larger diameter, this utilization should be dependent only on selected logs from well-managed plantations grown on premium sites.

In open market situations, it is recommended that only plywood suited to wet area (for example, concrete formwork) and specialized applications is produced from plantation-grown *Eucalyptus* logs. For more generic structural plywood, it is considered the competitive threat of OSB is simply too strong to justify new investment in its production.

However, it is important to note that there are limitations on the suitability of plantation-grown *Eucalyptus* wood for rotary veneer production. With many taxa, the wood's basic density is a deterrent to its use and, for wet area plywood applications; it is possible that taxa susceptible to collapse should be avoided. For many others...and this probably includes wood of *E. grandis*...growth stresses may cause production of thinner rotary veneer (for example, 1.2 mm sheets) to be uneconomic because of the prevalence of end splitting. Almost certainly, economic production of quality thin veneer will require installation of greater veneer drying capacity because it can only be dried successfully at relative slow rates. Additionally, different lathe set-ups may be required if splitting is to be minimized. For example, based on their trials, Raute Wood (of Finland, global suppliers of plant and equipment for several wood processing industries) recommends rotary nose-bars for thicker *Eucalyptus* veneers and fixed nose-bars for thinner, high-quality veneers.

Thus, the author's recommendation above is made with important caveats as to taxa suitability and to the amenability of the intended plywood product to make use of exclusively thicker veneers (say, a minimum of 2 mm). This suggests that LVL manufacture could be a preferred application for *Eucalyptus* rotary veneer.

Reluctantly, this author suggests an increasingly difficult competitive future for LVL, especially when based on a "*green-field*" production site. This suggestion is made in full recognition of authoritative forecasts of strong global prospects for production growth. This author's opinion is that LSL substitution for LVL may be more rapid than anticipated. This is more likely to occur when its production is more widely undertaken, geographically, in terms of raw material used and in terms of manufacturing ownership. This opinion is reinforced by the fact that the manufacture of LVL is...probably unavoidably...a more costly undertaking.

A more competitive future for LVL may lie in very large, engineered structural applications. However, this application only rarely creates a demand of sufficient volume to justify new investment in minimum scale production facilities.

6.3 *Reconstituted Products*

This author believes the increasingly competitive advantages of reconstituted products in open market situations are such that they should be an additional prime focus for higher-value utilization of plantation-grown *Eucalyptus*. The prospect for combined production of what might be termed "*sunrise*" products...OSB, LSL and, perhaps, even MDF...on a single production line with a continuous press has particular appeal.

The fact that reconstituted products manufacture can be dependent on pulpwood log raw materials grown on sites of lower inherent productivity than those premium sites suggested as required for sawlog and peeler log production is central to augmenting this competitive advantage. It is also highly probable that manufacture and properties of reconstituted products are less impacted by inherent defects of the wood of the genus *Eucalyptus*.

However, it is also true that the mill residues of sawmilling and plywood manufacture, if conducted at a scale capable of under-writing the required competitiveness for open markets, may represent an even more attractive raw material for production of reconstituted products. Perhaps, these residues are best converted to chips and MDF may be the more appropriate target for their utilization. However, boxed hearts of juvenile and knotty wood from the sawmill and peeler cores from the rotary veneer mill might be suited to strand production for OSB and LSL, augmenting log pulpwood in the furnish of raw materials for this production.

All this leads to interesting and challenging opportunities for integration of both different, site-specific, plantation management regimes and common-site forest products' manufacture.

桉树人工林的利用：新资源…新途径

EVAN D. SHIELD

Forestry / Forest Industry Consultant
3/115 Chester Road, Annerley
Q., 4103, Australia

桉树人工林中的林木，特别是幼年采伐的林木，其木材特性同澳大利亚本土同一树种的木材有明显差异，对这些木材的特性进行鉴定非常重要。若对木材的利用还有更高的要求，如用作锯材或胶合板材，可能在利用上还会有一定的困难。而木材特性的鉴定可以解释这些木材难以利用的原因。本文的目的不是详述这些问题的细节及其解决方法，而是注重于阐述通过利用先进技术来克服这些困难，进而生产新一代林木新产品的经济逻辑。故本文的重点在于如何利用桉树人工林中的林木去生产诸如中密度纤维板（MDF）、定向板条刨花板（OSB）和板条层积胶合板（LSL）之类的产品。这些产品被称为“日出”产品，从工艺方面讲，在许多用途上可以作为硬板材、刨花板、锯材和胶合板的替代产品。而且，这些产品在生产成本上比那些“日落”产品更具有竞争力。

GAINS AND LOSSES ON SAWN WOOD YIELD AND QUALITY THROUGH FOREST IMPROVEMENT, MANAGEMENT AND SAWING STRATEGIES

JOSÉ N. GARCIA

Wood Engineering Laboratory, University of São Paulo
ESALQ – LCF - IPEF - P.O. Box 9
13418-900 Piracicaba/SP, Brazil

The present work put together many results obtained from different research projects in order to give an overall view on the latest understanding on the improvement of wood yield and quality through forest improvement, management and sawing techniques. Sawn wood defects related traits like log end splitting, board end splitting and cupping are very heritable and selection of trees based on them brings enormous genetic gain. Therefore much of the gains provided by a well improvement and management program may get lost before or during sawing. Sawing strategy cannot be taken from one eucalypt species to another because the correlation between log end splitting and board end splitting depends on the forest stand performance and/or species. It is possible to predict board end splitting in *Eucalyptus saligna* and *E. urophylla* but not in most of the planted *E. grandis*. The system of sawing which permits doing balanced cuts in order to release growth stresses in an equilibrated way, allows for a higher productivity and better board thickness accuracy. Nevertheless, neither cant end splitting during the cut in the twin band saw nor end splitting in the board extremity during sawing in the gang saw can be avoided. Both board bow and crook cannot be avoided for any sawing strategies. No strong influence exists by the thinning intensity on those traits, which are smaller at the tree bottom.

1 Introduction

Since the early 70's, many forests have been established in Brazil as sources for energy or pulp production. The first action was planting forests with fast growing *Eucalyptus* species. The quality of the produced wood was not so important because the idea was oriented toward round wood volume production only. Forest companies realize that it is important to manage forests aiming at quality rather than at quantity. The objective of the present work is to present in a direct simple way how inhomogeneous *Eucalyptus* wood is in terms of sawing related properties, even when produced by apparently well managed homogeneous forests.

2 Method and Materials

This paper presents a brief overview of some research projects, which have been carried out in the Department of Forestry Sciences of the University of São Paulo. Those efforts have been directed at *Eucalyptus* wood processing procedures, focusing on the defects related to growth stresses in trees and on the possibility of contributing to the technological development of this important wood production genus. The studies reviewed in this paper were based either on the cut strategy shown in Figure 1, which is focused on a simple band saw as breakdown machine and the consequent distortion of the cant that occurs just after cut by this machine, or on the cut strategy shown in Figure 2 that emphasizes a twin band saw which produces straight cants. In both case the breakdown machine is followed by a gang circular saw.

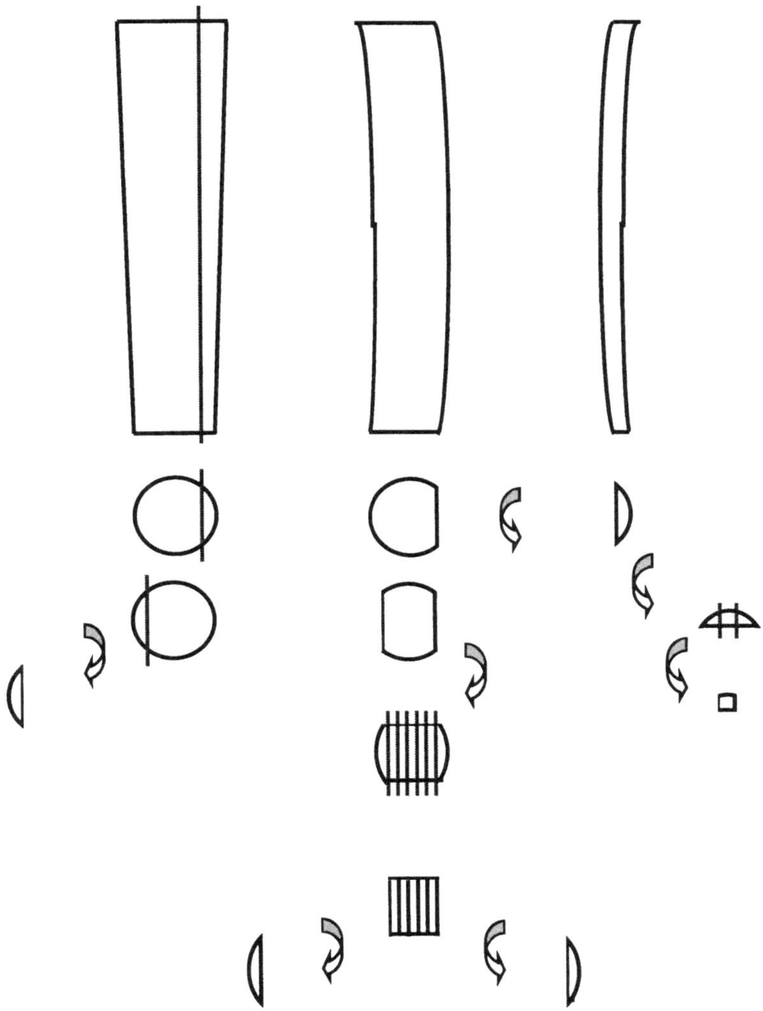

Figure 1. Sawing strategy in a simple band saw followed by a slab re-saw and a gang circular saw.

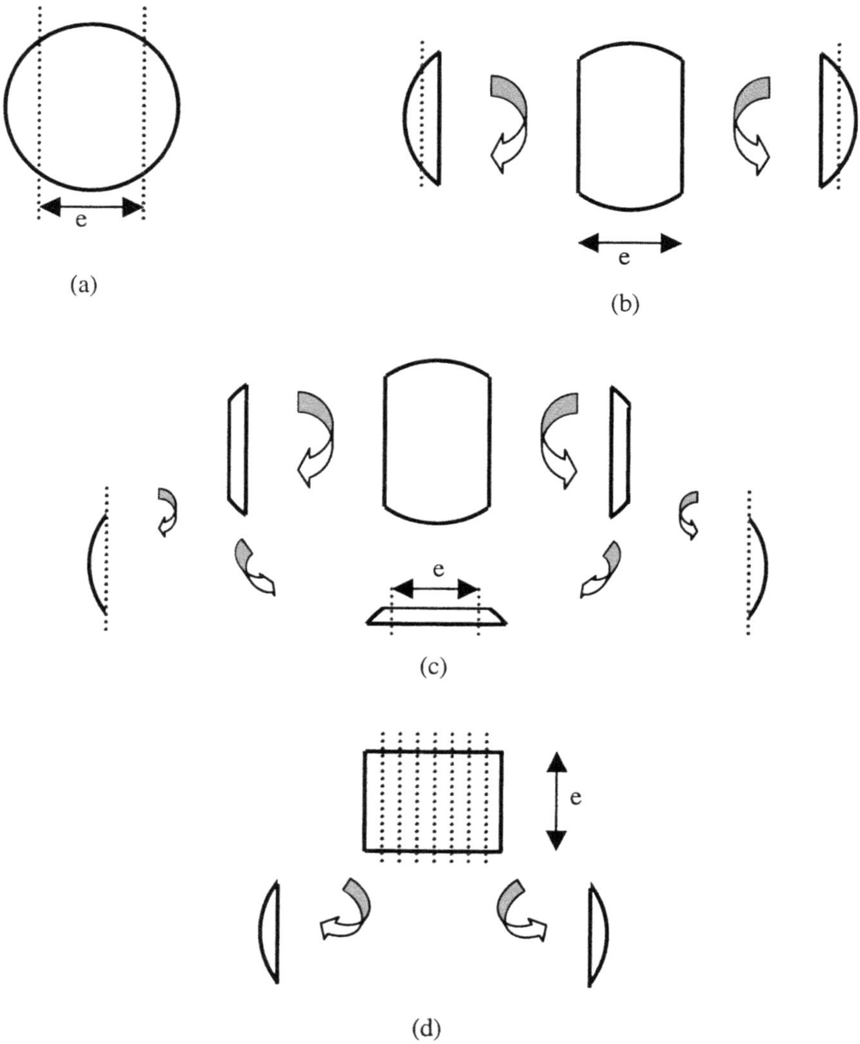

Figure 2. Sawing strategy in a layout composed by a twin band saw (a), slab re-saw (b), double edger (c) and a gang circular saw (d).

3 Results and Discussion

Even an apparently well managed forest as that one shown in Figure 3 will not give an exact idea of what is present inside the tree trunk in terms of yield and wood quality.

(c)

(a) (b)

Figure 3. Brazilian 50-year-old grown *Eucalyptus grandis.*

In the particular case of the *Eucalyptus* genus there are high levels of growth stresses in standing trees. These stresses are transformed into strains during cross cutting and/or sawing because a total release of those stresses occurs on the surface cut. Examples of defects that may occur when the strains are higher than the wood strength permits are shown in Figures 4 (adopted from Garcia et al. 2001) and 5.

(a)

(b)

Figure 4. Log end splitting (a) and board ending splitting (b) of 20- and 7-year-old grown *Eucalyptus grandis*, respectively.

(a) (b)

Figure 5. Eucalypts board bow (a) and board crook (b).

An extensive variation in the outside appearance of *Eucalyptus saligna* can be seen in Figure 6 (adopted from Lopes 2000). The tree on the left and the one on the right have barks that are smooth and rough, respectively. The type shown by the tree in the middle is probably a combination of both, and is called "half bark".

Figure 6. Phenotypic variation in bark of 22-year-old Brazilian *Eucalyptus saligna*.

Difference in some related properties that were observed among those types of trees (Lopes 2000), is summarized in Table 1. Choosing one of those types is always a matter of decision. It is important to balance the weight of each characteristic considered because certain expected gain may be offset by some unforeseen loss.

Table 1. Sawn wood related properties of 22-year-old grown *Eucalyptus saligna*.

Bark phenotype	LESI (%)	BESL (mm/m)	fc (MPa)	Basic density (g/cm^3)
Rough	30.9	104.6	63.0	0.68
Half bark	18.1	74.2	66.7	0.69
Smooth	22.3	94.3	58.4	0.62

It can be seen in Figure 4b that there are significant differences of board end splitting among trees, even when they are quite similar in their external appearance, as shown in Figure 3a. On the other hand, Table 1 also shows that trees that look quite different in the outside view (Figure 6) may exhibit about the same behaviour during the process of sawing.

So, it is important to keep in mind that trees, which look the same, may not be the same, while others, which look like different, may be the same.

There are many ways of evaluating the LESI. One of them is shown in eqn [1] obtained from Garcia and Lima (2000). Eqn [2] combined with Figure 7a shows an easy way to evaluate the BESL. The procedure, which has been used for evaluating cupping, either bow or crook, is shown in Figure 7b.

$$\text{LESI}=200\left(\sum_{i=1}^{n} a_i C_i\right) \qquad [1]$$

$$\text{BESI}=\frac{\sum_{i=1}^{n}(l_1+l_2)_i}{n} \qquad [2]$$

where a_i= split opening,
C_i=split length, and
n=number of boards obtained from the log.

Table 2 shows the heritability of growth stress related defects described above, obtained from a 9-year-old clonal trial of *E. urophylla*. The heritability changes with the number of clones tested, meaning that it is important to have good experiments in order to have precise and reproducible evaluation.

These h^2 values reported by Schacht et al. (1998a, 1998b) indicate that *E. urophylla* traits are genetically manageable. They can, therefore, be considered as good indicators for selection of superior trees in forest improvement programs for sawmill.

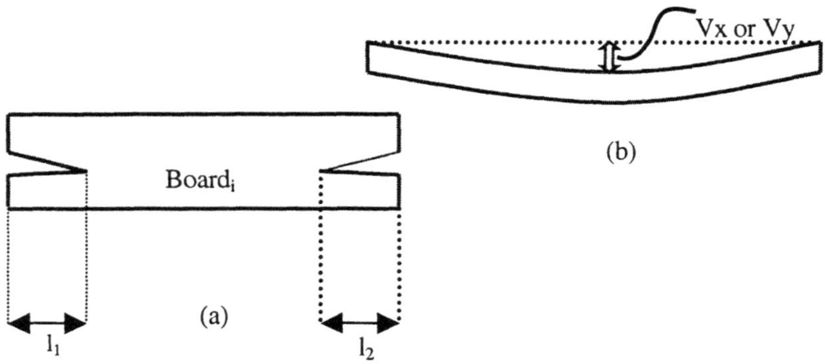

Figure 7. Procedure for measuring BESL (a) and cupping (b) either bow (Vx) or crook (Vy).

Table 2. Heritability of *E. urophylla* traits at 9-year-old.

Trait	$h^{2(1)}$	$h^{2(2)}$
LESI (log end splitting index)	0.83 (27 clones)	
	0.74 (6 clones)	0.50
BESL (board end splitting length)	0.74 (27 clones)	
	0.60 (10 clones)	0.46
SD (slab distortion)	0.83 (27 clones)	
SBD (slab based bean distortion)	0.83 (27 clones)	
BB (board bow)	0.83 (27 clones)	
	0.83 (10 clones)	0.76

(1) First log (bottom) of the tree only.
(2) First two logs of the tree, from the bottom.

Table 3 shows some data taken from Garcia and Lima (2000), which exhibit very high potential for improvement of eucalypts forests, either for LESI or BESL. High genetic gain is possible because eucalypts are still wild for those characteristics. Even if selection is not intensive, the genetic gain can still be very high when compared to those obtained for many other forestry traits. The gain for BESL is not as high as that estimated for LESI; but it is more useful for final sawn wood utilization.

Table 3. Genetic gain (ΔG) in log end splitting index (LESI) and board end splitting length (BESL) of 8-year-old *E. urophylla* at 20% of selection intensity.

	LESI	BESL
Average over all trees	1.20	35.19
Average over selected trees	0.01	14.54
DS[(1)]	1.19	20.65
ΔG	1.06	18.37
100ΔG/Average over all trees (%)	88	52

[(1)] DS=selection differential.

A good response can be obtained in selecting trees based on either LESI or BESL. But there may be some problems if the behaviour shown in Figure 8 is encountered (Garcia and Lima 2000). This figure shows that the correlation between LESI and BESL can be considered very poor in indicating two different meanings:

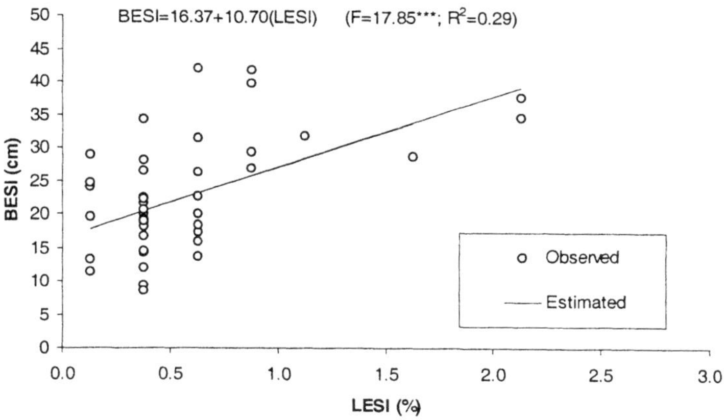

Figure 8. Phenotypic relationship between board end splitting length (BESL) and log end splitting index (LESI) of 18-year-old *E. grandis*.

1) Trees can be selected based on LESI without having much gain in BESL because the problem of board end splitting occurs mostly inside the sawmill, during log processing, as shown in Figure 9. This happens because the residual stresses that still exist in the log may be very high and provoke severe board end splitting when the ratio between board width and board thickness is very high.
2) Trees can be selected based on BESL, but probably having not much success, since the variability is only an environmental effect; which means that there is a strong effect of the sawing strategy as is shown in comparison of the LESI of Figure 10a to the BESL of Figure 10b. In this case it is important to check the h^2 values for the species and the sawing strategy you are dealing with. In the positive sense, however, it is still strongly recommended that trees are selected by using the BESL indicator.

Figure 9. *E. grandis* board end splitting during sawing.

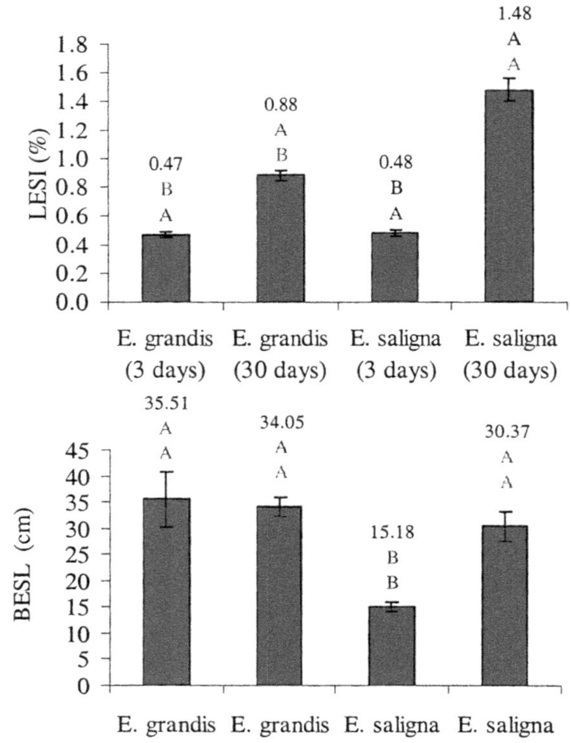

Figure 10. Log end splitting index (LESI) (a, up) and board end splitting length (BESL) (b, down) of 20-year-old *E. grandis* and *E. saligna*.

Figure 10 adopted from Garcia et al. (2001), shows that a respective increase about 100 % and 200% in the LESI for *E. grandis* and *E. saligna* can be expected if the time interval between cross cutting and sawing logs is large. In this instance the loss will outweigh any gain provided by any improvement program. This is confirmed by comparing Figures 10a and 10b, which shows that *E. grandis* splits more than *E. saligna* during sawing.

It is important to have this knowledge because it permits to conclude whether a set of clones is good or not in case of having no significant differences among them. If the correlation between BESL and LESI exists, there is a high probability that the clones are all good; otherwise, the clones are not so good because they react to the adopted sawing strategy.

In the case of *E. saligna*, the time interval from cross cutting at the felling time, to sawing affects both the LESI and the BESL. So, the relationship between those variables ($R^2 = 0.77$) is quite good in this species.

Figures 11 and 12, adopted from Garcia and Lima (2000), show the influence of thinning intensity on LESI and BESL, respectively. Thinning intensity only have a slight influence on the LESI but have no influence on BESL. In a broad sense we may consider that thinning has a slightly negative effect on LESI and a slightly positive effect on BESL. Both were a little higher at the intermediate thinning intensities, a result which cannot be reasonably explained yet. LESI was smaller in the bottom log but BESL was about the same along the tree trunk. As not influenced by either thinning intensity or log position of the tree trunk, BESL can, if heritable, be used as a pure indicator for selecting trees for sawmill, which is the matter of judgement.

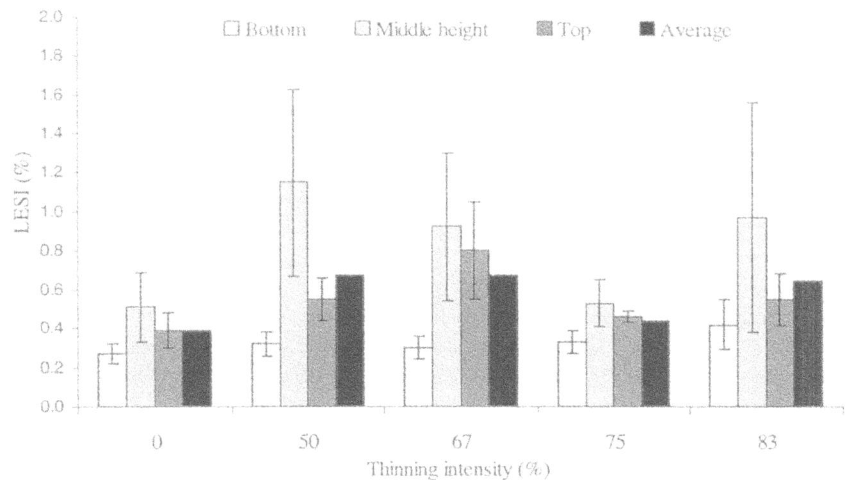

Figure 11. Log end splitting (LESI) of 18-year-old *E. grandis* at different log positions and thinning intensities.

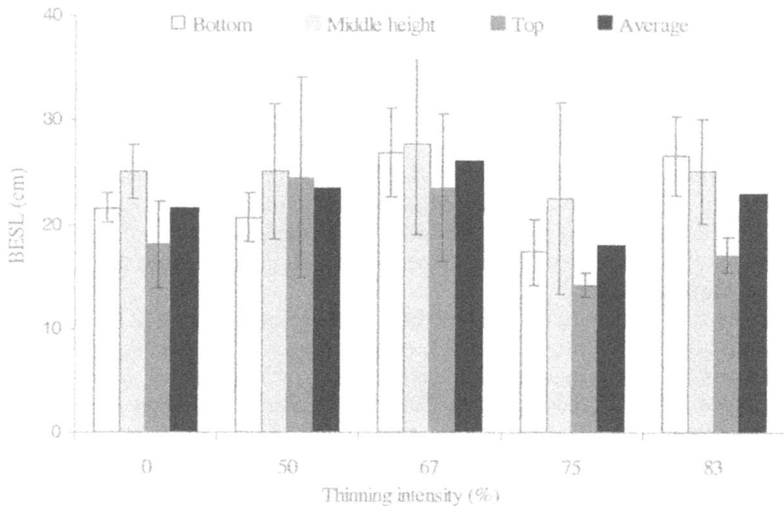

Figure 12. Green board end splitting (BESL) of 18-year-old *E. grandis* at different log positions and thinning intensities.

4 Conclusion

The correlation between BESL and LESI depends on the forest stand and/or species. A good sawing strategy for one case is not necessary good for another. There is no correlation between board end splitting and log end splitting in *E. grandis* because it splits during sawing. However board end splitting is predictable in *E. saligna.*

The system of sawing which permits doing balanced cuts in order to release growth stresses in an equilibrated way, allow for a higher productivity and better board thickness accuracy. Nevertheless, neither cant end splitting during cut in the twin band saw nor end splitting in the board extremity during sawing in the gang saw can be avoided. There is a strong effect of board width/thickness ratio, which has to be monitored.

Both board bow or crook cannot be avoided for any sawing strategies. They may be reduced by decreasing the growth stress in trees through forest improvement. It is important to select axis-symmetric trees in terms of wood properties.

Thinning intensity has no significant effect on either log end splitting or board end splitting.

References

Garcia, J.N. 1992. States of stresses in trees and strains in sawn wood pieces. PhD Thesis, University of São Paulo, Brazil, 240p.

Garcia, J.N. and Lima, I.L. 2000. Log end splitting and improvement in sawing *Eucalyptus grandis*. In: Proc. of IUFRO Conference "The Future of eucalypts for wood products", Launceston, Hobart, Forestry Tasmania, 19-24 March 2000, pp113-123.

Garcia, J.N. 1999. Theory and practice on the high quality *Eucalyptus* lumber production on forest improvement and sawmill techniques. In: Proc. of the Twelfth Silvotecna Eucalypt - Present and Future, Corma Concepción, Chile, 1999, pp1-22.

Schacht, L., Garcia, J.N., Dias, C.T. and Dos, S. 1998a. *Eucalyptus urophylla* log end splitting variation: a methodological investigation. In: "Variation of characters and its implication for the genetic improvement of *Eucalyptus urophylla* saw wood", L. Schacht, MSc Thesis, University of São Paulo, Brazil, pp17-25.

Schacht, L., Garcia, J.N. and Venkovsky, R. 1998b. Genetic variation of growth stress indicators in *Eucalyptus urophylla* clones. Sci. For., Piracicaba 54:55-68.

Garcia, J.N., Crêspo, E.A. and Baillères, H. 2001. Influence of log characteristics and machines performance on *Eucalyptus* board end splitting and board deflections. In: Proc. of 15th International Wood Machining Seminar, Wood Machining Institute, Berkeley, Los Angeles, USA, 20 July - 1 August 2001, pp163-171.

Lopes G.A. 2000. Wood quality of *Eucalyptus saligna* Smith from Itatinga, associated to the bark patterns observed in the population. MSc Thesis, University of São Paulo, 74p.

林分改良、管理和制材手段对锯材产量和质量得失的影响

JOSÉ N. GARCIA

Wood Engineering Laboratory, University of São Paulo
ESALQ – LCF - IPEF - P.O. Box 9
13418-900 Piracicaba/SP, Brazil

本文将不同研究项目的研究结果进行综合分析，目的是通过林分改良、经营管理和锯材技术改进等研究，全面了解木材产量和质量改良的最新动态。与锯材材性缺限相关的性状，如原木尾端开裂，板材尾端开裂曲翘等性状具有高度遗传性，据此对不同树种进行性状选择，将会产生巨大的遗传增益。多数通过改良和管理手段获得的增益在制材前或制材过程中可能会流失。桉树种间制材手段不能效仿，因为原木尾端开裂与板材尾端开裂的相关性依赖于林分或树种的表现。柳桉和尾叶桉人工林林木板材尾端开裂情况或许可以预测，而巨桉人工林板材的尾端开裂情况却无法预测。平衡切割系统可以均衡方式缓解林木生长胁迫产生的内应力，从而获得较多的锯材和较厚的板材。然而，在双带切割过程中，板材的尾端开裂和直锯制材过程中板材尾端的开裂都是不可避免的。对任何的制材手段和策略来说，板材的弯曲和翘裂都是不可避免的。这些尾端开裂性状在林木的根部表现较弱，疏伐强度对它们的影响不大。

SILVICULTURAL MANAGEMENT OF PRIVATE *EUCALYPTUS* PLANTATIONS FOR WOOD CHIPS IN THAILAND

BUNVONG THAIUTSA

Faculty of Forestry, Kasetsart University
Bangkok 10900, Thailand

Although the first *Eucalyptus* tree was introduced to Thailand and planted in Bangkok as an ornamental tree over one hundred years ago, a series of research on species and provenance trials for wood production was started in 1950. Most industrial plantations focusing on *Eucalyptus camaldulensis* for wood chips were begun less than twenty years ago. Demands of eucalypt log for wood-chip industry in the year 2002 for 3 main categories, i.e. pulp mills, chip mills and MDF, are 3,700,000 (57.86%), 2,015,000 (31.51%) and 680,000 tons (10.63%), respectively leading to the total demands of 6,395,000 GMT. Most private plantations set up rotation length at 5 years old with an average yield of about 75 tons/ha under proper establishment and maintenances. Thus, total plantation areas required for wood chips are 426,335 ha in order to be harvested 85,267 ha/yr.

In 1997, Thailand had 438,524 ha of eucalypt private plantations distributed 47.38 % in the Northeast, 28.73% in the East, 12.77% in the North, 11.07% in the Central Plains, and 0.05% in the South. New technologies were introduced to private plantations recently. Most planting stocks were produced from cutting, followed by tissue culture and seeds with the percentages of 80, 12, and 8, respectively. The spacing of 2 m x 3 m was a common practice planted after intensively mechanized site preparation. Some plantations preferred early dry-season planting in order to minimize the first-year weeding problems. Chemical fertilizers were applied periodically. Clearcutting with coppice system was carried out at 5 years old, followed by thinning the new shoots and remaining 2 –3 sprouts for the next rotation.

1 Production and Consumption of Wood Chips

It is obvious that woodpulp plays a major role of wood-chip consumption in Thailand. Only short-fiber pulp has been considered as domestic production, while long-fiber pulp is totally dependent upon importation. During the last three years annual export of the short-fiber pulp was about 277,530 tons (5,560 million Baht), while the long-fiber pulp import was about 422,533 tons/year, which was as high as 9,248 million Baht (Royal Forest Department 2001). Hence, woodpulp production mentioned in this paper is solely referred to the short-fiber pulp. Data from the Thai Pulp and Paper Industries Association showed that domestic pulp production increased dramatically, from 185,000 tons in 1991 to 316,000 tons in 1995, 756,000 tons in 1999, 764,000 tons in 2000 and 919,000 tons in 2001. Fast-growing trees recommended for wood chip and woodpulp manufactures are *E. camaldulensis, Acacia mangium*, and *Casuarina junghuhniana*. Due to annual rainfall requirement of *Acacia mangium* and poor coppicing ability of *Casuarina junghuniana,* however, only *E. camaldulensis* has been planted for such purposes in Thailand (Thaiutsa 1988).

Besides woodpulp, wood-chips also used as raw material for producing medium-density fiberboard (MDF) and hardboard as well as for exporting. In 1998 Thailand produced 2.972 million tons of wood chips and the figures became 3.121 million tons in 1999 and 3.317 million tons in 2000. It is of interest that percentages of wood chips used for woodpulp slightly decreased, from 80.87 in 1998 to 77.00 in 1999 and 72.45 in 2000, while the exports increased from 10.51% in 1998 to 12.50% in 1999 and 17.66% in 2000. Table 1 showed the figures of wood-chip exports during 1996 to 2000 and money gained from such exportation, which increased about 2.3 times within five years.

Table 1. Wood-chip export of Thailand during 1996-2000.

Year	Metric ton	1,000 Baht *	Baht/ton
1996	193,322	518,256	2,618
1997	260,314	714,188	2,744
1998	312,418	792,427	2,536
1999	390,055	958,399	2,457
2000	585,986	1,192,324	2,035

* US$1=Baht42.

Table 2 indicated that 72.45% of wood-chip production in Thailand was used for pulping. The second category next to pulping is export (17.66%), followed by hardboard (5.25%) and medium-density fiberboard (4.64%).

Table 2. Wood-chip consumption in Thailand in 2000.

Category	Metric Ton	%
Pulp	2,403,500	72.45
Export	585,986	17.66
Hardboard	174,000	5.25
MDF	153,760	4.64
Total	3,317,246	100.00

As mentioned earlier that nearly 100% of wood chips in Thailand came from *E. camaldulensis*, therefore, the Thai Pulp and Paper Industries Association estimated the demands of eucalypt logs for wood-chip industry in Thailand for the year 2002 (Table 3). Pulp mills would consume 57.86% of total demands, followed by chip mills (31.51%) and medium-density fiberboard (10.63%). Advance Agro and Phoenix seem to be the leading companies in demands for woodpulp; while Kittawee and Bangpakong (SHS) are leading in wood chips and Agro Line in medium-density fiberboard.

Table 3. Demands of eucalypt logs for wood-chip industry in Thailand in 2002.

Industry	Company	Province	Green Metric Ton
Pulp mill	Advance Agro	Prachinburi	1,800,000
	Phoenix	Khon Kaen	1,000,000
	Siam Cellulose	Kanchanaburi	600,000
	Panjapon	Ayutthaya	300,000
		Sub-Total	3,700,000 (57.86%)
Chip mill	Kittawee	Surin	600,000
	Bangpakong (SHS)	Chachoengsao	525,000
	STD	Chonburi	350,000
	Sriracha (SHS)	Chonburi	220,000
	Thai Wittawat	Surin	150,000
	KM 17	Buriram	100,000
	STD	Nakhonratchasima	70,000
		Sub-Total	2,015,000 (31.51%)
MDF, etc.	Agro Line	Chachoengsao	400,000
	Thai Pylwood	Saraburi	150,000
	Metro Fiber	Kanchanaburi	70,000
	Wanachai	Chonburi	60,000
		Sub-Total	680,000 (10.63%)
Total			6,395,000

Figure 1 expressed the distribution of pulp mill, chip mill, MDF and etc. in various provinces which mostly are in the central part of the country, possibly because of the main markets (Bangkok) and harbours (Bangkok and Chonburi).

Based on the demanding figure of 6,395,000 GMT/year of eucalypt logs for wood-chip industry, the plantation areas required can be estimated if the rotation length and target yields of wood are fixed. The common rotation length of *E. camaldulensis* for woodchip industry in Thailand is five years old with mean annual increment of about 15 tons per hectare or 75 tons/ha/5 yrs. Hence, the annual harvested areas should be about 85,267 hectares.

It means that Thailand should have wood-chip plantations at least 426,335 hectares; otherwise, wood import cannot be avoided.

2 Eucalypt as an Exotic Tree to Thailand

Eucalypt is commonly known as an Australian tree. The genus *Eucalyptus* consists almost 600 species. From economic points of view, it plays greater roles outside Australia rather than its natural range. Aside from *E. camaldulensis*, which is well known in Thailand, *E. citriodora*, *E. deglupta*, *E. grandis*, *E. globulus*, *E. tereticornis*, and *E. urophylla* are among principal species of commercial plantations in the Tropics.

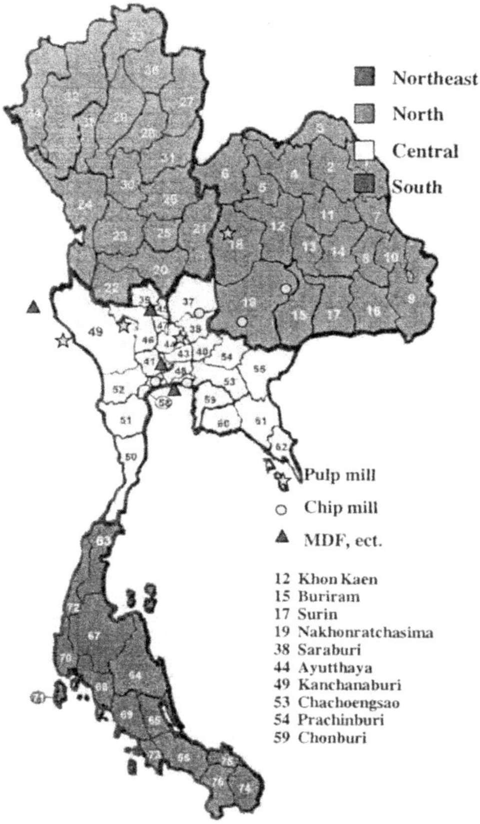

Figure 1. Distribution of eucalypt-wood-based mills in Thailand.

The earliest record on eucalypt introduction to Thailand might be referred to a huge standing tree in front of the Vimanmeak Palace in Bangkok. It is believed that this eucalypt was planted as an ornamental tree before 1900, during the construction of the Palace. Another record is a photograph of 5-year-old *E. camaldulensis* shown at Forestry Library of Kasetsart University. It revealed that Prof. Dr. Sa-ard Boonkird of Kasetsart University planted this eucalypt tree in Prae province in 1946. However, official information of the Royal Forest Department recorded that the first introduction of eucalypts from Australia to Thailand was made in 1950. Several *Eucalyptus* species including *E. alba*, *E. citriodora*, *E. paniculata*, *E. saligna* and *E. grandis* were introduced to Doi Suthep Experimental Plot of Chiangmai Province.

Through the RFD/FAO Project on surveying raw materials for pulp and paper industry, fifteen species of *Eucalyptus* were introduced from Queensland to be planted at Surathani in the South, Kanchanaburi in the West, Chiangmai in the North and Srisakaet in the Northeast in 1964. Species and provenance trials of *Eucalyptus* were repeated at those planting sites in the following years. *E. camaldulensis* was recorded to be included in the later trials. Another extensive *Eucalyptus* species trials were carried out in 1973 through Thai/Danish Tree Improvement Project on pines and fast-growing trees by introducing more than forty species of *Eucalyptus* to be planted at various sites.

In 1978, the Silvicutural Research Section of the RFD (Royal Forest Department) started a field research on agroforestry in the Northeast. The findings showed that *E. camaldulensis* grew faster than *Acacia auriculiformis, Leucaena leucocephala*, and *Peltophorum dasyrachis*. Yields of upland crops intercropped with *E. camaldulensis* were also found to be higher than those gained from the remaining treatments. Experiences gained from these long-term experiments made the Village Woodlot Project, a project financially supported by NEA/USAID aiming to overcome fuel wood shortage in the Northeast considered *E. camaldulensis* as a suitable tree to be planted. Since then, RFD has vigorously promoted *E. camaldulensis*. Due to its ability to grow successfully in a wide range of environments, including different altitudes, soil types, rainfall regimes, and terrain as well as ability to coppice with rapid growth. *E. camaldulensis* has commercially been planted by private sectors since 1978. However, the large-scaled commercial plantations did not started until 1983 (Thaiutsa and Taweesuk 1987). Then, strong controversies on ecological impacts of growing eucalypts, especially on soil, water and air were begun in 1986 resulting to the decline of planting areas for a certain period. Some chronological information on eucalypt introduction to Thailand could be summarized as follows:

- 1900 - Planting an eucalypt as an ornamental tree at Vimanmeak Palace in Bangkok;
- 1946 - Planting an eucalypt at a reforestation unit in Phrae province;
- 1950 - *Eucalyptus* species trials by RFD in the North (Chiangmai);
- 1964 - *Eucalyptus* species trials, including *E. camaldulensis* by RFD/FAO in the North (Chiangmai), South (Suratthani), Northeast (Srisakaet) and the West (Kanchanaburi);
- 1973 - *Eucalyptus* species and provenance trials by Thai/Danish Tree Improvement Project at Chiangmai;
- 1978 - Planting *E. camaldulensis* in agroforestry systems at Srisakaet by the Village Woodlot Project of RFD/NEA/USAID;
- 1979 - Beginning the private plantations of eucalypt;
- 1983 - Beginning the large-scaled commercial plantations of eucalypt;

Recently, Sunthornhao (1999) surveys the private-eucalypt plantations throughout the country by using questionnaires together with field investigation. Table 4 showed that there were 438,524 hectares of eucalypt plantations owned by private sectors in Thailand in 1997. The northeastern region had the highest amounts (47.38%), followed by the East (28.73%), the North (12.77%) and Central Plains (11.07%) while the southern region had only 236 hectares (0.05%) of private-eucalypt plantations. Distribution by province plantation areas is expressed in Figure 2.

Table 4. Areas of private-eucalypt plantations in Thailand in 1997.

Region	Area (ha)	%
Northeast	207,785	47.38
East	125,976	28.73
North	55,996	12.77
Central	48,531	11.07
South	236	0.05
Total	438,524	100.00

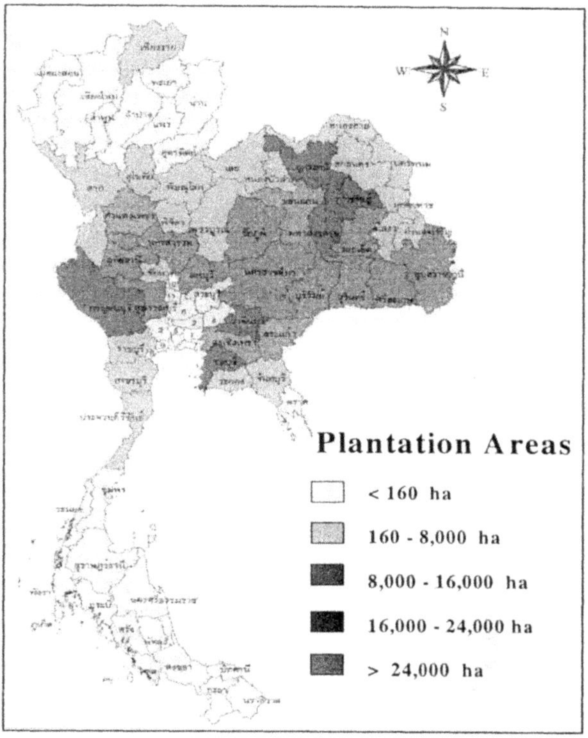

Figure 2. Distribution of private-eucalypt plantation by planting areas of each province.

Regarding to the farm sizes of eucalypt plantation, a survey in the West was carried by Cherdkiatikul (2002) and found that most large-scaled plantation had areas of about 32-80 hectares (30.4%), followed by less than 16 hectares (24.1%), 16-32 hectares (21.5%), 80-160 hectares (12.7%), and more than 160 hectares (11.4%). The findings suggested that some farmers could earn their main income from growing eucalypt, but at least one-third of them had too small farms and another sources of income would be necessary.

3 Silvicultural Management of Eucalypt Plantations

Silviculture is not only a tree planting, but also deals with the controlling of establishment, growth, composition and quality of the forest crops concerned. In case of eucalypt plantation for wood-chip industry, silvicultural management is crucially involved from tree planting to harvesting leading to the successful program of forest plantation. From a basic concept of biology, Phenotype = Genotype + Environment, performances or growth and yields of planted eucalypts vary with planting stocks, planting site and plantation management.

Thailand was begun with wrong information mentioning, "Eucalypt is a fast growing tree, is easy to grow and is a magic tree". Actually, eucalypt is just a normal tree species. In order to gain maximal profit, intensive and proper silvicultural treatments are crucially required from seedling through harvesting.

Although seedlings were produced from seeds in the early stage of eucalypt plantations in Thailand, most nursery stocks are currently produced from vegetative propagation. It is roughly estimated that 80% of eucalypt-planting stocks consumed by private plantation is propagated by rooted cutting. Tissue culture accounts for 12% and 8% comes from seeds through seedlings. Rooted cutting and tissue culture are vegetative propagation, thus, selection of elite trees is very important in order to maintain their superior characteristics. It is very risky to have narrow genetic bases of mother trees.

Regarding the preparation of planting site, most private plantations nowadays prepare site by ploughing surface soil with farm tractors. If the 75 tons per hectares of green-weight wood is a target yield for 5-year-old rotation, then the common spacing is 2 m x 3 m. Cost of plantation establishment and maintenances is about US$765 per hectare. Price of small eucalypt wood for wood chips is about US$17 per ton. So, the farmers may get net profit of about US$102 per hectare per year. However, if planting site is not ploughed, mean annual increment may not reach 15 tons per hectare resulting in lower net profit than such figures.

Aramsri (1999) did an experiment on growth analysis of 5-year-old *E. camaldulensis* clones planted at 2 m x 4 m spacing at three planting sites in the Northeast. Under the same manner of establishment and maintenances, significant clonal and site variations in green-weight yields were reported as expressed in Table 5. Clone No.5 showed superior performances to No.P7, No.S10 and seedlings produced from seeds at all sites. Yield from Khon Kaen planting site were only 50% or even one-third of those gained from Sra Kaew and Ubonratchathani, mainly because of infertile lateritic soil at Khon Kaen. The findings suggested that in order to maximize yields and of course income per hectare, the farmers must plant the suitable clones on proper sites. Furthermore, it was found that clonal plantation provided higher yields than the reproductive ones.

Table 5. Variation in green weight of eucalypt wood (tons per hectare) with genotypes and planting sites.

Clone	Sra Kaew	Ubon	Khon Kaen
T5	178	144	59
P7	128	114	54
S10	109	106	39
Seedling	68	60	29

Another silvicultural problem of eucalypt plantations in Thailand is disease. The most common fungal pathogen recently found is leaf and shoot blight caused by *Cryptosporiopsis eucalypti*. Degree of severity is dependent on soil and climatic factors. Eucalypt trees may show severe symptoms during late rainy season, especially for trees growing on high pH soil of poor drainage with annual rainfall of over 1,200 mm and more than 70% relative humidity. Base on a simple scoring index of 60 clones, it was found that 55% of eucalypt planted in the West on such environments would be considered resistant to leaf and shoot blight diseases, 11.67% moderately resistant, 10% moderately susceptible and 23.33% susceptible. It means that one-third of the trees is susceptible to *Crystosporiopsis eucalypti*. If it is not too severe, the trees may not die, but it affects growth rate. It is believed that the major cause of this disease is the narrow genetic base of *E. camaldulensis* planted in Thailand. Thus, introduction of more and wider seed sources is required.

Other silvicultural considerations of eucalypt plantations are weed control, early-season planting, fire protection, fertilizer application, stump management, and agroforestry systems with eucalypts. About 25% of successful forest plantation in the Tropics depends on weed control, especially during the early stage of the planted tree development. Forest vegetation might play the negative role as weeds during the rainy season, but they would be fuel in dry season. So, weeding is not only a method of weed control, but also a fire protection as well. Many private plantations satisfy with the result of early-season planting by preparing planting sites completely in early dry season, i.e. around early December after filling up water in the planting holes. Through these techniques, the planted trees could grow faster because of long growing periods and less weed problems during the first year. Forest fertilization seems to be rarely practiced in Thailand, but it is very necessary for such a short-rotation species as *E. camaldulensis* in order to replace nutrient losses due to crop removal. With an exception of soil/foliar analysis, nutrients leading to vegetative growth promotion should be applied both at the planting time and late rotation. Mycorrhizal inoculation should also be very fruitful, because mycorrhiza could increase phosphorus and water uptakes. For stump management, the number of shoots to be remained after the first-rotation coppicing should be about 2-3 shoots per stump. Shoot thinning should be done within few months after harvesting. Also, agroforestry systems should be widely practiced for private plantations, because they provide opportunity to plantation owners to earn money both from primary crops and intercrops. The species of agricultural crops combined vary with sites and regions, but mungbean, cassava, corn, upland rice and forage grasses are more common.

4 Conclusions

Data in the year 2002 shows that Thailand demands 6,395,000 green metric tons of eucalypt wood for wood-chip industry. Based on 75 metric tons per hectare for 5-year-old rotation, it is necessary to have at least 426,335 hectares of eucalypt plantations. In 1997, Thailand had private-eucalypt plantations of about 438,524 hectares. With an exception of 5-year differences, it seems to have production over consumption of about 12,189 hectares, equivalent to 914,175 tons or 182,835 tons per year. Actually, the country is now facing a shortage problem of eucalypt logs for wood chips, because most plantations have yields less than 75 tons per hectare at 5-year-old rotation. Moreover, those plantations are too far from the wood-chip manufactures which affect higher transportation costs fixed at a radius of about 100-120 kilometers from the markets. Hence, any silvicultural managements resulting in increasing growth and yields of eucalypt wood are crucially required by the private sectors.

References

Aramsri, C. 1999. Growth analysis of 5-year-old *Eucalyptus camaldulensis* Dehnh. clones. MS Thesis, Kasetsart University, Bangkok, 146p.

Cherdkiatikul, P. 2002. Stratergies to raise *Eucalyptus* plantations promoting capability of pulp and paper business in Thailand: A case study of large-scale plantations. MBA Thesis, Kasetsart University, Bangkok, 109p.

Royal Forest Department 2001. Forestry statistics of Thailand, 2001. Data Center, Information Office, RFD, Bangkok, 156 p.

Sunthornhao, P. 1999. Supply of *Eucalyptus* timber and financial analysis of Thailand private plantation in 1997. MS Thesis, Kasetsart University, Bangkok, 106p.

Thaiutsa, B. 1988. Industrial plantation of fast-growing trees. A paper submitted to TDRI, Bangkok, 24p.

Thaiutsa, B. and Taweesuk, S. 1987. *Eucalyptus* plantation in Thailand. Thai J. For. 6:437-443.

泰国私有桉树木片人工林的经营管理

BUNVONG THAIUTSA

Faculty of Forestry, Kasetsart University
Bangkok 10900, Thailand

泰国首次引进桉树的时间是在 100 多年前，引进的第一批桉树被作为观赏树种种植于曼谷，将桉树作为用材树种对其进行的系列树种和种源试验研究却开始于 1950 年。大约 20 年前，开始营造用于木片生产的桉树工业用材林，主要是赤桉林。2002 年对桉树木片材的需求主要集中在纸浆、木片和建筑用材这三个方面，需求量分别为 370(57.86%)、201.5(31.51%)和 68(10.63%)万吨，总需求量为 639.5 万吨。多数私有桉树人工林的轮伐期为 5 年，平均每公顷产量约 75 吨。可见，为使每年采伐 85,267 公顷桉树人工林，木片生产林分的总面积应达到 426,335 公顷。到 1997 年，泰国建立桉树私有林 438,524 公顷，其中东北部占 47.38%，东部 28.73%，北部 12.77%，中部平原 11.07%和南部 0.05%。最近，在私有林的营建中引入了新技术，种植的苗木 80%来自扦插，12%来自组培苗，8%来自实生苗。机械化整地后，最常用的株行距为 2 m × 3 m。为减少造林后第一年杂草的危害，通常在年初干旱季节造林，并阶段性地施用化学肥料。林分 5 年生时开始砍伐，保留树桩，对萌芽条进行疏伐，保留2-3 个萌芽条使之发展为第二代人工林。

LIST OF PARTICIPANTS

Arnold, Roger
Division of Forestry, CSIRO
PO Box E4008
Kingston ACT 2604
Australia
roger.arnold@csiro.au

Bai, Jiayu
Research Institute of Tropical Forestry
Chinese Academy of Forestry
Longdong, Guangzhou 510520
P.R. China
baijy@public.guangzhou.gd.cn

Borralho, Nuno
RAIZ
Centro de Investigacao Florestal Quinta da Torre Bela
Apartado 15
2065-999 Alcoentre, Portugal
nborralho@raiz-iifp.pt

Chen, Beiguang
South China Agricultural University
Wushan, Guangzhou 510642
P.R. China
bgchen@scau.edu.cn

Chen, Chongzheng
Guangxi Research institute of Forestry
23 Yongwu Road
Nanning 530001, Guangxi
P.R. China
chenchongzheng@lol365.com

Chen, Jianbo
Guangxi Research institute of Forestry
23 Yongwu Road, Nanning 530001
Guangxi
P.R. China
chenjb2005@mail.com

Chen, Shaoxiong
China Eucalypt Research Center
30 Mid Renmin Road
Zhanjiang 524022, Guangdong
P. R. China
sxchen@pub.zhanjiang.gd.cn

Chen, Yongling
Jiangxi Academy of Forestry
Maiyuan, Changbei Exploring District
Nanchang, Jiangxi
P.R. China
jxlky@pub.nc.jx.cn

Clegg, Barbara
RGM International
Forestry Service Division
Aek Nauli Tree Improvement Station
Tromolpos 11, Parapat 21174
Sumatra Utara
Indonesia

Clegg, Paul
RGM International
Forestry Service Division
Aek Nauli Tree Improvement Station
Tromolpos 11, Parapat 21174
Sumatra Utara
Indonesia
paclegg@indosat.net.id

Cossalter, Christian
Center for International Forestry Research
P.O.Box 6596 JKPWB
10065 Jakarta, Indonesia
c.cossalter@cgiar.org

Danusevicius, Darius
Lithuanian Forest Research Institute
Girionys, LT-4312, Kaunas Region
Lithuania
darius.danusevicius@takas.lt

De Lellis Garcia Bertolucci, Fernando
Research & Technology Center
Aracruz Celulose, Caixa Postal: 33.1011
Rodovia Aracruz/Barra do Riacho
Km 25, 29179-000 Aracruz-ES-Brazil
flb@aracruz.com.br

Dell, Bernie
School of Biological Sciences and Biotechnology, Murdoch University
South Street, Perth, WA6150, Australia
dell@central.murdoch.edu.au

Ding, Hongmei
China Green Times
18 East Hepingli Street
Beijing 100714, P.R. China
elphant@263.net

Endo, Masatoshi
Guangxi Oji Plantation Forest Co. Ltd
21/F., Wukuang Building
32 Yuanhu Road
Nanning 530022, Guangxi
P.R. China
forester@nn.gx.cninfo.net

Fryer, John
ACIAR, Forestry Program, Australian Centre for International Agricultural Research, GPO Box 1571
Canberra ACT 2601, Australia
fryer@aciar.gov.au

Fung, Jayco
Sino-Forest Corporation
3129-40, 31/F. Sun Hung Kai Centre
30 Harbour Road, Wanchai
Hong Kong
jayco-fung@sinoforest.com

Garcia, José Nivaldo
Wood Engineering Laboratory
University of São Paulo
ESALQ – LCF – IPEF - P.O. Box 9
13418-900 Piracicaba/SP
Brazil
jngarcia@esalq.usp.br

Halinen, Markku
UPM-Kymmene Forest
Niementie 1, P.O. Box 32
FIN-37601 Valkeakoski, Finland
markku.halinen@upm-kymmene.com

Hardy, Giles
School of Biological Sciences and Biotechnology, Murdoch University
South Street, Perth, WA6150
Australia
g-hardy@central.murdoch.edu.au

He, Xiang
UPM-Kymmene Group
Beijing Representative Office
Rm.1005, Kerry Center North Tower
1 Guang Hua Road, Beijing 100020
P.R. China
he.xiang@upm-kymmene.com

Henson, Michael
Eucalyptus.biz Limited
Rosemead
Polesden Lane, Ripley
Surrey, GU236DX, UK
michael@eucalyptus.biz

Hou, Yuanzhao
Research Institute of Forestry
Information
Chinese Academy of Forestry
Beijing 100091
P.R. China
houyuanzhao@163.net

Hu, Tianyu
Sichuan Academy of Forestry
18 Xing Hui Road
Chengdu 610081, Sichuan
P.R. China
asclky@shell.scsti.ac.cn

Huang, Baoqiang
Jiachang Forest Development Co. Ltd.
37 Jinwei Building, Erjing Road
Nanchang 330006, Jiangxi
P.R. China

Huang, Longbin
School of Environmental Science
Murdoch University
South Street, Perth, WA6150
Australia
huang@central.murdoch.edu.au

Huang, Shaowei
College of Forestry
South China Agricultural University
Wushan, Guangzhou 510642
P.R. China
shwhuang@scau.edu.cn

Huang, Wending
Stora-Enso, Asia Pacific
Room 1908 CITIC Plaza
233 North Tianhe Road
Guangzhou 510613, P.R. China
wending.huang@storaenso.com

Huang, Yu
Guangxi State-Owned Forest Farm
Development Corporation
107 Dongge East Road
Nanning 530022, Guangxi
P. R. China
htiger5@21cn.com

Kang, Lihua
Research Institute of Tropical Forestry
Chinese Academy of Forestry
Longdong, Guangzhou 510520
P.R. China
sbliao@pub.guangzhou.gd.cn

Li, Danqing
Yunnan Academy of Forestry Science
Heilongtan, Kunming 650204
Yunnan, P.R. China
ldq2001@163.com

Li, Guangyou
Research Institute of Tropical Forestry
Chinese Academy of Forestry
Longdong, Guangzhou 510520
P.R. China
leerwater3000@yahoo.com.cn

Li, Kaifu
College of Forestry
South China Agricultural University
Wushan, Guangzhou 510642
P.R. China
kfli@scau.edu.cn

Li, Mei
Research Institute of Tropical Forestry
Chinese Academy of Forestry
Longdong, Guangzhou 510520
P.R China
smggan@pub.guangzhou.gd.cn

Liang, Tailin
Jia-He Forestry Development Limited
40 South Huada Street
Heyuan 517000, Guangdong
P. R. China

Lin, Miranda
Sino-Forest Corporation
3129-40, 31/F. Sun Hung Kai Centre
30 Harbour Road, Wanchai
Hong Kong
miranda-lin@sinoforest.com

Lindgren, Dag
Department of Forest Genetics and
Plant Physiology, Swedish
University of Agricultural Sciences
SLU, SE-901 83, Umeå, Sweden
dag.lindgren@genfys.slu.se

Little, Keith
Institute for Commercial Forestry
Research, P.O. Box 100281
Scottsville, KwaZulu-Natal, 3209
Republic of South Africa
keith@icfr.unp.ac.za

Lu, Zhaohua
Research Institute of Tropical Forestry
Chinese Academy of Forestry
Longdong, Guangzhou 510520
P.R. China
luzhaohua@163.com

Martin, Bernard
Chargé de mission à l'ENGREF
14, rue Girardet
54042 NANCY cedex
France
betj.martin@9online.fr

Midgley, Stephen
Division of Forestry, CSIRO
PO Box E4008
Kingston, ACT 2604, Australia
stephen.midgley@csiro.au

Mo, Xiaoyong
Lei Zhou Forestry Bureau
Chengyue, Suixi 524348
Guangdong
P.R. China
mxyyxh@21cn.com

Morris, Jim
Forest Science Center
123 Brown Street, P.O. Box 137
Heidelberg, Victoria 3084, Australia
jim.morris@nre.vic.gov.au

O'Connel, Anthony
CSIRO Forestry and Forest Products
Private Bag No.5, Wembley
WA6014, Australia
t.oconnell@ccmar.csiro.au

Pan, Yongyan
Hainan Foresource Forestry Co. Ltd.
5/F., 1# Building
Country Community Road
Baishali Autonomy County
Hainan, P.R. China

Poon, Simon
Sino-Forest Corporation
3129-40, 31/F., Sun Hung Kai Centre
30 Harbour Road, Wanchai
Hong Kong
simon-poon@sinoforest.com

Qi, Shuxiong
Advisers Committee, Sino-Forest
Corporation, 33 Middle Renmin Road
Zhanjiang 524022, Guangdong
P.R. China

Saragih, Thomas
RGM International
Forestry Service Division
Aek Nauli Tree Improvement Station
Tromolpos 11, Parapat 21174
Sumatra Utara, Indonesia

Sartório, Robert Cardoso
Research & Technology Center
Aracruz Celulose, Caixa Postal:
33.1011
Rodovia Aracruz/Barra do Riacho
Km 25, 29179-000 Aracruz-ES-Brazil
rcsa@aracruz.com.br

Shield, Evan
Forestry/Forest Industry Consultant
3/115 Chester Road, Annerley
Queensland, 4103
Australia
efabne@optusnet.com.au

Simpson, John
Queensland Forest Research Institute
Locked Bag 16
M/S 483 Fraser Road, Gympie, 4570
Australia
john.simpson@dpi.qld.gov.au

Song, Yajie
School of Forestry and Environmental
Studies, Yale University
205 Prospect Street, New Haven
CT 06511, USA
songyaj@aol.com

Swain, Tammy
Institute for Commercial Forestry
Research, P.O. Box 204
Sabie, Scottville 1260
South Africa
tammy@icfrsabie.unp.ac.za

Thaiutsa, Bunvong
Faculty of Forestry
Kasetsart University
Bangkok 10900, Thailand
fforbvt@ku.ac.th

Tu, Farlay
Sino-Forest Corporation
3129-40, 31/F. Sun Hung Kai Centre
30 Harbour Road, Wanchai
Hong Kong
farlay-tu@sinoforest.com

Varghese, Mohan
Institute of Forest Genetics and
Tree Breeding, P.O.B. No.1061
Forest Campus
Coimbatore-641002, India
mvarghese1@rediffmail.com

Vuokko, Risto
Stora-Enso, Asia Pacific
Room 1908, CITIC Plaza
233 North Tianhe Road
Guangzhou 510613, P.R. China
risto.vuokko@storaenso.com

Wang, Guoxiang
Room 401, 27 Middle Renmin Road
Zhanjiang 524022
Guangdong
P.R. China
guoxwang@163.net

Wang, Huoran
Research Institute of Forestry
Chinese Academy of Forestry
Beijing 100091
P.R. China
wanghr@forestry.ac.cn

Wang, Meiyan
International Cooperation Division
Chinese Academy of Forestry
Beijing 100091
P.R. China
mwang@caf.forestry.ac.cn

Wei, Run-Peng
Sino-Forest Corporation
3129-40, 31/F., Sun Hung Kai Centre
30 Harbour Road, Wanchai
Hong Kong
runpeng-wei@sinoforest.com

Wei, Shanhua
Dongmen Forestry Center
Dongmen Town,
Fusui County Nanning 532108
Guangxi, P.R. China
wei-1.L@163.com

Wu, Zhongmin
Research Institute of Tropical Forestry
Chinese Academy of Forestry
Longdong, Guangzhou 510520
P.R. China
zhongminwu@pub.guangzhou.gd.cn

Xiang, Dongyun
Guangxi Research institute of Forestry
23 Yongwu Road
Nanning 530001, Guangxi
P.R. China
xiangdy@public.nn.gx.cn

Xie, Yaojian
China Eucalypt Research Center
30 Middle Renmin Road
Zhanjiang 524022, Guangdong
P.R. China
xieyj@pub.zhanjiang.gd.cn

Xu, Daping
Research Institute of Tropical Forestry
Chinese Academy of Forestry
Longdong, Guangzhou 510520
P.R. China
gzfsrd@public.guangzhou.gd.cn

Xu, Jianmin
Research Institute of Tropical Forestry
Chinese Academy of Forestry
Longdong, Guangzhou 510520
P.R. China
jianmxu@public.guangzhou.gd.cn

Yang, Yaoqing
Guijia Forestry Limited Co.
Room 1202A & B, 12/F. Ninghui
Building, 42 Binhu Road, Nanning
330006, Guangxi, P.R. China
yangyq0703@sino.com

Zeng, Linghai
Guangdong Academy of Forestry
Longdong, Shahe,
Guangzhou 510520
P.R. China

Zhang, Meiya
Hainan Foresource Forestry Company
Fengtai P.O.B. 7158
Beijing
P.R. China
mu681@yahoo.com

Zhang, Ningnan
Research Institute of Tropical Forestry
Chinese Academy of Forestry
Longdong, Guangzhou 510520
P.R. China
gzfsrd@pub.guangzhou.gd.cn

Zheng, Bai
Dongmen Forestry Center
Dongmen Town, Fusui 532108
Nanning, Guangxi
P.R. China

Zhou, Zaizhi
Research Institute of Tropical Forestry
Chinese Academy of Forestry
Longdong, Guangzhou 510520
P.C. China
zzzw@pub.guangzhou.gd.cn

www.ingramcontent.com/pod-product-compliance
Ingram Content Group UK Ltd.
Pitfield, Milton Keynes, MK11 3LW, UK
UKHW020402150926
8624IPUK00004B/301

9 789812 385574